The Institute of Mathematics
and its Applications
Conference Series

Volumes in the previous series were published by Academic Press to whom all enquiries should be addressed. The following and all forthcoming titles are published by Oxford University Press throughout the world.

NEW SERIES

1. *Supercomputers and parallel computation* Edited by D. J. Paddon
2. *The mathematical basis of finite element methods* Edited by David F. Griffiths
3. *Multigrid methods for integral and differential equations* Edited by D. J. Paddon and H. Holstein
4. *Turbulence and diffusion in stable environments* Edited by J. C. R. Hunt
5. *Wave propagation and scattering* Edited by B. J. Uscinski
6. *The mathematics of surfaces* Edited by J. A. Gregory
7. *Numerical methods for fluid dynamics II* Edited by K. W. Morton and M. J. Baines
8. *Analysing conflict and its resolution* Edited by P. G. Bennett
9. *The state of the art in numerical analysis* Edited by A. Iserles and M. J. D. Powell
10. *Algorithms for approximation* Edited by J. C. Mason and M. G. Cox
11. *The mathematics of surfaces II* Edited by R. R. Martin
12. *Mathematics in signal processing* Edited by T. S. Durrani, J. B. Abbiss, J. E. Hudson, R. N. Madan, J. G. McWhirter, and T. A. Moore
13. *Simulation and optimization of large systems* Edited by Andrzej J. Osiadacz
14. *Computers in mathematical research* Edited by N. M. Stephens and M. P. Thorne
15. *Stably stratified flow and dense gas dispersion* Edited by J. S. Puttock
16. *Mathematical modelling in non-destructive testing* Edited by Michael Blakemore and George A. Georgiou
17. *Numerical methods for fluid dynamics III* Edited by K. W. Morton and M. J. Baines
18. *Mathematics in oil production* Edited by Sir Sam Edwards and P. R. King
19. *Mathematics in major accident risk assessment* Edited by R. A. Cox
20. *Cryptography and coding* Edited by Henry J. Beker and F. C. Piper
21. *Mathematics in remote sensing* Edited by S. R. Brooks
22. *Applications of matrix theory* Edited by M. J. C. Gover and S. Barnett
23. *The mathematics of surfaces III* Edited by D. C. Handscomb
24. *The interface of mathematics and particle physics* Edited by D. G. Quillen, G. B. Segal, and Tsou S. T.
25. *Computational methods in aeronautical fluid dynamics* Edited by P. Stow
26. *Mathematics in signal processing II* Edited by J. G. McWhirter
27. *Mathematical structures for software engineering* Edited by Bernard de Neumann, Dan Simpson, and Gil Slater
28. *Computer modelling in the environmental sciences* Edited by D. G. Farmer and M. J. Rycroft
29. *Statistics in medicine* Edited by F. Dunstan and J. Pickles
30. *The mathematical revolution inspired by computing* Edited by J. H. Johnson and M. J. Loomes
31. *The mathematics of oil recovery* Edited by P. R. King
32. *Credit scoring and credit control* Edited by L. C. Thomas, J. N. Crook, and D. B. Edelman
33. *Cryptography and coding* Edited by Chris Mitchell
34. *The dynamics of numerics and numerics of dynamics* Edited by D. S. Broomhead and A. Iserles
35. *The unified computation laboratory: modelling, specifications, and tools* Edited by Charles Rattray and Robert G. Clark
36. *Mathematics in the automotive industry* Edited by James R. Smith
37. *The mathematics of control theory* Edited by N. K. Nichols and D. H. Owens
38. *Mathematics in transport planning and control* Edited by J. D. Griffiths
39. *Computational ordinary differential equations* Edited by J. R. Cash and I. Gladwell

Continued overleaf

40. *Waves and turbulence in stably stratified flows* Edited by S. D. Mobbs and J. C. King
41. *Robotics: applied mathematics and computational aspects* Edited by Kevin Warwick
42. *Mathematical modelling for materials processing* Edited by Mark Cross, J. F. T. Pittman, and Richard D. Wood
43. *Wavelets, fractals, and Fourier transforms* Edited by M. Farge, J. C. R. Hunt, and J. C. Vassilicos
44. *Computational aeronautical fluid dynamics* Edited by J. Fezoui, J. Hunt, and J. Periaux
45. *Cryptography and coding III* Edited by M. J. Ganley
46. *Parallel computation* Edited by A. E. Fincham and B. Ford
47. *Aerospace vehicle dynamics* Edited by M. V. Cook and M. J. Rycroft
48. *Computer-aided surface geometry and design: mathematics of surfaces IV* Edited by A. Bowyer
49. *Mathematics in signal processing III* Edited by J. G. McWhirter
50. *Design and application of curves and surfaces: mathematics of surfaces V* Edited by R. B. Fisher
51. *Artificial intelligence in mathematics* Edited by J. H. Johnson, S. McKee, and A. Vella
52. *Stably stratified flows: flow and dispersion over topography* Edited by I. P. Castro and N. Rockliffe
53. *Computational learning theory: EuroCOLT 93* Edited by J. S. Shawe-Taylor and M. Antony
54. *Complex stochastic systems and engineering* Edited by D. M. Titterington
55. *Mathematics of dependable systems* Edited by C. J. Mitchell and V. Stavridou
56. *The mathematics of deforming surfaces* Edited by D. G. Dritschel and R. J. Perkins
57. *Mathematical education of engineers* Edited by L. R. Mustoe and S. Hibberd
58. *The mathematics of surfaces VI* Edited by G. Mullineux
59. *Applications of finite fields* Edited by Dieter Gollmann
60. *Applications of combinatorial mathematics* Edited by C. J. Mitchell
61. *Image processing: mathematical methods and applications* Edited by J. M. Blackledge
62. *Flow and dispersion through groups of obstacles* Edited by R. J. Perkins and S. E. Belcher
63. *The state of the art in numerical analysis* Edited by I. S. Duff and G. A. Watson
64. *Mathematics of dependable systems II* Edited by V. Stavridou
65. *Approximations and numerical methods for the solution of Maxwell's equations* Edited by F. El Dabaghi, K. Morgan, A. K. Parrott and J. Periaux
66. *Mathematics of heat transfer* Edited by G. E. Tupholme and A. S. Wood
67. *Mathematics in signal processing IV* Edited by J. G. McWhirter and I. K. Proudler

Mathematics in Signal Processing IV

Based on the proceedings of a conference on Mathematics in Signal Processing. Organized by the Institute of Mathematics and its Applications and held at the University of Warwick in December 1996.

Edited by

J. G. McWHIRTER

Head of Signal Processing Group (Senior Fellow)
Defence Evaluation and Research Agency, Malvern

and

I. K. PROUDLER

Defence Evaluation and Research Agency, Malvern

CLARENDON PRESS • OXFORD • 1998

Oxford University Press, Great Clarendon Street, Oxford OX2 6DP

Oxford New York
Athens Auckland Bangkok Bogota Buenos Aires Calcutta
Cape Town Chennai Dar es Salaam Delhi Florence Hong Kong Istanbul
Karachi Kuala Lumpur Madrid Melbourne Mexico City Mumbai
Nairobi Paris São Paolo Singapore Taipei Tokyo Toronto Warsaw

and associated companies in
Berlin Ibadan

Oxford is a registered trade mark of Oxford University Press

Published in the United States by
Oxford University Press Inc., New York

A catalogue record for this book is available from the British Library

Library of Congress Cataloging in Publication Data
Data applied for
ISBN 0 19 850202 8 (hbk)

Typeset using LaTeX

Printed in Great Britain by
Bookcraft (Bath) Ltd,
Midsomer Norton, Avon

PREFACE

The fourth IMA conference on "Mathematics in Signal Processing" was held at the University of Warwick from 17th to 19th December 1996. This book comprises a selection of the papers presented at that meeting and provides a representative cross-section of the many topics discussed. Unfortunately, it was only possible to include about 50% of all papers presented at the conference and many good papers, worthy of publication had to be omitted. A selection was made by the editors giving priority to papers with a strong mathematical content, clear relevance to signal processing and tutorial value. We would like to offer our sincere apologies to the authors of papers which could not be included in this book.

The opening session of the conference was devoted to three invited talks on the theme of alternative algebras for signal processing. It got off to a great start with the keynote paper on Multilinear Algebra presented in a very lively manner by Professor Bart de Moor from the Katholieke Universiteit in Leuven. Multilinear algebra is already making a very significant impact in the context of signal processing based on Higher Order Statistics. The potential application of Geometric Algebra and Gröbner Bases, as discussed in the other two papers, is much more speculative but fascinating none the less.

The other conference sessions covered a wide range of topics from Array Processing and Digital Communications to Wavelets and Nonlinear Signal Processing. These, in turn, encompassed a rich variety of mathematical techniques including Padé Approximation, Convex Optimisation and Generalised Eigenvalue Decomposition. We hope that the reader will enjoy this intriguing blend of theory and application as much as we did.

Finally, we would like to thank: all members of the conference organising committee for their help in planning a very successful event and their contribution to evaluating the written papers for inclusion in this book; Pamela Bye, Debbie Brown and staff at the Institute for their assistance with the conference organisation and in the preparation of this volume; all authors who submitted papers to the conference.

Editors:
J.G. McWhirter and I.K. Proudler
Defence Evaluation and Research Agency, Malvern

Members of the Organising Committee:
Professor J.G. McWhirter F.Eng. (Chairman) (DERA, Malvern); Dr. I.K. Proudler (DERA, Malvern); Professor C. Budd (University of Bath); Professor O.R. Hinton (University of Newcastle); Dr. J.E. Hudson (Nortel, Harlow); Dr. M. McLeod (University of Cambridge); Professor M. Sandler (King's College, London); Dr. M. Moonen (Katholieke Universiteit, Leuven)

ACKNOWLEDGEMENTS

The Institute of Mathematics and its Applications would like to thank the following who contributed to the organisation of the conference and the production of the proceedings:

Professor John McWhirter (DERA) Editor of the proceedings and Chairman of the Conference Organising Committee, Dr. Ian Proudler who co-edited the proceedings and ran the conference web site, the members of the Conference Organising Committee, the authors of the papers, Mrs. Kay Pitt who provided the Editor with administrative support, and the following IMA staff for the organisation of the conference: Dr. Adrian Lepper (Executive Secretary) and Mrs. Pamela Bye (Conference Officer), and for the production of the proceedings: Mrs. Hilary Hill (Deputy Secretary), Miss Debbie Brown (Publications Officer), Mr. Simon Byles (Publications Officer) and Miss Karen Jenkins (Proceedings Officer).

CONTENTS

CONTRIBUTORS

P. BARONE; Istituto per le Applicazioni del Calcolo, CNR, Viale del Policlinico 137, 00161 Roma, Italy.

R.A. BATES; Department of Engineering, University of Warwick, Coventry, CV4 7AL.

D.S. BROOMHEAD; Department of Mathematics, University of Manchester Institute of Science and Technology, P.O. Box 88, Manchester, M60 1QD.

A. CANTONI; Australian Telecommunications Research Institute and Cooperative Research Centre for Broadband Telecommunications and Networking, Curtin University of Technology, Bentley, WA 6102, Australia.

J-F. CARDOSO; Départemen Signal, Ecole Nationale Supérieure des Télécommunications, Paris, France.

M.J. CHAPPELL; Department of Engineering, University of Warwick, Coventry, CV4 7AL.

L. DE LATHAUWER; Department of Electrical Engineering, University of Leuven, K. Mercierlaan 94, B-3001 Leuven, Belgium.

B. DE MOOR; Department of Electrical Engineering, University of Leuven, K. Mercierlaan 94, B-3001 Leuven, Belgium.

G.D. DE VILLIERS; E505, DERA Malvern, St. Andrews Road, Malvern, Worcestershire, WR14 3PS.

J-P. DELMAS; Départemen Signal et Image, Institut Nationale des Télécommunications, 91011 Evry, France.

H. FASSBENDER; Fachbereich 3 Mathematik und Informatik, Zentrum für Technomathematik, Universität Bremen, 28334 Bremen, Germany.

P. FITZPATRICK; Department of Mathematics, National University of Ireland, Cork, Ireland.

N. GARCÍA; Grupo de Tratamiento de Imágenes, E.T.S. Ingenieros de Telecomunicación, Universidad Politécnica de Madrid, E-28040 Madrid, Spain.

K.R. GODFREY; Department of Engineering, University of Warwick, Coventry, CV4 7AL.

S.J. GODSILL; Signal Processing and Communications Group, Department of Engineering, University of Cambridge, Trumpington Street, Cambridge, CB2 1PZ.

M. HARTENECK; Signal Processing Division, Department of Electrical and Electronic Engineering, University of Strathclyde, Glasgow, Scotland, G1 1XW.

S.D. HAYWARD; DERA Malvern, St. Andrews Road, Malvern, Worcestershire, WR14 3PS.

M.F. HILTON; Department of Respiratory Physiology, Birmingham Heartlands Hospital, Birmingham, B9 5SS.

J.P. HUKE; Department of Mathematics, University of Manchester Institute of Science and Technology, P.O. Box 88, Manchester, M60 1QD.

T. KAISER; Department of Communication Engineering, Gerhard-Mercator-Universität Duisburg, FB 9, Fachgebiet Nachrichtentechnik, Bismarckstrasse 81, BA 249, 47048 Duisburg, Germany.

M. KIRBY; Department of Mathematics, Colorado State University, Fort Collins, CO 80523, USA.

J. LASENBY; Department of Engineering, University of Cambridge, Trumpington Street, Cambridge, CB2 1PZ.

R.C. LE BORNE; Department of Mathematics, University of Tennessee at Chattanooga, Chattanooga, TN 37403-2598, USA.

T.S.T. LEUNG; Institute of Sound and Vibration, University of Southampton, Highfield, Southampton, SO17 1BJ.

I. LEVY; Department of Computer Science, University of Warwick, Coventry, CV4 7AL.

F.T. LUK; Department of Computer Science, Rensselaer Polytechnic Institute, Troy, New York 12180, USA.

R. MARCH; Istituto per le Applicazioni del Calcolo, CNR, Viale del Policlinico 137, 00161 Roma, Italy.

G.J. MARSEILLE; Department of Applied Physics, Delft University of Technology, P.O. Box 5046, 2600 GA Delft, The Netherlands.

R. MARTIN; GEC-Marconi Research Centre, Elstree Way, Borehamwood, Hertfordshire, WD6 1RX.

A. McLACHLAN; Neural Computing Research Group, Department of Computing Science and Applied Mathematics, Aston University, Birmingham, B4 7ET.

B. McNALLY; Department of Physics, King's College London, Strand, London, WC2R 2LS.

J.G. McWHIRTER; DERA Malvern, St. Andrews Road, Malvern, Worcestershire, WR14 3PS.

J.M. MENÉNDEZ; Grupo de Tratamiento de Imágenes, E.T.S. Ingenieros de Telecomunicación, Universidad Politécnica de Madrid, E-28040 Madrid, Spain.

B.G. MERTZIOS; Department of Electrical and Computer Engineering, Democritus University of Thrace, 67 100 Xanthi, Hellas, Greece.

P.R. MEULEMANS; Department of Computer Science, University of Warwick, Coventry, CV4 7AL.

M. MOONEN; ESAT - Katholieke Universiteit Leuven, K. Mercierlaan 94, 3001 Heverlee, Belgium.

C. MOSQUERA; Departamento Tecnologías de las Comunicaciones, ETSI Telecomunicación, Universidad de Vigo, 36200 Vigo, Spain.

M.R. MULDOON; Department of Mathematics, University of Manchester Institute of Science and Technology, P.O. Box 88, Manchester, M60 1QD.

F. PÉREZ GONZÁLEZ; Departamento Tecnologías de las Comunicaciones, ETSI Telecomunicación, Universidad de Vigo, 36200 Vigo, Spain.

E.R. PIKE; Department of Physics, King's College London, Strand, London, WC2R 2LS.

I.K. PROUDLER; DERA Malvern, St. Andrews Road, Malvern, Worcestershire, WR14 3PS.

S. QIAO; Department of Computer Science and Systems, McMaster University, Hamilton, Ontario, Canada, L8S 4K1.

R. RABENSTEIN; Lehrstuhl für Nachrichtentechnik I, Universität Erlangen-Nürnberg, Cauerstraße 7, D-91058 Erlangen, Germany.

E. RENDÓN; Grupo de Tratamiento de Imágenes, E.T.S. Ingenieros de Telecomunicación, Universidad Politécnica de Madrid, E-28040 Madrid, Spain.

L. SALGADO; Grupo de Tratamiento de Imágenes, E.T.S. Ingenieros de Telecomunicación, Universidad Politécnica de Madrid, E-28040 Madrid, Spain.

I.W. SELESNICK; Department of Electrical Engineering, Polytechnic University, 6 Metrotech Center, Brooklyn, NY 11201, USA.

I.M. SPILIOTIS; Department of Electrical and Computer Engineering, Democritus University of Thrace, 67 100 Xanthi, Hellas, Greece.

A.X.S. STEVENSON; Department of Engineering, University of Cambridge, Trumpington Street, Cambridge, CB2 1PZ.

R.W. STEWART; Signal Processing Division, Department of Electrical and Electronic Engineering, University of Strathclyde, Glasgow, Scotland, G1 1XW.

K.L. TEO; Australian Telecommunications Research Institute and Cooperative Research Centre for Broadband Telecommunications and Networking, Curtin University of Technology, Bentley, WA 6102, Australia.

J. TIAN; Computational Mathematics Laboratory, Rice University Houston, Texas 77005-1892, USA.

P.T. TROUGHTON; Signal Processing and Communications Group, Department of Engineering, University of Cambridge, Trumpington Street, Cambridge, CB2 1PZ.

P. VANDAELE; ESAT - Katholieke Universiteit Leuven, K. Mercierlaan 94, 3001 Heverlee, Belgium.

A-J. VAN DER VEEN; Department of Electrical Engineering, Delft University of Technology, Mekelweg 4, 2628 CD Delft, The Netherlands.

C. VAN MEIJEREN; Department of Applied Physics, Delft University of Technology, P.O. Box 5046, 2600 GA Delft, The Netherlands.

D. VAN ORMONDT; Department of Applied Physics, Delft University of Technology, P.O. Box 5046, 2600 GA Delft, The Netherlands.

B. VO; Australian Telecommunications Research Institute and Cooperative Research Centre for Broadband Telecommunications and Networking, Curtin University of Technology, Bentley, WA 6102, Australia.

R.O. WELLS, JR.; Computational Mathematics Laboratory, Rice University Houston, Texas 77005-1892, USA.

P.R. WHITE; Institute of Sound and Vibration, University of Southampton, Highfield, Southampton, SO17 1BJ.

R. WILSON; Department of Computer Science, University of Warwick, Coventry, CV4 7AL.

Z. ZANG; Australian Telecommunications Research Institute and Cooperative Research Centre for Broadband Telecommunications and Networking, Curtin University of Technology, Bentley, WA 6102, Australia.

From Matrix to Tensor: Multilinear Algebra and Signal Processing

L. De Lathauwer and B. De Moor

Department of Electrical Engineering, University of Leuven, Belgium

Abstract

In recent years research on Higher-Order Statistics has raised some fundamentally new insights in the area of mathematical signal processing. From an algebraic point of view this is reflected by an emerging interest in the assets of multinear algebra, and the development of numerical "tensor" tools, complementary to the arsenal of existing matrix techniques. We give an overview of some important tensor algebraic concepts, and some of their implications in signal processing.

The algebraic track of our argument starts from the question how the concepts of Eigenvalue Decomposition (EVD) and Singular Value Decomposition (SVD) could be generalized to the higher-order case. It appears that the rank properties of higher-order tensors, which are very different from their matrix counterparts, do not allow for an easy extension: instead, the generalization of different EVD/SVD-properties can lead to different multilinear decompositions. As far as orthogonal decompositions are concerned, we discuss the transformations to "all-orthogonality" resp. "maximal diagonality". In addition, we investigate the expansion of a higher-order tensor in a sum of non-orthogonal rank-1 components.

These multilinear techniques find a major application in the problem of Blind Source Separation, also known as Independent Component Analysis. Orthogonal tensor decompositions can be used to obtain the solution from the covariance matrix and a higher-order cumulant tensor of the data. As an alternative, non-orthogonal tensor decompositions allow to solve the problem by resorting only to the higher-order cumulant, in a way that is conceptually blind to additive Gaussian noise. A second basic type of problem that can be tackled with multilinear algebra, is the Blind Factor Analysis of multiway datasets. The signal processing part of the exposition is illustrated with real-life examples.

1 Introduction

Multilinear algebra is the algebra of higher-order tensors. In this paper an $(I_1 \times I_2 \times \ldots \times I_N)$-tensor is intuitively considered as an ordered $(I_1 \times I_2 \times \ldots \times I_N)$-table of numerical values, given with respect to the bases chosen in $\mathbb{R}^{I_1}$, $\mathbb{R}^{I_2}$, ..., $\mathbb{R}^{I_N}$. More formally, tensors can be defined as multilinear mappings over a set of vector spaces [1].

In the past, famous scientists like Gauss, Kronecker, Cayley, Weyl and Hilbert have indirectly contributed to the development of tensor algebra through research on homogeneous polynomials: their results can be casted in a tensor framework via the isomorphism between homogeneous polynomials and super-symmetric tensors (which are invariant under arbitrary index permutations), in the same way as quadratic forms and symmetric matrices are related. An interesting discussion of the early literature can be found in [2]. Recently, (numerical) multilinear algebra has received a great deal of attention through research in Higher-Order Statistics (HOS)(for a fairly recent overview of workshops, tutorials, etc. on this topic, we refer to [3]), where the basic quantities (higher-order moments, cumulants and spectra) of a stochastic vector are super-symmetric higher-order tensors [4].

Literature in fundamental multilinear algebra is often based on the "index-notation" and "Einstein summation convention" (for example [5], Chapters 1-3). It is a matter of practice to get acquainted with this notation, and to learn to benefit from its assets. For this introductory paper however, we have chosen to use a matrix-like notation.

In Section 2 we briefly introduce some basic material, such as the definition of scalar product and outer product, the concept of tensor rank, and the way in which a tensor changes by multiplication with a matrix. Sections 3 and 4 deal with orthogonal transformations of a tensor to make it "all-orthogonal" or "maximally diagonal" respectively. Section 5 discusses a non-orthogonal tensor transformation, used to obtain an expansion in a minimal number of rank-1 components. Section 6 explains how these tensor decompositions can be used in Independent Component Analysis (ICA), which is a refinement of the concept of Principal Component Analysis (PCA). In Section 7 we describe a second fundamentally new signal processing tool, namely the Blind Factor Analysis (BFA) of multiway datasets. The exposition is intended to give a very elementary introduction to the basic aspects of the topics that we have mentioned; for more details the reader is invited to consult the references.

All the concepts described in this paper, are formulated in terms of real-valued tensors. The generalization to the complex case is always possible, but from time to time it is more involved from a notational as well as computational point of view.

2 Basic definitions

First, the definition of an outer product generalizes expressions of the type AB^T, in which A and B are vectors:

Definition 1. (Outer product) *The outer product $\mathcal{A} \circ \mathcal{B}$ of a tensor $\mathcal{A} \in \mathbb{R}^{I_1 \times I_2 \times \ldots \times I_P}$ and a tensor $\mathcal{B} \in \mathbb{R}^{J_1 \times J_2 \times \ldots \times J_Q}$, is defined by*

$$(\mathcal{A} \circ \mathcal{B})_{i_1 i_2 \ldots i_P j_1 j_2 \ldots j_Q} \stackrel{\text{def}}{=} a_{i_1 i_2 \ldots i_P} b_{j_1 j_2 \ldots j_Q}$$

for all values of the indices.

Next, we give straightforward generalizations of the scalar product, orthogonality and Frobenius-norm:

Definition 2. (Scalar product) *The scalar product* $\langle \mathcal{A}, \mathcal{B} \rangle$ *of tensors* $\mathcal{A}, \mathcal{B} \in \mathbb{R}^{I_1 \times I_2 \times \ldots \times I_N}$ *is defined as*

$$\langle \mathcal{A}, \mathcal{B} \rangle \stackrel{\text{def}}{=} \sum_{i_1} \sum_{i_2} \cdots \sum_{i_N} b_{i_1 i_2 \ldots i_N} a_{i_1 i_2 \ldots i_N}.$$

Definition 3. (Orthogonality) *Tensors of which the scalar product equals* 0, *are mutually orthogonal.*

Definition 4. (Frobenius-norm) *The Frobenius-norm of a tensor* $\mathcal{A}$ *is given by*

$$||\mathcal{A}|| \stackrel{\text{def}}{=} \sqrt{\langle \mathcal{A}, \mathcal{A} \rangle}.$$

Multiplications of a higher-order tensor with a matrix can be defined as follows:

Definition 5. *The mode-n product of a tensor* $\mathcal{A} \in \mathbb{R}^{I_1 \times I_2 \times \ldots \times I_N}$ *by a matrix* $\mathbf{U} \in \mathbb{R}^{J_n \times I_n}$, *denoted by* $\mathcal{A} \times_n \mathbf{U}$, *is an* $(I_1 \times I_2 \times \ldots \times I_{n-1} J_n \times I_{n+1} \ldots \times I_N)$*-tensor defined by*

$$(\mathcal{A} \times_n \mathbf{U})_{i_1 i_2 \ldots j_n \ldots i_N} = \sum_{i_n} a_{i_1 i_2 \ldots i_n \ldots i_N} u_{j_n i_n}$$

for all index values.

In this notation, the matrix product $\mathbf{A} = \mathbf{U}^{(1)} \cdot \mathbf{B} \cdot \mathbf{U}^{(2)^T}$ takes the form of the "symmetric" expression $\mathbf{A} = \mathbf{B} \times_1 \mathbf{U}^{(1)} \times_2 \mathbf{U}^{(2)}$, reflecting the fact that $\mathbf{U}^{(2)}$ acts in exactly the same way on the columns of $\mathbf{B}$ as $\mathbf{U}^{(1)}$ does for the rows. The mode-n product allows to express the effect of a basis transformation in $\mathbb{R}^{I_n}$ on the tensor $\mathcal{A}$.

In tensor terminology, column vectors, row vectors, ... will be denoted as mode-1 vectors, mode-2 vectors, etc. The definitions of column rank, row rank and rank of a matrix can then be generalized as follows:

Definition 6. (Mode-n rank) *The mode-n vectors of a tensor* $\mathcal{A} \in \mathbb{R}^{I_1 \times I_2 \times \ldots \times I_N}$ *are the* I_n*-dimensional vectors obtained from* $\mathcal{A}$ *by varying the index* i_n *and keeping the other indices fixed. The mode-n rank of* $\mathcal{A}$, *denoted by* $R_n = \mathtt{rank_n}(\mathcal{A})$, *is the dimension of the vector space generated by the mode-n vectors.*

Definition 7. (Rank-1 tensor) *An* N*th-order tensor* $\mathcal{A}$ *has rank* 1 *when it equals the outer product of* N *vectors* $U^{(1)}$, $U^{(2)}$, ..., $U^{(N)}$:

$$\mathcal{A} = U^{(1)} \circ U^{(2)} \circ \ldots \circ U^{(N)}.$$

Definition 8. (Rank) *The rank of an arbitrary Nth-order tensor* $\mathcal{A}$, *denoted by* $R = \texttt{rank}(\mathcal{A})$, *is the minimal number of rank-1 tensors that yield* $\mathcal{A}$ *in a linear combination.*

Higher-order rank-related properties are thoroughly different from their matrix counterparts. The various mode-n ranks of a given tensor are not necessarily the same, and can also be different from the rank. From the definitions we have that always $R_n \leq R$. Some examples can for instance be found in [6].

3 The Higher-Order SVD

A first way to generalize the SVD of matrices, is as follows:

Definition 9. (HOSVD) *The Higher-Order Singular Value Decomposition (HOSVD) of a real* $(I_1 \times I_2 \times \ldots \times I_N)$*-tensor* $\mathcal{A}$ *is given by:*

$$\mathcal{A} = \mathcal{S} \times_1 \mathbf{U}^{(1)} \times_2 \mathbf{U}^{(2)} \ldots \times_N \mathbf{U}^{(N)}, \tag{3.1}$$

in which:

- $\mathbf{U}^{(n)} = \left[U_1^{(n)} U_2^{(n)} \ldots U_{I_n}^{(n)}\right]$ *is an orthogonal* $(I_n \times I_n)$*-matrix* $(1 \leq n \leq N)$.
- $\mathcal{S}$ *is a real* $(I_1 \times I_2 \times \ldots \times I_N)$*-tensor of which the subtensors* $\mathcal{S}_{i_n=\alpha}$, *obtained by fixing the nth index to* α, *have the properties of:*
  - *all-orthogonality: two subtensors* $\mathcal{S}_{i_n=\alpha}$ *and* $\mathcal{S}_{i_n=\beta}$ *are orthogonal for all possible values of* n, α *and* β *subject to* $\alpha \neq \beta$:

    $$\langle \mathcal{S}_{i_n=\alpha}, \mathcal{S}_{i_n=\beta} \rangle = 0 \quad \text{when} \quad \alpha \neq \beta$$

  - *ordering:*

    $$||\mathcal{S}_{i_n=1}|| \geq ||\mathcal{S}_{i_n=2}|| \geq \ldots \geq ||\mathcal{S}_{i_n=I_n}|| \geq 0$$

    for all possible values of n.

The Frobenius-norms $||\mathcal{S}_{i_n=i}||$, *symbolized by* $\sigma_i^{(n)}$, *are called mode-n singular values of* $\mathcal{A}$ *and the vector* $U_i^{(n)}$ *is an ith mode-n singular vector. The decomposition is visualized for third-order tensors in Figure 1.*

This decomposition can be obtained by computing any $\mathbf{U}^{(n)}$ as the left singular matrix of a matrix $\mathbf{A}_{(n)}$, in which all the mode-n vectors are stacked as columns; the tensor $\mathcal{S}$ then follows from Equation (3.1). A consequence of this strong link with the matrix decomposition, is that many properties of the tensor decomposition show a strong analogy with the matrix case as well (uniqueness, link with EVD, perturbation properties, ...) For this reason, it was proposed to denote the decomposition as *the* HOSVD in [7].

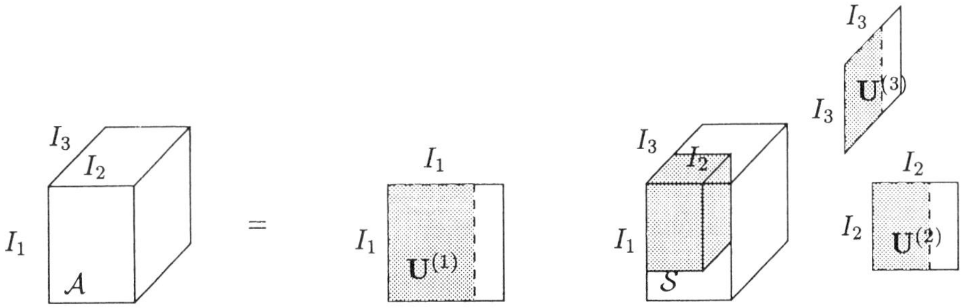

Figure 1. Visualization of the HOSVD for a third-order tensor

Imposing the condition of all-orthogonality, instead of diagonality, on tensor $\mathcal{S}$, implies that the HOSVD is always defined. As a matter of fact, $\mathcal{S}$ cannot be diagonal in general, which means that the HOSVD does not necessarily reveal the rank of $\mathcal{A}$: in the cases where $\mathcal{S}$ is diagonal, the orthogonality of the matrices of mode-n singular vectors implies that

$$\mathcal{A} = \sum_{i_n}^{R_n} s_{i_n i_n \ldots i_n} U_{i_n}^{(1)} \circ U_{i_n}^{(2)} \circ \ldots \circ U_{i_n}^{(N)}$$

is a decomposition in a minimal number of rank-1 terms. On the other hand, the number of non-zero (significant) mode-n singular values corresponds to the mode-n rank (in a numerical sense) of $\mathcal{A}$.

4 "Maximal diagonality"

Another way to deal with the fact that a generic tensor cannot be diagonalized by orthogonal transformations, when looking for a tensorial kind of SVD, could be to claim that the transformed tensor is "as diagonal as possible" (in least-squares sense):

Definition 10. *For every real $(I_1 \times I_2 \times \ldots \times I_N)$-tensor $\mathcal{A}$, the decomposition*

$$\mathcal{A} = \mathcal{S} \times_1 \mathbf{U}^{(1)} \times_2 \mathbf{U}^{(2)} \ldots \times_N \mathbf{U}^{(N)}, \tag{4.1}$$

in which:

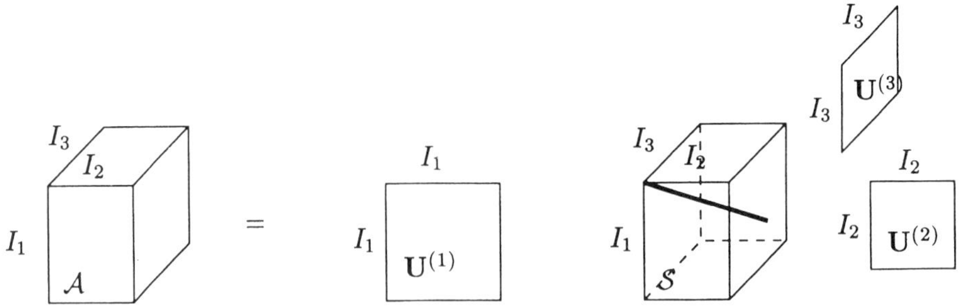

Figure 2. Visualization of the optimal diagonalization, in least-squares sense, of a third-order tensor by orthogonal transformations

- $\mathbf{U}^{(n)} = \left[U_1^{(n)} U_2^{(n)} \ldots U_{I_n}^{(n)} \right]$ *is an orthogonal* $(I_n \times I_n)$*-matrix* $(1 \leq n \leq N)$.
- $\mathcal{S}$ *is a real* $(I_1 \times I_2 \times \ldots \times I_N)$*-tensor with the following properties:*
  - *maximal diagonality (in least-squares sense): for all* $\mathcal{S}'$*, obtained by an orthogonal decomposition of* $\mathcal{A}$*, we have that* $\sum_i |s_{ii\ldots i}|^2 \geq |s'_{ii\ldots i}|^2$,
  - *ordering:*
  $$|s_{11\ldots 1}| \geq |s_{22\ldots 2}| \geq \ldots \geq |s_{II\ldots I}|,$$
  in which $I = \mathtt{min}\{I_1, I_2, \ldots, I_N\}$,

defines a generalized SVD. The diagonal entries $s_{ii\ldots i}$ *and the vectors* $U_i^{(n)}$ *can be considered as higher-order singular values respectively singular vectors of* $\mathcal{A}$*. The decomposition is visualized for third-order tensors in Figure 2.*

The computation of the decomposition, for super-symmetric higher-order tensors, was first studied by Comon, who proposed a Jacobi-type iteration algorithms [8]. Related computation schemes have been explored in [9] and [10].

5 Canonical Decomposition

An expansion of a tensor $\mathcal{A}$ as a sum of $\mathtt{rank}(\mathcal{A})$ rank-1 terms, generically involves non-orthogonal components:

Definition 11. (CANDECOMP) *A Parallel Factors (PARAFAC) Decomposition or Canonical Decomposition (CANDECOMP) of a real* $(I_1 \times I_2 \times \ldots \times I_N)$*-tensor* $\mathcal{A}$ *is a decomposition of* $\mathcal{A}$ *as a linear combination of a minimal number of rank-1 terms:*

$$\mathcal{A} = \sum_r^R \lambda_r \, U_r^{(1)} \circ U_r^{(2)} \circ \ldots \circ U_r^{(N)}. \tag{5.1}$$

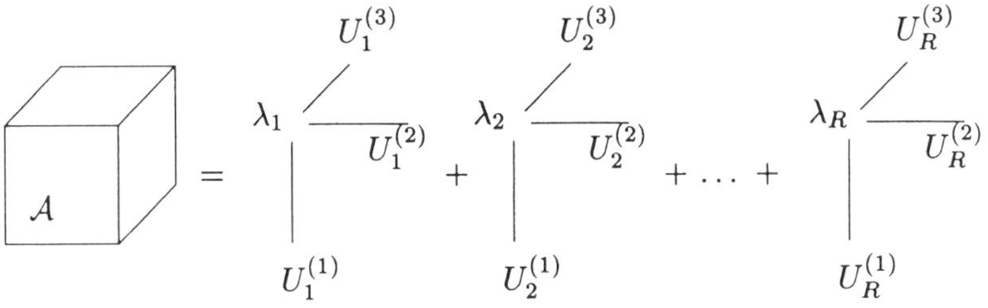

Figure 3. Visualization of the CANDECOMP for a third-order tensor

The decomposition is visualized for third-order tensors in Figure 3.

Even if non-orthogonal rank-1 terms are allowed, the minimal number of terms is in general not bounded by $\texttt{min}\{I_1, I_2, \ldots, I_N\}$. For generic rank values and rank upperbounds we refer to [2] (the paper addresses the case of super-symmetric tensors, but many theorems can easily be generalized for unsymmetric tensors). For computational aspects we refer to [13, 14, 15, 6].

The uniqueness properties of the CANDECOMP are very different from (and also much more complicated than) their matrix equivalents. Where a decomposition of a matrix $\mathbf{A}$ as a sum of $\texttt{rank}(\mathbf{A})$ rank-1 terms is usually made unique - up to some trivial indeterminacies - by imposing orthonormality constraints, the uniqueness of the CANDECOMP depends on conditions of linear independence. As an example, we assume the special but important case of an $(I \times I \times I)$-tensor $\mathcal{A}$ with rank $R = I$. It can be verified that the CANDECOMP of $\mathcal{A}$ is essentially unique if for instance (a) the set $\{U_r^{(1)}\}_{(1 \leq r \leq R)}$ is linearly independent, (b) the set $\{U_r^{(2)}\}_{(1 \leq r \leq R)}$ is linearly independent, and (c) the set $\{U_r^{(3)}\}_{(1 \leq r \leq R)}$ does not contain collinear vectors [16]. [17] contains a more general study of the uniqueness in the third-order case.

6 Independent Component Analysis

Where PCA relies on second-order statistical information, ICA involves the use of HOS. In a first subsection we introduce some basic elements of HOS. The second subsection deals with the topic of ICA itself.

6.1 Moments and cumulants of a stochastic vector

Moment tensors of a real stochastic vector are defined as follows:

Definition 12. (Moment) *The Nth order moment tensor $\mathcal{M}_X^{(N)}$ of a real stochastic vector X is defined by the element-wise equation:*

$$(\mathcal{M}_X^{(N)})_{i_1 i_2 \dots i_N} = \mathtt{Mom}(X_{i_1}, X_{i_2}, \dots, X_{i_N}) \stackrel{\text{def}}{=} \mathtt{E}\{X_{i_1} X_{i_2} \dots X_{i_N}\}. \tag{6.1}$$

$\mathtt{E}$ denotes the expectation. The first-order moment is the mean of the stochastic vector. The second-order moment is the correlation matrix.

On the other hand, cumulants of a real stochastic vector are defined as follows:

Definition 13. (Cumulant) *The Nth order cumulant tensor $\mathcal{C}_X^{(N)}$ of a real stochastic vector X is defined by the element-wise equation [18]:*

$$(\mathcal{C}_X^{(N)})_{i_1 i_2 \dots i_N} = \mathtt{Cum}(X_{i_1}, X_{i_2}, \dots, X_{i_N})$$

$$\stackrel{\text{def}}{=} \sum (-1)^N (N-1)!\, \mathtt{E}\{\prod_{i \in \mathtt{A}_1} X_i\}\, \mathtt{E}\{\prod_{i \in \mathtt{A}_2} X_i\} \dots \mathtt{E}\{\prod_{i \in \mathtt{A}_K} X_i\} \tag{6.2}$$

where the summation involves all possible partitions $\{\mathtt{A}_1, \mathtt{A}_2, \dots, \mathtt{A}_K\}$ $(1 \leqslant K \leqslant N)$ of the integers $\{i_1, i_2, \dots, i_N\}$. For a real zero-mean stochastic vector X the cumulants up to order four are explicitly given by:

$$(C_X)_i = \mathtt{Cum}(X_i) \stackrel{\text{def}}{=} \mathtt{E}\{X_i\} \tag{6.3}$$

$$(\mathbf{C}_X)_{i_1 i_2} = \mathtt{Cum}(X_{i_1}, X_{i_2}) \stackrel{\text{def}}{=} \mathtt{E}\{X_{i_1} X_{i_2}\} \tag{6.4}$$

$$(\mathcal{C}_X^{(3)})_{i_1 i_2 i_3} = \mathtt{Cum}(X_{i_1}, X_{i_2}, X_{i_3}) \stackrel{\text{def}}{=} \mathtt{E}\{X_{i_1} X_{i_2} X_{i_3}\} \tag{6.5}$$

$$\begin{aligned}(\mathcal{C}_X^{(4)})_{i_1 i_2 i_3 i_4} = \mathtt{Cum}(X_{i_1}, X_{i_2}, X_{i_3}, X_{i_4}) \stackrel{\text{def}}{=}\ & \mathtt{E}\{X_{i_1} X_{i_2} X_{i_3} X_{i_4}\} \\ & - \mathtt{E}\{X_{i_1} X_{i_2}\} \mathtt{E}\{X_{i_3} X_{i_4}\} \\ & - \mathtt{E}\{X_{i_1} X_{i_3}\} \mathtt{E}\{X_{i_2} X_{i_4}\} \\ & - \mathtt{E}\{X_{i_1} X_{i_4}\} \mathtt{E}\{X_{i_2} X_{i_3}\}.\end{aligned} \tag{6.6}$$

For every component X_i of X that has a non-zero mean, X_i has to be replaced in these formulas, except Equation (6.3), by $X_i - \mathtt{E}\{X_i\}$.

Again, the first-order cumulant is the mean of the stochastic vector. The second-order cumulant is the covariance matrix.

Some crucial properties are [4, 19]:

- *Super-symmetry:* moments and cumulants are symmetric in their arguments, i.e.

$$(\mathcal{M}_X^{(N)})_{i_1 i_2 \dots i_N} = (\mathcal{M}_X^{(N)})_{\mathtt{P}(i_1 i_2 \dots i_N)} \tag{6.7}$$

$$(\mathcal{C}_X^{(N)})_{i_1 i_2 \dots i_N} = (\mathcal{C}_X^{(N)})_{\mathtt{P}(i_1 i_2 \dots i_N)} \tag{6.8}$$

  in which $\mathtt{P}$ is an arbitrary permutation of the indices.

- *Multilinearity:* if a real stochastic vector X is transformed into a stochastic vector $\tilde{X}$ by a matrix multiplication $\tilde{X} = \mathbf{A} \cdot X$, with $\mathbf{A} \in \mathbb{R}^{J \times I}$, then we have:

$$\mathcal{M}_{\tilde{X}}^{(N)} = \mathcal{M}_X^{(N)} \times_1 \mathbf{A} \times_2 \mathbf{A} \ldots \times_N \mathbf{A} \tag{6.9}$$
$$\mathcal{C}_{\tilde{X}}^{(N)} = \mathcal{C}_X^{(N)} \times_1 \mathbf{A} \times_2 \mathbf{A} \ldots \times_N \mathbf{A}. \tag{6.10}$$

- *Partitioning of independent variables:* if a subset of I stochastic variables $X_1, \ldots, X_I$ is independent of the other variables, then we have:

$$\mathtt{Cum}(X_1, X_2, \ldots, X_I) = 0. \tag{6.11}$$

  This property does not hold in general for moments. A consequence of this property is that a higher-order cumulant of a stochastic vector with mutually independent components, is a diagonal tensor.

- *Sum of independent variables:* if the stochastic variables $X_1, X_2, \ldots, X_I$ are mutually independent from the stochastic variables $Y_1, Y_2, \ldots, Y_I$, then we have:

$$\mathtt{Cum}(X_1 + Y_1, X_2 + Y_2, \ldots, X_k + Y_k) = \mathtt{Cum}(X_1, X_2, \ldots, X_k) + \mathtt{Cum}(Y_1, Y_2, \ldots, Y_k). \tag{6.12}$$

  The cumulant tensor of a sum of independent random vectors is the sum of the individual cumulants. This property does not hold for moments either; as a matter of fact, it explains the term "cumulant".

- *Non-Gaussianity:* if Y is a Gaussian variable with the same mean and variance as a given stochastic variable X, then holds for $N \geq 3$:

$$\mathcal{C}_X^{(N)} = \mathcal{M}_X^{(N)} - \mathcal{M}_Y^{(N)}. \tag{6.13}$$

  Higher-order cumulants of a Gaussian variable are 0.

6.2 Independent Component Analysis

Assume the following basic linear statistical model:

$$Y = \mathbf{M}X + N \tag{6.14}$$

in which $Y \in \mathbb{R}^I$ is denoted as the *observation vector*, $X \in \mathbb{R}^J$ is called the *source vector* and $N \in \mathbb{R}^I$ represents additive *noise*. $\mathbf{M} \in \mathbb{R}^{I \times J}$ is the *mixing matrix.*

The concept of Independent Component Analysis can be formulated as follows:

The goal of Independent Component Analysis (ICA), or Blind Source Separation (BSS), consists of the estimation of the transfer matrix $\mathbf{M}$ *and/or the corresponding realizations of the source vector* X, *given only realizations of the output vector* Y, *under the following assumptions:*

- *The columns of* $\mathbf{M}$ *are linearly independent.*
- *The components of* X *are mutually statistically independent, as well as statistically independent from the noise components.*

The second assumption is the key ingredient for ICA. It is a very strong hypothesis, but also quite natural in lots of applications. From an algebraic point of view it implies that:

$$\mathbf{C}_Y = \mathbf{C}_X \times_1 \mathbf{M} \times_2 \mathbf{M} + \mathbf{C}_N \tag{6.15}$$

$$\mathcal{C}_Y^{(n)} = \mathcal{C}_X^{(n)} \times_1 \mathbf{M} \times_2 \mathbf{M} \ldots \times_n \mathbf{M} + \mathcal{C}_N^{(n)} \tag{6.16}$$

in which $\mathbf{C}_X$ and $\mathcal{C}_X^{(n)}$ are diagonal, and in which $\mathcal{C}_N^{(n)}$ vanishes if the noise is Gaussian.

It is impossible to determine the norm of columns of $\mathbf{M}$ in Equation (6.14), since a rescaling of these vectors can be compensated by the inverse scaling of the source signal values. Similarly the ordering of the source signals, having no physical meaning, cannot be identified. For non-Gaussian sources, these indeterminacies are the only way in which an ICA-solution is not unique [8, 20].

The ICA-assumptions do not allow to distinguish between the signal and the noise term in Equation (6.14). Hence the source signals will be estimated as $\hat{X}$, by a simple matrix multiplication:

$$\hat{X} = \mathbf{W}^T Y. \tag{6.17}$$

As an example, $\mathbf{W}^T$ can take the form of $\hat{\mathbf{M}}^\dagger$, in which $\hat{\mathbf{M}}$ is an estimate of the transfer matrix. More generally various beamforming strategies [21] can be applied.

Equation (6.15) leads to a classical *Principal Component Analysis* (PCA), which only allows to estimate the sources as well as the mixing matrix up to an orthogonal transformation. To illustrate this, let us assume that the sources have unit variance. Then we have (we omit the noise term at this point, for clarity):

$$\mathbf{C}_Y = \mathbf{M}\mathbf{M}^T. \tag{6.18}$$

Substitution of the SVD of the mixing matrix $\mathbf{M} = \mathbf{U}\mathbf{S}\mathbf{V}^T$ shows that the EVD of the observed covariance, as in PCA, allows to estimate the column space of $\mathbf{M}$ while the factor $\mathbf{V}$ remains unknown:

$$\mathbf{C}_Y = \mathbf{U}\mathbf{S}^2\mathbf{U}^T = (\mathbf{U}\mathbf{S})(\mathbf{U}\mathbf{S})^T. \tag{6.19}$$

Substitution of the PCA-subresults in Equation (6.16) leads to an orthogonal transformation of a higher-order cumulant tensor into a diagonal tensor:

$$\mathcal{C}_Z^{(n)} = \mathcal{C}_X^{(n)} \times_1 \mathbf{V}^T \times_2 \mathbf{V}^T \ldots \times_n \mathbf{V}^T \tag{6.20}$$

(up to the noise term), where $Z \stackrel{\text{def}}{=} \mathbf{S}^\dagger \mathbf{U}^T Y$. Theoretically the unknown orthogonal factor can thus be estimated via an HOSVD (diagonality is a special

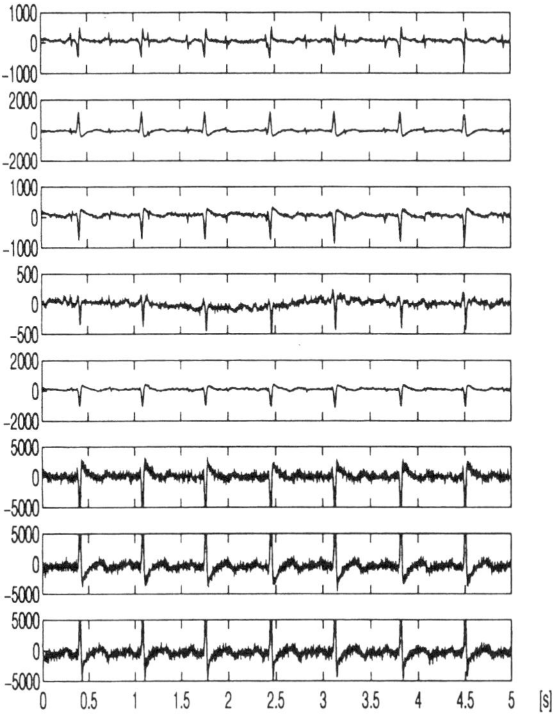

Figure 4. 8-channel set of cutaneous potential recordings

case of all-orthogonality) [22]. However, when dealing with sample statistics, and taking the noise terms into account, the "maximal diagonality" approach is more robust [23, 8].

On the other hand, up to the noise term, Equation (6.16) is a CANDECOMP of the cumulant of Y, which leads to higher-order-only solutions to the ICA-problem, conceptually blind for additive Gaussian noise [13].

It is clear that ICA is a very basic problem; its application range is very wide. The most well-known application is perhaps blind spatial multiplexing in telecommunications (see for example [24], in which an efficient algorithm based on simultaneous matrix diagonalization is presented). Figures 4 and 5 give a biomedical example [25]. Figure 4 displays 8 channels of cutaneous potential signals of a pregnant woman. The large pulses correspond to the mother electrocardiogram (ECG); channels 1 to 3 clearly contain an additional weaker signal, which corresponds to the fetal ECG. Figure 5 displays the sources estimated with the ICA-algorithm of [8]. Channels 1 to 3, and to some extent also channel 4, contain the components of the mother ECG. Channels 6 and 8 show the fetal

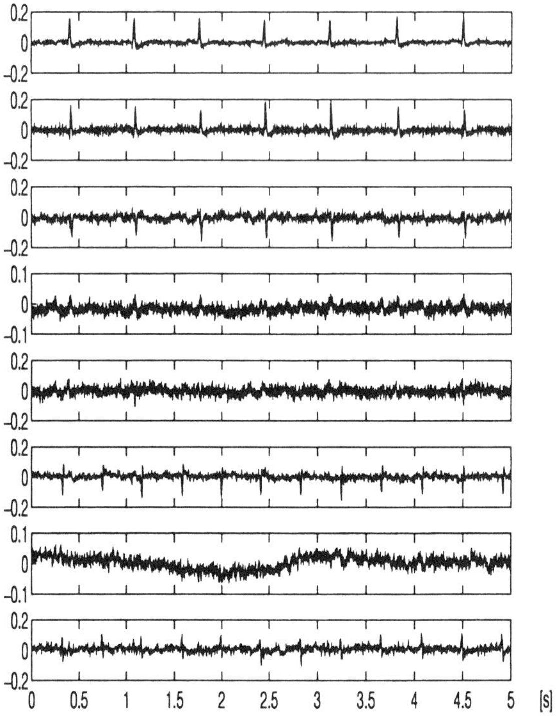

Figure 5. Source estimates obtained via ICA

ECG. The fact that the mixing matrix has been estimated as well, could be used to find better electrode locations, in this application.

7 Blind Factor Analysis

The decomposition of a matrix dataset as a sum of rank-1 terms is sometimes denoted as the *Factor Analysis* problem. With the decomposition, one aims at relating the different rank-1 terms to the different "physical mechanisms" that have contributed to the dataset. However, the diagonalization of a matrix by equivalence is, as such, essentially underdetermined. The extra conditions (maximal variance, orthonormality, ...) that are usually imposed to guarantee uniqueness, are not seldom physically irrelevant, or require a priori knowledge. In a wide range of parameters, this is not the case for the higher-order equivalent problem, i.e. the CANDECOMP of a higher-order dataset; the weaker conditions of linear independence to ensure uniqueness, often have a physical meaning. This makes the CANDECOMP of higher-order tensors to an interesting signal processing tool.

We illustrate the technique by means of an application in Chemometrics [26, 27]. In laboratoria for chemical analysis, one makes frequently use of bilinear equipment (tandem massa spectrometry, 2D-NMR, ...): the output of the device is a matrix, formed by rank-1 contributions of the various chemical components. For example for emission-excitation fluorescence data, each row is an emission spectrum and each column an excitation scan or vice-versa. Theoretically these row- and column-vectors are linear combinations of the emission and excitation spectra of the components, which have to be determined. At a matrix-level however, an infinite number of matrix decompositions is possible, and the chemical analysis cannot be realized without a priori knowledge. On the other hand, it is often necessary to examine several samples. If its identifiability conditions are satisfied, then the CANDECOMP of the corresponding third-order dataset allows to determine the components and their concentrations in a blind way.

8 Conclusion

In 1995, there was an LA-NET discussion on the question whether numerical linear algebra was dead (it wasn't). We hope to have demonstrated that numerical multinear algebra at least, is a new challenging and important field of research. In this summary paper we have briefly proposed three tensor generalizations of the SVD of matrices, each focusing on a different aspect of the matrix decomposition. Independent Component Analysis and Blind Factor Analysis are two examples of new signal processing methodologies based on tensor techniques.

Acknowledgements

This research has been supported by the Flemish Government (Concerted Research Action MIPS), the Fund for Scientific Research - Flanders (G.0292.95), the Belgian Government (Interuniversity Attraction Poles IUAP-17 and IUAP-50), the European Commission (Human Capital and Mobility Network SIMONET) and the European Community Research program ESPRIT, Basic Research Working Group nr. 6620 (ATHOS) and the Flemish Institute for Support of Scientific-Technological Research in Industry (I.W.T). L. De Lathauwer is a Research Assistant supported by the I.W.T.; B. De Moor is a Research Associate of the F.W.0. and an Associate Professor at the K.U.Leuven.

Bibliography

1. Lang, S. (1984). *Algebra*, Addison Wesley, Reading.

2. Comon, P. and Mourrain, B. (1996). Decomposition of quantics in sums of powers of linear forms. *Signal Processing*, Special Issue *Higher Order Statistics*, **53**, Nos. 2-3, 93-108.

3. Swami, A., Giannakis, G. (1996) Editorial. Higher-order statistics. *Signal Processing*, Special Issue *Higher Order Statistics*, **53**, Nos. 2-3, 89-91.

4. McCullagh, P. (1987). *Tensor Methods in Statistics*, Chapman and Hall, London.

5. Kay, D.C. (1988). *Theory and Problems of Tensor Calculus*, McGraw-Hill.

6. De Lathauwer, L. (1997). *Signal Processing with Multilinear Algebra*, Ph.D. thesis, K.U.Leuven, E.E.Dept.-ESAT, Belgium (in preparation).

7. De Lathauwer, L., De Moor, B. and Vandewalle, J. (1994). A singular value decomposition for higher-order tensors. ESAT/SISTA report 94-31, K.U.Leuven, submitted to: *SIAM J. Matrix Anal. Appl.*.

8. Comon, P. (1994). Independent component analysis, a new concept? *Signal Processing*, Special Issue *Higher Order Statistics*, **36**, No. 3, 287-314.

9. De Lathauwer, L., Comon, P., De Moor, B. and Vandewalle, J. (1995). Higher-order power method - Application in independent component analysis. *Proc. 1995 Int. Symp. on Nonlinear Theory and Its Applications, NOLTA'95*, **1**, Las Vegas, USA, Dec. 10-14, 91-96.

10. De Lathauwer, L., De Moor, B. and Vandewalle, J. (1996). Blind source separation by simultaneous third-order tensor diagonalization. *Proc. EUSIPCO-96*, **3**, Trieste, Italy, Sept. 10-13, 2089-2092.

11. Carroll, J.D. and Pruzansky, S. (1984). The CANDECOMP-CANDELINC family of models and methods for multidimensional data analysis. *Research Methods for Multimode Data Analysis*, Editors: H.G. Law, C.W. Snyder, J.A. Hattie and R.P. McDonald, Praeger, N.Y., 372-402.

12. Harshman, R.A. and Lundy, M.E. (1984). The PARAFAC model for three-way factor analysis and multidimensional scaling. *Research Methods for Multimode Data Analysis*, Editors: H.G. Law, C.W. Snyder, J.A. Hattie and R.P. McDonald, Praeger, N.Y., 122-215.

13. De Lathauwer, L., De Moor, B. and Vandewalle, J. (1996). Independent component analysis based on higher-order statistics only. *Proc. 8th IEEE SP Workshop on Stat. Signal and Array Process., SSAP-96*, Corfu, Greece, June 24-26, 356-359.

14. Cardoso, J.-F. (1992). Iterative techniques for blind source separation using only fourth-order cumulants. *Proc. EUSIPCO-92*, **2**, Brussels, Aug. 24-27, 739-742.

15. Cardoso, J.-F. (1991). Super-symmetric decomposition of the fourth-order cumulant tensor. Blind identification of more sources than sensors. *Proc. ICASSP-91*, **5**, 3109-3112.

16. Leurgans, S.E., Ross, R.T., and Abel, R.B. (1993). A decomposition for three-way arrays. *SIAM J. Matrix Anal. Appl.*, **14**, No. 4, 1064-1083.

17. Kruskal, J.B. (1977). Three-way arrays: rank and uniqueness of trilinear decompositions, with application to arithmetic complexity and statistics. *Lin. Alg. Appl.*, **18**, 95-138.

18. Leonov, V.P. and Shiryaev, A. (1959). On the technique of computing semi-invariants. *Theory of Probability Applications*, **4**, 319-329.

19. Nikias, C.L. and Mendel, J.M. (1993). Signal processing with higher-order spectra. *IEEE Signal Proc. Mag.*, 10-37.

20. Tong, L. et al. (1991). Indeterminacy and identifiability of blind identification. *IEEE Trans. on Circuits and Systems*, **38**, No. 5, 499-509.

21. Van Veen, B.D. and Buckley, K.M. (1988). Beamforming: a versatile approach to spatial filtering. *IEEE ASSP Magazine*, April, 4-24.

22. De Lathauwer, L., De Moor, B. and Vandewalle, J. (1994). Blind source separation by higher-order singular value decomposition. *Proc. EUSIPCO-94*, **1**, Edinburgh, Scotland, U.K., Sept. 13-16, 175-178.

23. Lacoume, J.L. and Ruiz, P. (1992). Separation of independent sources from correlated inputs. *IEEE Trans. on Signal Process.*, **40**, No. 12, 3074-3078.

24. Cardoso, J.-F. and Souloumiac, A. (1994). Blind beamforming for non-Gaussian signals. *IEE Proc.-F*, **140**, No. 6, 362-370.

25. De Lathauwer, L., De Moor, B. and Vandewalle, J. (1995). Fetal electrocardiogram extraction by source subspace separation. *Proc. IEEE SP / ATHOS Workshop on HOS*, Girona, Spain, June 12-14, 134-138.

26. Sanchez, E. and Kowalski, B.R. (1988). Tensorial calibration: II. Second-order calibration. *J. Chemometrics*, **2**, 265-280.

27. Sanchez, E. and Kowalski, B.R. (1990). Tensorial resolution: a direct trilinear decomposition. *J. Chemometrics*, **4**, 29-45.

The Analysis of Linear Functions using Geometric Algebra – an Alternative to Matrices and Tensors

J. Lasenby*, **A.X.S. Stevenson*** **and I.K. Proudler****

**Department of Engineering, University of Cambridge, and **DERA, Malvern*

Abstract

In this paper we discuss a mathematical system based on the algebras of Grassmann and Clifford [1, 5], called geometric algebra [7, 8]. It is shown how geometric algebra can be used to carry out, in a simple manner, various complex manipulations relevant to matrix and tensor-based problems, including that of optimization. In particular it will be shown how differentiation of certain matrix functions with respect to the matrix, can easily be achieved. The encoding of structure into such problems will be discussed and various examples will be given. In particular we will look at a multi-source signal separation problem and the problems of estimating the fundamental matrix and the trilinear tensor in computer vision.

1 Introduction

Let $\mathcal{G}_n$ denote the geometric algebra of n-dimensions – this is a linear space in which the elements are graded (see later). As well as vector addition and scalar multiplication we have a non-commutative product which is associative and distributive over addition – this is the **geometric** or **Clifford** product. A further distinguishing feature of the algebra is that any vector squares to give a scalar. The geometric product of two vectors $\boldsymbol{a}$ and $\boldsymbol{b}$ is written $\boldsymbol{ab}$ and can be expressed as a sum of its symmetric and antisymmetric parts

$$\boldsymbol{ab} = \boldsymbol{a}\cdot\boldsymbol{b} + \boldsymbol{a}\wedge\boldsymbol{b}. \tag{1.1}$$

We are therefore able to define the inner product $\boldsymbol{a}\cdot\boldsymbol{b}$ and the outer product $\boldsymbol{a}\wedge\boldsymbol{b}$ in terms of the more fundamental geometric product as follows;

$$\mathbf{a}\cdot\mathbf{b} = \frac{1}{2}(\mathbf{ab} + \mathbf{ba}) \qquad \mathbf{a}\wedge\mathbf{b} = \frac{1}{2}(\mathbf{ab} - \mathbf{ba}). \tag{1.2}$$

The inner product of two vectors is the standard *scalar* or *dot* product and produces a scalar. The outer or wedge product of two vectors is a new quantity we call a **bivector**. We think of a bivector as a directed area in the plane containing $\boldsymbol{a}$ and $\boldsymbol{b}$, formed by sweeping $\boldsymbol{a}$ along $\boldsymbol{b}$ – see Figure 1.

Thus, $\boldsymbol{b}\wedge\boldsymbol{a}$ will have the opposite orientation making the wedge product anticommutative. The outer product is immediately generalizable to higher dimensions – for example, $(\boldsymbol{a}\wedge\boldsymbol{b})\wedge\boldsymbol{c}$, a **trivector**, is interpreted as the oriented

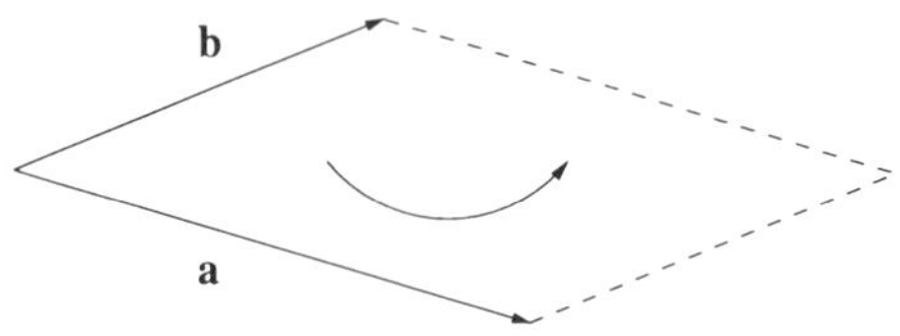

Figure 1. The directed area, or bivector, $\boldsymbol{a}\wedge\boldsymbol{b}$

volume formed by sweeping the area $\boldsymbol{a}\wedge\boldsymbol{b}$ along vector $\boldsymbol{c}$. The outer product of k vectors is a k-vector (or more correctly, a k-blade), and has *grade* k. A general element of the geometric algebra of n-dimensions is a *multivector*, which is a linear combination of objects of any grade. If a multivector possesses only terms of a single grade it is termed *homogeneous*. The geometric algebra provides a means of manipulating multivectors which allows us to keep track of different grade objects simultaneously. We are already familiar with such a process in dealing with complex numbers: there one has two different types of object (real and imaginary), but the algebra is such that we can manipulate the complex number in a way which gives us the correct behaviour in the real and imaginary domains. In a space of 3 dimensions we can construct a trivector $\boldsymbol{a}\wedge\boldsymbol{b}\wedge\boldsymbol{c}$, but no 4-vectors exist since there is no possibility of sweeping the volume element $\boldsymbol{a}\wedge\boldsymbol{b}\wedge\boldsymbol{c}$ over a 4th dimension. The highest grade element in a space is called the **pseudoscalar**. The unit pseudoscalar is denoted by I, or i in 2 and 3 dimensions.

We can generalize the definitions of inner and outer products given in Equation (1.2). For two homogeneous multivectors A_r and B_s (i.e. multivectors of grades r and s respectively), we define the inner and outer products as

$$A_r\cdot B_s = \langle A_r B_s\rangle_{|r-s|} \tag{1.3}$$

$$A_r\wedge B_s = \langle A_r B_s\rangle_{r+s}. \tag{1.4}$$

Where $\langle M\rangle_t$ denotes the t-grade part of the multivector M. Thus, the inner product produces an $|r-s|$-vector – which means it effectively reduces the grade of B_s by r; and the outer product gives an $(r+s)$-vector, therefore increasing the grade of B_s by r. This is an extension of the general principle that dotting with a vector lowers the grade of a multivector by 1 and wedging with a vector raises the grade of a multivector by 1. In the following sections we will frequently evaluate a vector dotted with a bivector; according to the above this produces a vector and is explicitly given by

$$\boldsymbol{a}\cdot(\boldsymbol{b}\wedge\boldsymbol{c}) = (\boldsymbol{a}\cdot\boldsymbol{b})\boldsymbol{c} - (\boldsymbol{a}\cdot\boldsymbol{c})\boldsymbol{b}. \tag{1.5}$$

Another concept which will be used elsewhere in this paper is that of the *reciprocal frame*. Given a set of linearly independent vectors $\{\boldsymbol{e}_i\}$ (where no assumption of orthonormality is made), we can form a **reciprocal frame**, $\{e^i\}$, which is such that

$$e^i \cdot e_j = \delta_{ij}. \tag{1.6}$$

For details of the explicit construction of such a reciprocal frame in n-dimensions see [8]. In three dimensions this is a very simple operation and the reciprocal frame vectors for a linearly independent set of vectors $\{e_i\}$, $i = 1,..,3$, are as follows

$$\begin{aligned} e^1 &= \frac{1}{\alpha} i e_2 \wedge e_3 \\ e^2 &= \frac{1}{\alpha} i e_3 \wedge e_1 \\ e^3 &= \frac{1}{\alpha} i e_1 \wedge e_2, \end{aligned} \tag{1.7}$$

where $i\alpha = e_3 \wedge e_2 \wedge e_1$.

The idea of *reversion* will also be needed in the paper: this operation reverses the order of vectors in any multivector. The reverse of A is written as $\tilde{A}$, and for a product we have

$$(AB)\tilde{} = \tilde{B}\tilde{A}. \tag{1.8}$$

In the following sections the summation convention will be used unless otherwise stated, i.e. repeated indices are summed over.

2 Linear algebra and multivector calculus

Geometric algebra has very powerful associated linear algebra and calculus frameworks and is therefore a natural language for the study of linear functions and non-orthonormal frames. Consider a linear function $f(a)$ which maps vectors to vectors in the same space. It is then possible to extend f to act linearly on multivectors. This extension of f (the outermorphism) is written as $\underline{f}$ and is given by

$$\underline{f}(a_1 \wedge a_2 \wedge \ldots \wedge a_r) \equiv f(a_1) \wedge f(a_2) \wedge \ldots \wedge f(a_r). \tag{2.1}$$

$\underline{f}$ is thus grade-preserving. The **adjoint** to $\underline{f}$ is written as $\overline{f}$ and satisfies the relation $\underline{f}(a) \cdot b = \overline{f}(b) \cdot a$ for any vectors a and b; note that it is also possible to define the adjoint in a frame-independent manner by using derivatives. If F is a matrix with elements F_{ij}, we can write these elements in terms of a linear function $\underline{f}$ as

$$F_{ij} = \sigma_i \cdot \underline{f}(\sigma_j) \tag{2.2}$$

where the σ_i are unit vectors for the reference frame implied by the matrix. Since $\sigma_i \cdot \underline{f}(\sigma_j) = \overline{f}(\sigma_i) \cdot \sigma_j$ we see that the adjoint corresponds to the transpose of the matrix.

As the outermorphism preserves grade, we know that the pseudoscalar of the space must be mapped onto some multiple of itself. The scale factor in this mapping is the **determinant** of $\underline{f}$;

$$\underline{f}(I) = \det(\underline{f})I. \tag{2.3}$$

This is much simpler than many definitions of the determinant. Using this definition, most properties of determinants can be established with little effort. A simple example of this is seen by considering the determinant of the product of two matrices A and B. Suppose A and B are represented by the linear functions $\underline{f}$ and $\underline{g}$;

$$\begin{aligned} \det(\underline{fg})I &= (\underline{fg})(I) \\ &= \underline{f}[\underline{g}(I)] \\ &= \det(\underline{g})\underline{f}(I) \\ &= \det(\underline{g})\det(\underline{f})I. \end{aligned} \tag{2.4}$$

We therefore see that the result $\det(AB) = \det(A)\det(B)$ drops out trivially from our definition of the determinant.

In addition, it is simple to obtain expressions for the inverse of a function (if it exists) and the inverse of an adjoint [3];

$$\underline{f}^{-1}(A) = \det(f)^{-1}I\overline{f}(I^{-1}A) \tag{2.5}$$

$$\overline{f}^{-1}(A) = \det(f)^{-1}I\underline{f}(I^{-1}A) \tag{2.6}$$

where A is any arbitrary multivector.

Within geometric algebra it is possible to differentiate with respect to any multivector quantity [8] – in practice this turns out to be very useful in many optimization processes in signal analysis. Here we will give a brief description of multivector differentiation and a few of the standard results.

If X is a mixed-grade multivector, $X = \sum_r X_r$, and $F(X)$ is a general multivector-valued function of X, then the derivative of F in the A "direction" (where A has the same grades as X) is written as $A * \partial_X F(X)$ (here we use $*$ to denote the scalar part of the product of two multivectors, i.e. $A * B \equiv \langle AB \rangle$, where $\langle\,\rangle$ is shorthand for the scalar part, $\langle\,\rangle_0$), and is defined as

$$A * \partial_X F(X) \equiv \lim_{\tau \to 0} \frac{F(X + \tau A) - F(X)}{\tau}. \tag{2.7}$$

We impose the constraint that A must have the same grades as X so that the differentiation in terms of the limit makes some physical sense. If X contains no terms of grade-r and A_r is a homogeneous multivector, then we define $A_r * \partial_X = 0$. This definition of the derivative also ensures that the operator $A * \partial_X$ is a scalar operator and satisfies all of the usual partial derivative properties. We can now use the above definition of the directional derivative to formulate a general expression for the multivector derivative ∂_X without reference to one particular direction. This is accomplished by introducing an arbitrary frame $\{e_j\}$ and extending this to a basis (vectors, bivectors, etc.) for the entire algebra, $\{e_J\}$. Then ∂_X is defined as

$$\partial_X \equiv \sum_J e^J e_J * \partial_X, \tag{2.8}$$

where $\{e^J\}$ is an extended basis built out of the reciprocal frame. The directional derivative, $e_J * \partial_X$, is only non-zero when e_J is one of the grades contained in X (as previously discussed) so that ∂_X inherits the multivector properties of its argument X. Although we have here defined the multivector derivative using an extended basis, it should be noted that the sum over all the basis ensures that ∂_X is independent of the choice of $\{e_j\}$ and so all of the properties of ∂_X can be formulated in a frame-free manner. One of the most useful results concerning multivector derivatives is

$$\partial_X \langle XB \rangle = \mathrm{P}_X(B), \tag{2.9}$$

where $\mathrm{P}_X(B)$ is the projection of the multivector B onto the grades contained in X. We can see this as follows. Since

$$\partial_X \langle XB \rangle = \sum_J e^J e_J * \partial_X \langle XB \rangle \tag{2.10}$$

and $e_J * \partial_X = 0$ if e_J is not a grade of X, we see that

$$\begin{aligned} \partial_X \langle XB \rangle &= \sum_{J'} e^{J'} \lim_{\tau \to 0} \frac{\langle (X + \tau e'_J)B \rangle - \langle XB \rangle}{\tau} \\ &= \sum_{J'} e^{J'} \langle e_{J'} B \rangle \\ &= P_X(B) \end{aligned} \tag{2.11}$$

where the sum over J' is over those grades contained in X. Using this result the following relations can be shown to hold

$$\partial_X \langle \tilde{X} B \rangle = \mathrm{P}_X(\tilde{B}) \tag{2.12}$$

$$\partial_{\tilde{X}} \langle \tilde{X} B \rangle = \mathrm{P}_{\tilde{X}}(B) = \mathrm{P}_X(B) \tag{2.13}$$

$$\partial_\psi \langle M \psi^{-1} \rangle = -\psi^{-1} \mathrm{P}_\psi(M) \psi^{-1} \text{ for general multivectors } \psi \text{ and } M \tag{2.14}$$

It is often convenient to indicate, via an overdot, the quantity on which ∂_X operates. For example, $\dot{\partial}_X A\dot{B}$ means that the derivative part of ∂_X acts on B. A complete discussion of the geometric calculus is given in [8], where the results in Equations (2.12) and (2.13) are proved. The result in Equation (2.14) is discussed in [2].

One often wants to take the derivative, ∂_a, with respect to a vector quantity a. If we replace A by a, e^J by e^j and e_J by e_j in Equation (2.8) and use the definition in Equation (2.7) we see that the differential operator ∂_a can be written as

$$\partial_a = e^i \frac{\partial}{\partial a^i} \qquad \text{where} \qquad a = a^i e_i. \tag{2.15}$$

This will be used several times in the following sections. Note that we will not write vectors in bold when they appear as subscripts in the vector derivative. At this point it is appropriate to give some explanation of the expansion of $\boldsymbol{a}$ used above, namely $\boldsymbol{a} = a^i \boldsymbol{e}_i$. Recall that our basis $\{\boldsymbol{e}_i\}$ was not necessarily an orthonormal basis and that we were able to define a reciprocal basis $\{\boldsymbol{e}^i\}$, such that $\boldsymbol{e}^i \cdot \boldsymbol{e}_j = \delta_{ij}$. Consider the component of the vector $\boldsymbol{a}$ in the $\boldsymbol{e}^j$ direction – this is given by $\boldsymbol{a} \cdot \boldsymbol{e}^j$ and we will call this component a^j, where the upstairs index tells us that it is the component of $\boldsymbol{a}$ in the direction of the jth basis vector of the *reciprocal* frame. Similarly, the component of $\boldsymbol{a}$ in the $\boldsymbol{e}_j$ direction is given by $\boldsymbol{a} \cdot \boldsymbol{e}_j$ which we call a_j. We note the useful identities

$$\boldsymbol{a} = (\boldsymbol{a} \cdot \boldsymbol{e}_j)\boldsymbol{e}^j \qquad \text{and} \qquad \boldsymbol{a} = (\boldsymbol{a} \cdot \boldsymbol{e}^j)\boldsymbol{e}_j. \tag{2.16}$$

Of course, if the basis is an orthonormal basis, say $\{\sigma_i\}$, then $\sigma^i = \sigma_i$ and we would, without ambiguity, write $\boldsymbol{a} = a_i \sigma_i$. This ability to differentiate with respect to any multivector is often very useful – for example, one can easily differentiate with respect to rotations (represented by a scalar plus a bivector) in order to find the optimal motion in a given problem.

To complete our calculus it is now necessary to look at the formulation of functional differentiation in GA. In [3] it is shown that we can differentiate with respect to a linear function according to the following rule:

$$\partial_{\underline{f}(a)} \{\underline{f}(\boldsymbol{b}) \cdot \boldsymbol{c}\} = (\boldsymbol{a} \cdot \boldsymbol{b})\boldsymbol{c} \tag{2.17}$$

for any vectors $\boldsymbol{a}, \boldsymbol{b}, \boldsymbol{c}$. This is the basic result and can be extended to the case where we have general multivectors as follows:

$$\partial_{\underline{f}(a)} \langle \underline{f}(A) B \rangle = \sum_r \langle \underline{f}(\boldsymbol{a} \cdot A_r) B_r \rangle_1 \tag{2.18}$$

where the right hand side represents a sum of vector parts.

Using this functional differentiation we can differentiate expressions which would be considerably harder to deal with in the usual matrix formulation – an example of this is the differentiation of $\det A$ with respect to a matrix A. This is rather hard to do conventionally but reasonably straightforward in GA. If the function $\underline{f}$ represents the matrix A, then from the definition of the determinant in Equation (2.3) we see that

$$\partial_{\underline{f}(a)} \det(\underline{f}) = \partial_{\underline{f}(a)} \underline{f}(I) I^{-1}. \tag{2.19}$$

From the formula for functional differentiation in Equation (2.18) and from the definition of the inverse given in Equation (2.5), we can evaluate the above to give

$$\begin{aligned}\partial_{\underline{f}(a)} \det(\underline{f}) &= \underline{f}(\boldsymbol{a}\cdot I)I^{-1} \\ &= \det(\underline{f})\overline{f}^{-1}(\boldsymbol{a}),\end{aligned} \tag{2.20}$$

which agrees with the standard result.

3 Optimization and encoding matrix structure

Equipped with the basic results of functional differentiation and some results which tell us how to implement the chain rule, it is possible to apply these techniques to more complicated problems. The aim in this paper will be to use this framework to optimize expressions containing structured matrices. Firstly we consider a simple example: find the matrix R, such that R is cyclic Toeplitz, which maximizes

$$p(\boldsymbol{x}) = \frac{1}{(2\pi)^{\frac{n}{2}}}|R|^{-\frac{1}{2}} \exp\{-\frac{1}{2}\boldsymbol{x}^T R^{-1}\boldsymbol{x}\}. \tag{3.1}$$

For data $\boldsymbol{x}$ coming from an underlying zero-mean Gaussian process we expect R to be the estimate of the autocorrelation matrix. Here we want to optimize only over the space of *allowed* R's, i.e. those that are cyclic Toeplitz. An $n \times n$ cyclic Toeplitz matrix R can be written in terms of a generating vector $\boldsymbol{a} = a_1 e_1 + a_2 e_2 + \ldots + a_n e_n$, where $\{e_i\}$ is a basis for the n-dimensional space – for example:

$$R = \begin{bmatrix} a_1 & a_2 & a_3 & \cdots & a_n \\ a_n & a_1 & a_2 & \cdots & a_{n-1} \\ \vdots & \vdots & \vdots & \vdots & \vdots \\ a_2 & a_3 & a_4 & \cdots & a_1 \end{bmatrix}. \tag{3.2}$$

Consider another set of basis vectors $\{\sigma_i\}$ such that $\sigma_i \cdot \sigma_j = \delta_{ij}$ and $\sigma_i \cdot e_j = 0$ for all i, j – we then think of the linear function, f, representing R^{-1}, as mapping from σ-space to e-space (for convenience we will not write the basis vectors $\{\sigma_i\}$ and $\{e_i\}$ in bold). In the geometric algebra formulation of Hamiltonian dynamics one uses quantities called *doubling bivectors*, and it turns out that similar quantities can be used here to encode the Toeplitz nature of the function – i.e. the fact that each row of the matrix is just a shifted version of the first. We define $J_k = e_k \wedge \sigma_k$ and $\boldsymbol{J} = \sum_k J_k$, so that the effect of $\boldsymbol{J}$ is to take e_k to σ_k and vice versa;

$$e_k \cdot \boldsymbol{J} = \sigma_k, \quad \sigma_k \cdot \boldsymbol{J} = -e_k, \tag{3.3}$$

which can be verified using the result in Equation (1.5). By modifying $\boldsymbol{J}$ we can produce a shifted vector for use in describing R^{-1}. If we define

$$S_k = \sum_{j=1}^{n} e_j \wedge \sigma_{[j+k-2]_n+1} \tag{3.4}$$

(where $[..]_n$ indicates modulo-n) then each row of the matrix can be written as $\boldsymbol{a}\cdot S_k$ – in terms of the adjoint of f this gives

$$\overline{f}(e_k) = \boldsymbol{a}\cdot S_k. \tag{3.5}$$

With this characterization of R^{-1} we can differentiate Equation (3.1) straightforwardly with respect to $\boldsymbol{a}$ and set this to zero to give

$$\begin{aligned} \partial_a p(\boldsymbol{x}) &= \partial_{\underline{f}(\sigma_k)} p(\boldsymbol{x}) \cdot \partial_a \underline{f}(\sigma_k) \\ &= \frac{1}{(2\pi)^{\frac{n}{2}}} (\det f)^{\frac{1}{2}} \exp\{-\frac{1}{2}\tilde{\boldsymbol{x}}\cdot\underline{f}(\boldsymbol{x})\} \times \\ &\quad \{\underline{f}^{-1}(e_k)\cdot S_k - (e_k\cdot\tilde{\boldsymbol{x}})\boldsymbol{x}\cdot S_k\} \\ &= 0. \end{aligned} \tag{3.6}$$

In the above expression $\tilde{\boldsymbol{x}} = x_j e_j$ and $\boldsymbol{x} = x_j \sigma_j$, and we have used the *chain rule* in the differentiation. Here, the chain rule is applied in a straightforward manner – i.e. we know that the derivative with respect to $\boldsymbol{a}$ produces a vector; since the derivative of $p(\boldsymbol{x})$ wrt $\underline{f}$ gives a vector and the derivative of $\underline{f}(\sigma_k)$ wrt $\boldsymbol{a}$ will give a bivector, it follows that we must dot these together to produce a vector as required. Thus we see that the solution to this is given by

$$\underline{f}^{-1}(e_k)\cdot S_k = (e_k\cdot\tilde{\boldsymbol{x}})\boldsymbol{x}\cdot S_k. \tag{3.7}$$

Since $\underline{f}$ was constructed to represent R^{-1}, $\underline{f}^{-1}$ (mapping from e-space to σ-space) will represent R. The above then reduces to

$$a_j = \frac{1}{n}(e_k\cdot\tilde{\boldsymbol{x}})\boldsymbol{x}\cdot(e_j\cdot S_k) = \frac{1}{n} x_k x_{[j+k-2]_n+1}. \tag{3.8}$$

This indeed gives the required estimate of the autocorrelation matrix. Carrying out a similar operation with matrices is possible but harder. A general $m \times n$ Toeplitz matrix can be similarly characterized by its two generating vectors and two doubling bivectors and it is therefore possible to use the above techniques to form derivatives of much more complicated expressions.

Once a given characteristic has been encoded in terms of linear functions – as in Equation (3.5) for the cyclic Toeplitz case – the procedures of differentiation and functional differentiation are straightforward and offer an attractive new technique for dealing with structure in matrices. However, we note here that it is the encoding of the structure into the linear function which is the hardest stage of the problem and that there is no set recipe for doing this. However, we also note that such structure-encoding is also far from easy in conventional matrix algebra.

4 Application to signal separation

In order to illustrate further the above techniques we will look at one particular formulation of a signal separation problem. Suppose we have n sources, represented by the time series $s_i(t)$, $i = 1, .., n$, and N measured estimates, r_i, $i = 1, ..., N$ – below, the estimates will be a sequence of n^2 correlations. The process of obtaining the estimates can be modelled as passing the source signals through a *mixer* followed by an *estimator*. The mixer is modelled by an $n \times n$ matrix so that

$$\begin{bmatrix} x_1(t) \\ x_2(t) \\ \vdots \\ x_n(t) \end{bmatrix} = \begin{bmatrix} a_{11} & a_{12} & \cdots & a_{1n} \\ a_{21} & a_{22} & \cdots & a_{2n} \\ \vdots & \vdots & \vdots & \vdots \\ a_{n1} & a_{n2} & \cdots & a_{nn} \end{bmatrix} \begin{bmatrix} s_1(t) \\ s_2(t) \\ \vdots \\ s_n(t) \end{bmatrix} \tag{4.1}$$

which can also be written as $\boldsymbol{x}(t) = A\boldsymbol{s}(t)$. If we then take expectation values of the matrix $\boldsymbol{x}(t)\boldsymbol{x}(t+q)^T$ and make the assumption that the signals are uncorrelated, i.e. that the expectation of $\boldsymbol{s}(t)\boldsymbol{s}(t+q)^T$ has all off-diagonal terms zero, it is possible to write the measured correlations as

$$\boldsymbol{r}_q = \tilde{A}\boldsymbol{s}_q. \tag{4.2}$$

In this, $\boldsymbol{r}_q = [r_{11}(q), r_{12}(q), \ldots, r_{nn}(q)]^T$ (an $n^2 \times 1$ vector), and $r_{ij}(q) = E[x_i(t)x_j(t+q)]$. The vector $\boldsymbol{s}_q$ is given by $[s_{11}(q), s_{22}(q), \ldots, s_{nn}(q)]^T$ (an $n \times 1$ vector) with $s_{ii}(q) = E[s_i(t)s_i(t+q)]$ (no summation convention). The matrix $\tilde{A}$ is given by

$$\tilde{A} = [\boldsymbol{a}_1 \otimes \boldsymbol{a}_1, \boldsymbol{a}_2 \otimes \boldsymbol{a}_2, ...] \tag{4.3}$$

where $\boldsymbol{a}_i$ is the vector formed from the ith column of matrix A and $\otimes$ is the Kronecker product, such that $\boldsymbol{a}_1 \otimes \boldsymbol{a}_1 = [a_{11}a_{11}, a_{11}a_{21}, ..., a_{21}a_{11}, a_{21}a_{21}, ..., a_{n1}a_{n1}]^T$. We are now able to formulate this in the form of a least squares problem: suppose our observations are correlations over m lags, we would then want to minimize

$$\mathcal{E} = \sum_{q=1}^{m} \left(\boldsymbol{r}_q - \tilde{A}\boldsymbol{s}_q\right)^2, \tag{4.4}$$

i.e. find the $\tilde{A}$ and $\boldsymbol{s}_q$ which minimize $\mathcal{E}$ such that $\tilde{A}$ has the structure given in Equation (4.3). To do this our approach is to differentiate with respect to the $\{\boldsymbol{s}_q\}$ and the $\{\boldsymbol{a}_i\}$ and for this we need to express the matrix $\tilde{A}$ as a linear function. Writing the vector $\boldsymbol{a}_i$ as $\boldsymbol{a}_i = a_{1i}\sigma_i + a_{2i}\sigma_2 + \ldots + a_{ni}\sigma_n$, we then want to look for a linear function, $\underline{f}$, representing $\tilde{A}$, such that $\underline{f}$ maps σ-space onto e-space, as before. It is reasonably straightforward to see that we can write the ith column of the matrix as

$$\begin{aligned} \underline{f}(\sigma_i) &= (\sigma_{\bar{j}} \cdot \boldsymbol{a}_i)(\sigma_{\{j\}_n} \cdot \boldsymbol{a}_i)e_j \\ &= a_{\bar{j}i}a_{\{j\}_n i}e_j, \end{aligned} \tag{4.5}$$

where $\tilde{j} = int[\frac{j-1}{n}] + 1$ (*int* refers to *integer part*), $\{j\}_n = [j-1]_n + 1$ and j runs from 1 to n^2. Equation (4.4) can then be written as

$$\mathcal{E} = \sum_{q=1}^{m} (\boldsymbol{r}_q - \underline{f}(\boldsymbol{s}_q))^2, \tag{4.6}$$

where we now think of $\boldsymbol{s}_q$ as the vector $s_{11}(q)\sigma_1 + s_{22}(q)\sigma_2 + \ldots + s_{nn}(q)\sigma_n$, and $\boldsymbol{r}_q$ as the vector $r_{11}(q)e_1 + r_{12}(q)e_2 + \ldots + r_{nn}(q)e_{n^2}$ – so that $\underline{f}$ takes σ-space onto e-space. Differentiating with respect to $\boldsymbol{s}_q$ is straightforward and gives the standard solution

$$\boldsymbol{s}_q = (\overline{f}\underline{f})^{-1}\overline{f}(\boldsymbol{r}_q). \tag{4.7}$$

Differentiating with respect to the vectors $\boldsymbol{a}_i$ is achieved by first evaluating the expression obtained from the chain rule

$$\partial_{a_i}\mathcal{E} = (\partial_{\underline{f}(\sigma_j)}\mathcal{E})\cdot(\partial_{a_i}\underline{f}(\sigma_j)) = 0. \tag{4.8}$$

Evaluating the derivatives gives us the following equations (one for each $i = 1, ..., n$)

$$\partial_{a_i}\mathcal{E} = 2\left\{\sum_{q=1}^{m}(e_i\cdot\boldsymbol{s}_q)(\underline{f}(\boldsymbol{s}_q) - \boldsymbol{r}_q)\right\}\cdot B_i = 0, \tag{4.9}$$

where B_i is a bivector given by

$$a_{\{k\}_n i}\sigma_{\tilde{k}}\wedge e_k + a_{\tilde{k}i}\sigma_{\{k\}_n}\wedge e_k. \tag{4.10}$$

We can now try to solve Equations (4.7) and (4.10) or use the derivatives in a gradient-based optimization scheme. Having analytic derivatives is useful in many schemes for several reasons: firstly we may be able to solve the resulting equations, if not directly then often iteratively, and secondly, they provide more robustness in search methods than do differencing techniques.

5 Applications in computer vision

If we have a moving camera or several cameras placed at different positions, the data we obtain when viewing a scene is a number of 2D images. For many applications we would like to be able to estimate the relative positions or motion of the camera as well as the scene structure (i.e. distances from the cameras to the objects). Of course, the situation of a single camera and a moving object is an equivalent set-up. If the intrinsic parameters of the cameras are known (i.e. the focal length and how the image coordinates relate to the projection of the world coordinates onto the image plane) the camera is said to be *calibrated* and there are a variety of techniques for recovering the structure and motion [12, 11, 10]. However, in recent years there has been an increased interest in dealing with the case of *uncalibrated* cameras – so that we would like to recover as much information as possible when all we have are the various images and know nothing about the cameras they were taken with. Assume that we are able to find matching points in each of our images, i.e. points which correspond to the same world point. In Figure 2 we see that, assuming a simple pinhole camera model, the optical centres of the cameras, $\boldsymbol{a}_0$ and $\boldsymbol{b}_0$, and the projections of world point $\mathbf{X}$ in the images, $\boldsymbol{a}'$ and $\boldsymbol{b}'$, lie on a plane.

Suppose that matching points in the two images, (u_i, v_i) and (u'_i, v'_i), are written as 3D-vectors $\boldsymbol{x} = [u, v, 1]^T$ and $\boldsymbol{x}' = [u', v', 1]^T$; it is then possible to show that this coplanarity condition can be written as

$$\boldsymbol{x}^T F \boldsymbol{x}' = 0 \tag{5.1}$$

where F is a rank 2, 3×3 matrix known as the *fundamental matrix*. For many purposes we would like to be able to estimate F accurately given a set of point matches in the two images. One obvious method of doing this is via a linear algorithm. In the noiseless case, each pair of matching points, $(u_i, v_i, 1)$ and $(u'_i, v'_i, 1)$, satisfies a linear equation of the form

$$u_i u'_i F_{11} + u_i v'_i F_{12} + u_i F_{13} + v_i u'_i F_{21} + v_i v'_i F_{22} + v_i F_{23} + u'_i F_{31} + v'_i F_{32} + F_{33} = 0 \tag{5.2}$$

where F_{ij} is the ij^{th} element of F. Thus if we have n matching pairs we can form the equation

$$A\boldsymbol{y} = 0 \tag{5.3}$$

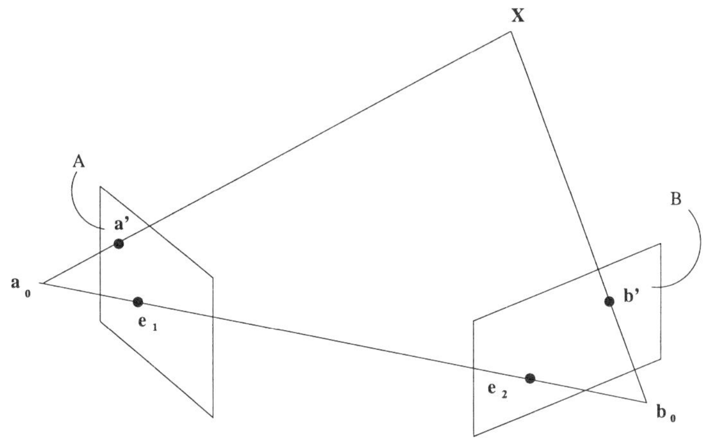

Figure 2. Two images planes showing projected points and epipoles

where $\boldsymbol{y} = [F_{11}, F_{12}, F_{13}, F_{21}, F_{22}, F_{23}, F_{31}, F_{32}, F_{33}]^T$ and A is an $n \times 9$ matrix made up of functions of the data points. One way of solving this is to minimize $\boldsymbol{y}^T B \boldsymbol{y}$ subject to $\boldsymbol{y}$ being non-zero, where $B = A^T A$. This is a standard problem which can be solved easily by finding the eigenvector corresponding to the least eigenvalue of B.

However, the solution found in this way does not impose the correct structure and the F which results will not in general have $\det F = 0$. There are a variety of techniques which can be used to adapt this linear algorithm to give a better answer – the most common of which is to project the resultant F onto the "*closest*" matrix which has rank 2 [6].

Our approach to this problem is to treat F as a linear function, $\underline{f}$, which takes two vectors (in practice, one from each image) and maps them onto a scalar; the vectors are mapped onto zero if they are projections of the *same* world point in the two images, for example

$$\begin{aligned} \underline{f}(\boldsymbol{a}, \boldsymbol{b}) &= \alpha \quad \alpha \text{ a real number} \\ \underline{f}(\boldsymbol{a}, \boldsymbol{b}) &= 0 \quad \text{if } \boldsymbol{a}, \boldsymbol{b} \text{ formed by same world point.} \end{aligned} \tag{5.4}$$

From Figure 2 we can see that the two points where the line joining the optical centres intersects the image planes, will be important in any discussions involving matched points. These points are called the ***epipoles*** and are denoted by $\boldsymbol{e}_1$ and $\boldsymbol{e}_2$ in Figure 2. The epipoles satisfy $F\boldsymbol{e}_2 = 0$ and $F^T\boldsymbol{e}_1 = 0$ – i.e. $\boldsymbol{e}_1$ and $\boldsymbol{e}_2$ are the left and right zero-eigenvectors of the matrix F.

Suppose we take $\boldsymbol{e}_1, \boldsymbol{e}_2$ and $\boldsymbol{e}_3 = \boldsymbol{e}_1 \times \boldsymbol{e}_2$ (cross product) as a basis for our 3D space and consider $\underline{f}$ in terms of this basis. Let $F_{ij} = \underline{f}(\boldsymbol{e}_i, \boldsymbol{e}_j)$; we then know, from a knowledge of how F acts on the epipoles, that

$$\underline{f}(\boldsymbol{e}_1, \boldsymbol{b}) = 0 \quad \forall \boldsymbol{b} \tag{5.5}$$

$$\underline{f}(\boldsymbol{a}, \boldsymbol{e}_2) = 0 \quad \forall \boldsymbol{a}. \tag{5.6}$$

If $\boldsymbol{b} = b^i \boldsymbol{e}_i$ then (5.5) tells us that $b^i F_{1i} = 0 \ \forall \boldsymbol{b}$, which implies $F_{11} = F_{12} = F_{13} = 0$. Similarly, (5.6) tells us that $F_{12} = F_{22} = F_{32} = 0$. Of the 9 elements of F we therefore see that only 4 are non-zero in this basis. The total number of degrees of freedom (dof) in the problem is thus $(4 + 2 + 2) = 8$; 4 from F and 2 from each of the epipoles (they can be normalized in some way). However, F is only determined up to an overall scale factor so we can subtract 1 dof leaving us with 7. The plan will therefore be to optimize a given cost function over the 7 parameters (4 from the epipoles and 3 from F in the epipole basis). Given n sets of matching points in the image $(\boldsymbol{x}_i, \boldsymbol{x}'_i)$ $i = 1, .., n$, we will minimize

$$S = \sum_{i=1}^{n} \frac{[\underline{f}(\boldsymbol{x}_i, \boldsymbol{x}'_i)]^2}{{\sigma_i}^2} \tag{5.7}$$

where the ${\sigma_i}^2$ are weighting factors to be determined. If the coordinates of the points that we actually measure in the two image planes are (u_i, v_i) and (u'_i, v'_i)

then we have a number of options for the 3D-vectors $\boldsymbol{x}_i$ and $\boldsymbol{x}'_i$. The simplest is to place a 1 in the 3rd coordinate as given previously. However, we are free to choose some other normalization and here we will choose to use

$$\boldsymbol{x}_i = [(1 - u_i - v_i), u_i, v_i]^T \qquad \boldsymbol{x}'_i = [(1 - u'_i - v'_i), u'_i, v'_i]^T \tag{5.8}$$

so that the coordinates sum to 1. The F which results from using such a normalization is the F which is used in a particular method of forming 3D projective invariants [9]. While this will not be discussed further here it will be assumed that from now on we will work with the $\boldsymbol{x}_i$s and $\boldsymbol{x}'_i$s as given in Equation (5.8).

The mechanics of optimizing F are fairly straightforward, involving the use of the reciprocal frame $[\boldsymbol{e}^1, \boldsymbol{e}^2, \boldsymbol{e}^3]$. We write

$$\begin{aligned} \underline{f}(\boldsymbol{x}, \boldsymbol{x}') &= \underline{f}(x^j \boldsymbol{e}_j, x^{k\prime} \boldsymbol{e}_k) \\ &= (\boldsymbol{x} \cdot \boldsymbol{e}^j)(\boldsymbol{x}' \cdot \boldsymbol{e}^k) F_{jk}. \end{aligned} \tag{5.9}$$

Setting (for example) F_{21} to be unity (as F is only defined up to an overall scale factor) leads us to the following equation which enables us to express each term in the cost function in terms of the image points we observe and the parameters we vary (the reciprocal frame is calculated from Equation (1.8))

$$\underline{f}(\boldsymbol{x}, \boldsymbol{x}') = (\boldsymbol{x} \cdot \boldsymbol{e}^2)(\boldsymbol{x}' \cdot \boldsymbol{e}^1) + (\boldsymbol{x} \cdot \boldsymbol{e}^2)(\boldsymbol{x}' \cdot \boldsymbol{e}^3) F_{23} + (\boldsymbol{x} \cdot \boldsymbol{e}^3)(\boldsymbol{x}' \cdot \boldsymbol{e}^1) F_{31} + (\boldsymbol{x} \cdot \boldsymbol{e}^3)(\boldsymbol{x}' \cdot \boldsymbol{e}^3) F_{33}. \tag{5.10}$$

The weighting of each term, σ_i^2, must be worked out by considering the noise on the image points. If we assume that the noise on each of the u_i, v_i, u'_i and v'_i coordinates has the same variance and is independent, we can evaluate the variance of $\epsilon_i \equiv \underline{f}(\boldsymbol{x}_i, \boldsymbol{x}'_i)$ as follows. Since the ϵ_i are zero mean we have

$$\mathrm{Var}(\epsilon) = \mathrm{E}[(\epsilon - \bar{\epsilon})^2] = \mathrm{E}(\epsilon^2) \tag{5.11}$$

where E denotes expectation value. If $\overline{\boldsymbol{x}}$ and $\overline{\boldsymbol{x}}'$ are the true values of the matching points such that $\underline{f}(\overline{\boldsymbol{x}}, \overline{\boldsymbol{x}}') = 0$, we can expand ϵ using the Taylor expansion about these true values. If $\boldsymbol{x} = \overline{\boldsymbol{x}} + \Delta\boldsymbol{x}$ and $\boldsymbol{x}' = \overline{\boldsymbol{x}}' + \Delta\boldsymbol{x}'$, to first order in $\Delta\boldsymbol{x}$ and $\Delta\boldsymbol{x}'$ we have

$$\epsilon \approx \Delta\boldsymbol{x} \cdot \partial_x \underline{f} + \Delta\boldsymbol{x}' \cdot \partial_{x'} \underline{f}. \tag{5.12}$$

Squaring the above, taking expectation values and noting that $\mathrm{E}[\Delta\boldsymbol{x}\Delta\boldsymbol{x}'^T] = 0$ since the coordinates are all assumed independent, gives (reverting to matrix notation)

$$\mathrm{E}[\epsilon^2] = (\partial_x \epsilon)^T \mathrm{E}[\Delta\boldsymbol{x}\Delta\boldsymbol{x}^T](\partial_x \epsilon) + (\partial_{x'} \epsilon)^T \mathrm{E}[\Delta\boldsymbol{x}'\Delta\boldsymbol{x}'^T](\partial_{x'} \epsilon). \tag{5.13}$$

Now, noting that $\Delta \boldsymbol{x} = [(-\Delta u - \Delta v), \Delta u, \Delta v]^T$ and similarly for $\Delta \boldsymbol{x}'$, we see that, if we assume $\mathrm{E}[\Delta u^2] = \mathrm{E}[\Delta v^2] = \sigma^2$ and $\mathrm{E}[\Delta u \Delta v] = 0$, the covariance matrices reduce to

$$\mathrm{E}[\Delta \boldsymbol{x} \Delta \boldsymbol{x}^T] = \mathrm{E}[\Delta \boldsymbol{x}' \Delta \boldsymbol{x}'^T] = \sigma^2 \begin{bmatrix} +2 & -1 & 0 \\ -1 & 1 & 0 \\ 0 & 0 & 1 \end{bmatrix}. \tag{5.14}$$

The derivatives in Equation (5.13) can be found by differentiating Equation (5.10) with respect to $\boldsymbol{x}_i$ and $\boldsymbol{x}_i'$:

$$\partial_{x_i} \epsilon_i = \boldsymbol{e}^2[(\boldsymbol{x}_i' \cdot \boldsymbol{e}^1) + (\boldsymbol{x}_i' \cdot \boldsymbol{e}^3) F_{23}] + \boldsymbol{e}^3[(\boldsymbol{x}_i' \cdot \boldsymbol{e}^1) F_{31} + (\boldsymbol{x}_i' \cdot \boldsymbol{e}^3) F_{33}] \tag{5.15}$$

and

$$\partial_{x_i'} \epsilon_i = \boldsymbol{e}^1[(\boldsymbol{x}_i \cdot \boldsymbol{e}^2) + (\boldsymbol{x}_i \cdot \boldsymbol{e}^3) F_{31}] + \boldsymbol{e}^3[(\boldsymbol{x}_i \cdot \boldsymbol{e}^2) F_{23} + (\boldsymbol{x}_i \cdot \boldsymbol{e}^3) F_{33}]. \tag{5.16}$$

We now have all quantities necessary to form the cost function S that we wish to minimize.

5.1 Results

The above procedure was implemented in MATLAB using a minimization routine which did not require analytic gradients (although these are easily available) and which minimized S with respect to the parameters $[F_{23}, F_{31}, F_{33}, \boldsymbol{e}_1 \cdot \sigma_2, \boldsymbol{e}_1 \cdot \sigma_3, \boldsymbol{e}_2 \cdot \sigma_2, \boldsymbol{e}_2 \cdot \sigma_3]$. The routine was given an initial F calculated by Hartley's linear method (with normalization and projection, see [6]). Twenty six sets of matching points in the two images were generated and gaussian noise was added to 20 of these point pairs. The matrix F was then estimated from these 20 noisy matches. The F obtained was then tested on the 6 noiseless matching pairs $(\tilde{\boldsymbol{x}}_i, \tilde{\boldsymbol{x}}_i'), i = 1, .., 6$. Figure 3 plots Δ^2 against the noise standard deviation σ, where

$$\Delta^2 = \frac{1}{6} \sum_{i=1}^{n} \underline{f}(\tilde{\boldsymbol{x}}_i, \tilde{\boldsymbol{x}}_i')^2. \tag{5.17}$$

Δ^2 gives one measure of how good the estimated F is, since $\underline{f}(\tilde{\boldsymbol{x}}_i, \tilde{\boldsymbol{x}}_i')$ should be identically zero for each of the noiseless matched pairs. For each σ the value of F was estimated by averaging over 50 realizations of the noise. The solid line shows Δ^2 for the F calculated using the non-linear method described here and the dashed line shows Δ^2 calculated using Hartley's linear method. It should be noted here that we can make various other measures of how good our estimate of F is. For example, one might measure how "far" the estimated epipoles are from the true epipoles. In general, while it is important to establish the correct non-linear method by which to estimate F, it is generally the case that for cases of moderate noise, the approximate linear methods actually do very well.

We can see that if performance is measured in this way the non-linear method gives significantly better results. It should be noted here that this problem, an

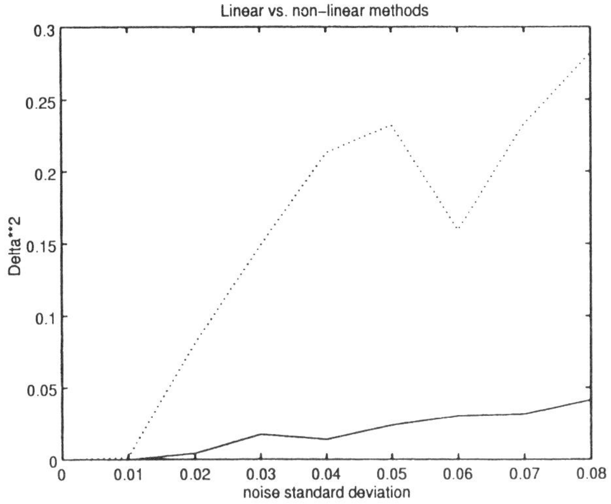

Figure 3. Comparison of the non-linear and linear methods of determining F

important one in computer vision, has been investigated in many different ways and the method described here is essentially similar to the non-linear technique described in [4]. However, we can now use exactly the same techniques to extend the discussion to 3 views where the aim is to recover a structured $3 \times 3 \times 3$ tensor, T_{ijk}, known as the *trilinear tensor*. This tensor has 27 components but is known to have at most 18 degrees of freedom – the problem of determining a T with the correct structure is currently of much interest, but conventionally there is, as yet, no convincing way of imposing the constraints T satisfies. Here we propose proceeding in the same way as for bilinear case and claim that this method, whereby the tensor is regarded as a linear function and the constraints are encoded by considering the action of the function on the epipoles (there are now six – two in each image plane) provides a very natural way of determining a T with the correct structure. A full account of this will be presented elsewhere.

6 Conclusions

The aim of this paper has been to outline the general principles of encoding structure in signal processing problems using linear functions and the geometric algebra framework instead of matrix calculus. We have also looked at the optimization of such structured problems. In general, once we have successfully encoded a desired structure into a linear function the subsequent differentiation is straightforward; it is this encoding of structure that varies from problem to problem and for which there is no easy recipe – this is, however, also a problem in conventional matrix algebra. We have given several examples of problems which can be addressed by these techniques, firstly a simple example of 1D zero-mean Gaussian data, for which the answer was known. Secondly, the question of multi-source signal separation was put into a form to which these methods could be applied. Lastly, the problems of estimating structured matrices and tensors in computer vision were addressed using this linear function approach. There are many other instances in signal processing where the encoding of structure may be of great use. Further areas we intend to investigate are power spectrum estimation in the case where we have missing data and various array processing tasks – both problems will involve the manipulation of complicated block-Toeplitz matrices, so that the starting point will be the encoding of such block-Toeplitz nature into a linear function.

Bibliography

1. Clifford, W.K. 1878. Applications of Grassmann's extensive algebra. *Am. J. Math.* 1: 350–358.
2. Doran, C.J.L. 1994. Geometric Algebra and its Applications to Mathematical Physics. Ph.D. Thesis, University of Cambridge.
3. Doran, C.J.L., Lasenby, A.N. and Gull, S.F. 1996. Linear Algebra. In W.E. Baylis, editor, *Clifford (Geometric) Algebras: with applications in physics, mathematics and engineering.* Birkhauser Boston, pp65-79.
4. Luong, Q-T and Faugeras, O.D. 1996. The Fundamental Matrix: Theory, Algorithms, and Stability Analysis. *International Journal of Computer Vision*, 17, 43-75.
5. Grassmann, H. 1877. Der Ort der Hamilton'schen Quaternionen in der Ausdehnungslehre. *Math. Ann.*, 12: 375.
6. Hartley, R.S. 1995. In defence of the eight-point algorithm *Proceedings of the 5th International Conference on Computer Vision*, IEEE Computer Society Press, Boston, MA, pp1064-1070.
7. Hestenes, D. 1986. New Foundations for Classical Mechanics. *D. Reidel*, Dordrecht.
8. Hestenes, D. and Sobczyk, G. 1984. Clifford Algebra to Geometric Calculus: A unified language for mathematics and physics. *D. Reidel*, Dordrecht.
9. Lasenby, J., Bayro-Corrochano, E., Lasenby, A.N. and Sommer, G. 1996. A new methodology for computing invariants in computer vision. Proceedings of ICPR'96, Vienna, Vol.I, pp334-338.
10. Lasenby, J., Lasenby, A.N., Doran, C.J.L. and Fitzgerald, W.J. 1998. New Geometric Methods in Computer Vision: an Application to Structure and Motion Estimation. to appear in *International Journal of Computer Vision.*
11. Sabata, B. and Aggarwal, J.K. 1991. Estimation of motion from a pair of range images: a review. *CVGIP: Image Understanding*, 54(3): 309–324.

12. Weng, J., Huang, T.S. and Ahuja, N. 1989. Motion and Structure from Two Perspective Views: Algorithms, Error Analysis and Error Estimation. *IEEE Trans.Pattern Anal.Mach.Intelligence*, 11(5): 451–476.

Rational Approximation using Gröbner Bases: Some Numerical Results

Patrick Fitzpatrick[1]

Department of Mathematics, University College Cork, Ireland

Abstract

The classical problem of approximating an unknown function by a rational function is considered. In the most general case we allow the specification of the unknown function and one or more of its derivatives at an arbitrary number of points. A new technique is given for determining a complete parametrisation of the set of all solutions satisfying a natural minimality condition, under the assumption that the arithmetic is exact. Results of some numerical experiments are presented indicating that the method may also be applied in the presence of roundoff.

1 Introduction

Let $f(x)$ be an unknown function which we want to approximate using a rational function $a(x)/b(x)$ where a, b are polynomials. Classically, there are two situations in which this problem is addressed. In the first place we may specify the values taken by f at a sequence of nodes (or knots) $\alpha_1, \alpha_2, \ldots, \alpha_N$; this is the (*scalar*) *rational interpolation* problem. Second we may specify the values of the derivatives $f^{(i)}, i = 0, 1, \ldots, N$ at a single node α, in which case we have the (*partial*) *realization problem.* These problems have a long history going back to Cauchy, Jacobi, Chebyshev, Frobenius, and Padé. In the engineering literature they appear, in particular, in the context of the design of IIR filters. In a different direction both problems have arisen in decoding algorithms for Reed–Solomon error correcting codes: the partial realization problem in Berlekamp's "key equation", solved by the Berlekamp–Massey algorithm (1969), and the interpolation problem in formulation and solution of the decoding problem by Welch and Berlekamp (1983).

The most general problem of this type (known in the context of approximation by polynomials as *Hermite interpolation*) is defined by the specification not only of the values of f at the nodes but also of one or more derivatives at each node. Moreover, rather than the determination of a single solution to the problem, we seek a parametrisation of the set of all solutions satisfying a minimality condition on the degrees of the polynomials a, b.

[1]Financial assistance from the Faculty of Arts, University College Cork, is gratefully acknowledged.

We begin therefore with an arbitrary sequence of points

$$(\alpha_1, \beta_1), (\alpha_2, \beta_2), \ldots, (\alpha_N, \beta_N) \tag{1.1}$$

where α_J, β_J lie in a field F. Let $\tau_I, I = 1, \ldots, L$ be the distinct first components and define the corresponding subsequences

$$(\tau_I, \sigma_{I,0}), (\tau_I, \sigma_{I,1}), \ldots, (\tau_I, \sigma_{I,n_I-1}), \; I = 1, \ldots, L \tag{1.2}$$

where $\sum_{I=1}^{L} n_I = N$ and $\sigma_{I,k} = \beta_J$ for some J. When F has characteristic zero the rational function $y = a/b$ is called an interpolation of (1.1) if $a \neq 0, b(\tau_I) \neq 0$ for all I, and

$$\frac{d^k y}{dx^k}|_{\tau_I} = \sigma_{I,k}, \; I = 1, \ldots, L, k = 0, \ldots, n_I - 1. \tag{1.3}$$

Alternatively, the Taylor series expansions of y about the τ_I satisfy

$$y = \frac{1}{0!}\sigma_{I,0} + \frac{1}{1!}\sigma_{I,1}(x - \tau_I) + \cdots + \frac{1}{(n_I - 1)!}\sigma_{I,n_I-1}(x - \tau_I)^{n_I-1} + \cdots \tag{1.4}$$

When F has non-zero characteristic these are replaced by formal power series expansions in powers of $x - \tau_I$ and y is an interpolation of (1.1) if

$$y = \sigma_{I,0} + \sigma_{I,1}(x - \tau_I) + \cdots + \sigma_{I,n_I-1}(x - \tau_I)^{n_I-1} + \cdots \tag{1.5}$$

For any pair of polynomials (a, b) let $m_r(a, b) = \max\{\delta a, \delta b + r\}$ where δ denotes degree and r is an integer parameter. We define the general rational approximation problem as the determination of all interpolations a/b of the given sequence (1.1) satisfying

CONDITION $\mathbf{A}_r$: $m_r(a, b)$ as small as possible.

The classical interpolation problem, in which all the α_J are distinct, corresponds to $n_I = 1$ for all I, and the partial realisation problem, in which all the α_J are equal (to zero here), corresponds to $n_1 = N$. Antoulas and Anderson [1] have given a solution which determines interpolations of minimum McMillan degree ($r = 0$).

Following Blackburn [2] we consider the *weak* rational approximation problem as follows. Let $h_I \in A = F[x]$ be given by

$$h_I = \sigma_{I,0} + \sigma_{I,1}(x - \tau_I) + \cdots + \sigma_{I,n_I-1}(x - \tau_I)^{n_I-1}. \tag{1.6}$$

A pair $(a, b) \in F[x]^2$ is a weak interpolation for the h_I if

$$a \equiv bh_I \bmod (x - \tau_I)^{n_I}, \quad I = 1, \ldots, L. \tag{1.7}$$

We are especially interested in non-zero weak interpolations (i.e. $(a,b) \neq (0,0)$) satisfying $\mathbf{A}_r$. All formal power series interpolations for (1.1) are weak interpolations for the h_I. On the other hand, a Taylor series interpolation satisfying (1.3) is a formal power series interpolation – and therefore a weak interpolation – for the modified set of polynomials with $\sigma_{I,k}$ replaced by $\frac{1}{k!}\sigma_{I,k}$.

In applications to the decoding problem for RS codes the underlying field of coefficients is finite so there are none of the usual difficulties associated with round-off. The Berlekamp–Massey and Welch–Berlekamp algorithms make use of polynomial algebra (as opposed to matrix techniques), where it is essential, in particular, that each polynomial have a well-defined degree. This is not usually regarded as appropriate for polynomials with real or complex coefficients: for example, what is the degree of the polynomial $10^{-10}x^2 + x + 1$ in arithmetic to 8 decimal places?

In this paper we outline a Gröbner basis approach that extends the algorithm of [3] to the determination of a parametrised solution of the general rational approximation problem. On the way we give a complete parametrisation of the solution set of the weak approximation problem[2]. Full details will appear elsewhere [4]. Our methods also depend on polynomial algebra and were originally derived with particular attention to the finite field case, using exact arithmetic in the coefficient domain. Of course, they also apply equally well when the coefficients are rational numbers and the computations are exact to arbitrary precision (for the convenience of the reader, our examples are given in this context).

Because of the problems associated with polynomial computations in the presence of roundoff, it is not at all clear whether the techniques might be adapted to a finite precision environment. In this paper, however, we present the results of some numerical experiments over the real numbers indicating that the algorithm can also perform successfully in this situation. Further experiments and analysis are currently being undertaken to make this rather surprising observation more precise.

2 Parametrisation of weak interpolations

Here condition $\mathbf{A}_r$ is to be interpreted as applying to (non-zero) weak interpolations (a,b). The set M of all pairs $(a,b) \in A^2$ satisfying (1.7) is closed under componentwise addition and under multiplication by elements of A defined as $f(a,b) = (fa, fb)$. Thus M is an A–submodule of A^2. The reader is referred to [4, 5] for the theory of Gröbner bases of submodules in A^2.

---

[2]Our algorithm for the weak interpolation problem is essentially the same as that derived by Blackburn [2], using different techniques, for the purpose of constructing a single solution rather than a parametrisation of all solutions.

The elements of the form $(x^i, 0), (0, x^j)$ in A^2 are called terms and each element of A^2 can be written as a (scalar[3]) linear combination of such terms. For example, $(3x^2 + x - 1, 2x + 1) = 3(x^2, 0) + (x, 0) - (1, 0) + 2(0, x) + (0, 1)$. In order, to recover as much of the structure of polynomial algebra as possible we need to order the terms. The term order $<_r$ is defined, with reference to the integer parameter r, by $(x^i, 0) <_r (x^j, 0)$ for $i < j$, $(0, x^i) <_r (0, x^j)$ for $i < j$, and $(x^i, 0) <_r (0, x^j)$ for $i \leq j + r$, With $r = 0$ we find that in descending order of terms $(3x^2 + x - 1, 2x + 1) = 3(x^2, 0) + 2(0, x) + (x, 0) + (0, 1) - (1, 0)$. In this way we can refer to the leading term and the corresponding leading coefficient of an element (a, b) : in the example $(x^2, 0)$ is the leading term, and 3 is the leading coefficient. In any submodule the elements can be classified according to their leading terms and, since the leading terms are well–ordered, there exist elements with least leading term. In fact, such a *minimal* element is uniquely defined up to scalar multiples (for if there were two distinct minimal elements then their leading terms would be the same and so, by subtracting a scalar multiple of one element from the other, we would obtain an element with lesser leading term, contrary to minimality). If the leading term of (a, b) has the form $(x^i, 0)$ we say that it is on the left while if it has the form $(0, x^j)$ we say that it is on the right.

Relative to $<_r$ the submodule M has a subset, of the form $\mathcal{B} = \{(a_1, b_1), (a_2, b_2)\}$, again uniquely defined up to scalar multiples, called a *reduced Gröbner basis*. Each element of M can be expressed as an A–linear combination $f_1(a_1, b_1) + f_2(a_2, b_2)$, and one or other of the two elements in the reduced Gröbner basis is the minimal element of M. It can be shown that $\mathcal{B}$ is characterised by the following properties

RGB1 $\mathcal{B}$ consists of two elements $(a_1, b_1), (a_2, b_2)$ with leading terms relative to $<_r$ on opposite sides (we may assume, without loss of generality, that (a_1, b_1) is the minimal element of M);

RGB2 if the leading term of (a_1, b_1) is on the left, then $\delta a_2 < \delta a_1, \delta b_1 < \delta b_2$, while if the leading term of (a_1, b_1) is on the right then $\delta a_1 < \delta a_2, \delta b_2 < \delta b_1$.

A set of two elements of M that satisfies only condition RGB1 (or, more generally, any set containing a subset satisfying RGB1) is called a Gröbner basis of M. It is easy to show that any Gröbner basis can be converted to a reduced Gröbner basis.

Example 1. The sequence $(0, 0), (1, 1), (0, 1), (1, 1/2), (1, -1/6)$ corresponds to the congruences

$$\begin{aligned} h_1 &\equiv x \bmod x^2 \\ h_2 &\equiv 1 + \frac{1}{2}x - \frac{1}{6}x^2 \bmod (x-1)^3. \end{aligned}$$

[3]Although in the context of polynomial modules the word "scalar" normally refers to the polynomial multiplier in expressions such as $f(a, b)$, we find it convenient to restrict its use elements of the field F (denoted throughout by lower case Greek letters).

It is easy to verify directly that the two elements of the set

$$\mathcal{B} = \{(x(-1+7x), -1+4x+x^2), (x(3x-1)(x-4), 4-10x)\}$$

are weak interpolations and that $\mathcal{B}$ satisfies properties RGB1 and RGB2 relative to $<_0$. It follows that $\mathcal{B}$ is the reduced Gröbner basis of M for this sequence of points. Note that only the first basis element (a_1, b_1) satisfies condition $\mathcal{A}_0$ and that $m_0(a_1, b_1) = 2$. As a consequence we deduce that the only elements of M satisfying $\mathcal{A}_0$ are the scalar multiples of (a_1, b_1). In other words, there is essentially only one weak interpolation satisfying $\mathcal{A}_0$.

Example 2. The sequence $(-1,2), (0,1), (1,-1)$ corresponds to the congruences

$$\begin{aligned} h_1 &\equiv 2 \bmod (x+1) \\ h_2 &\equiv 1 \bmod x \\ h_3 &\equiv -1 \bmod (x-1). \end{aligned}$$

Again, it can be verified that

$$\mathcal{B} = \{(-3+5x, -3+x), (-4, -4+3x+5x^2)\}$$

is the reduced Gröbner basis relative to $<_{-1}$ for this sequence, that the first element is the minimal element in M, and that both elements satsify condition $\mathcal{A}_{-1}$ with $m_{-1}(a, b) = 1$. Here there are infinitely many weak interpolations comprising the set of all scalar linear combinations $\gamma_1(a_1, b_1) + \gamma_2(a_2, b_2), \gamma_1, \gamma_2 \in F$.

These examples illustrate the two essentially distinct possibilities: either the minimal element in M is the only one that satisfies $\mathbf{A}_r$ or both elements of the reduced Gröbner basis satisfy $\mathbf{A}_r$ and then so also does any scalar linear combination of them. The following theorem makes this more precise.

Theorem 3. *Let $\{(a_1, b_1), (a_2, b_2)\}$ be the reduced Gröbner basis of M relative to $<_r$ with (a_1, b_1) minimal. Then*

1. *(a_1, b_1) satisfies $\mathbf{A}_r$.*
2. *One of the following holds:*

   I *(unique solution) $m_r(a_1, b_1) < m_r(a_2, b_2)$ and (a_1, b_1) is the unique element of M satisfying $\mathbf{A}_r$.*

   II *(parametrised solution) $m_r(a_1, b_1) = m_r(a_2, b_2)$ and the set of $(a, b) \in M$ satisfying $\mathbf{A}_r$ is $\{\gamma_1(a_1, b_1) + \gamma_2(a_2, b_2), \gamma_1, \gamma_2 \in F\}$.*

3 The algorithm

We define the sequence of submodules M_k, $k = 0, 1, \ldots, N$ where $M_0 = A^2, \ldots$, $M_N = M$, and M_k is the solution module that corresponds to the system of congruences analogous to (1.7) arising from the first k points in (1.1). More precisely, let τ_I appear k_I times as a first component in the first k points, so $\sum_{I=1}^{L} k_I = k$, and let M_k be the set of common solutions of the congruences

$$a \equiv bh_I^{(k_I)} \bmod (x - \tau_I)^{k_I}, \quad I = 1, \ldots, L \tag{3.1}$$

where

$$h_I^{(k_I)} = \sigma_{I,0} + \sigma_{I,1}(x - \tau_I) + \cdots + \sigma_{I,k_I-1}(x - \tau_I)^{k_I-1}$$

and initially $k_I = 0, h_I^{(k_I)} = 0$. To fix notation let the $(k+1)$ th point have first component τ_J.

The algorithm constructs a sequence of Gröbner bases $\{(a_1, b_1), (a_2, b_2)\}$ of the modules M_k according to the following theorem. Let $\mathcal{B} = 0\{(a_1, b_1), (a_2, b_2)\}$ be the reduced Gröbner basis of M_k (beginning with $\{(1,0), (0,1)\}$ for M_0) and suppose that (a_1, b_1) is minimal. Define scalars $\nu_i, i = 1, 2$ by the equations

$$a_i - b_i h_J^{(k_J)} \equiv \nu_i (x - \tau_J)^{k_J} \bmod (x - \tau_J)^{k_J+1} \tag{3.2}$$

and define the updated set $\mathcal{B}' = \{(a_1', b_1'), (a_2', b_2')\}$ by

$$\begin{aligned} &\text{if} \quad \nu_1 = 0 \quad \text{then} \begin{cases} (a_1', b_1') = (a_1, b_1) \\ (a_2', b_2') = ((x - \tau_J)a_2, (x - \tau_J)b_2) \end{cases} \\ &\text{if} \quad \nu_1 \neq 0 \quad \text{then} \begin{cases} (a_1', b_1') = ((x - \tau_J)a_1, (x - \tau_J)b_1) \\ (a_2', b_2') = \left(a_2 - \dfrac{\nu_2}{\nu_1}a_1, b_2 - \dfrac{\nu_2}{\nu_1}b_1\right). \end{cases} \end{aligned} \tag{3.3}$$

Generalising [3, Theorem 4.2] we have

Theorem 4. *The updated set $\mathcal{B}'$ is a Gröbner basis of M_{k+1} relative to $<_r$.*

In the algorithm (presented in Table 1) subscripts on the basis elements are denoted by the boolean variable $i = 1, 2$ which identifies the minimal element of the current basis (with bar denoting complement). The algorithm stores and updates a table each row of which contains values of $\tau_I, h_I^{(k_I)}, a_1, b_1, \nu_1, a_2, b_2, \nu_2$. At each step a new point (α, β) is accepted as input. If $\tau_J = \alpha$ has already appeared as a first component then the appropriate polynomial h is updated; otherwise a new row of the table is constructed. Next the values of ν_1, ν_2 are computed (at (1)) and the basis pairs $(a_1, b_1), (a_2, b_2)$ in the row corresponding to τ_J are updated accordingly.

Table 1. Weak interpolation algorithm

INPUT
$(\alpha_1, \beta_1), (\alpha_2, \beta_2), \ldots, (\alpha_N, \beta_N), r$

INITIALISATION
$(a_1, b_1) := (1, 0)$; $(a_2, b_2) = (0, 1)$; $k := 0$
IF $r \geq 0$ THEN
  $i := 0$; $d := 1 + r$
ELSE
  $i := 1$; $d := -r$
FI

WHILE $k \leq N$ DO
  $k := k + 1$; $(\alpha_k, \beta_k) \longrightarrow (\tau_J, \nu_i, \nu_{\bar{i}})$ (1)
  IF $\nu_i \neq 0$ THEN
    $(a_{\bar{i}}, b_{\bar{i}}) := \left(a_{\bar{i}} - \frac{\nu_{\bar{i}}}{\nu_i} a_i, b_{\bar{i}} - \frac{\nu_{\bar{i}}}{\nu_i} b_i\right)$
    $(a_i, b_i) := ((x - \tau) a_i, (x - \tau) b_i)$
    $d := d - 1$
    IF $d = 0$ THEN
      $i := \bar{i}$; $d := 1$
    FI
  ELSE
    $(a_{\bar{i}}, b_{\bar{i}}) := ((x - \tau) a_{\bar{i}}, (x - \tau) b_{\bar{i}})$
    $d := d + 1$
  FI
OD

UPDATE table (2)

EXTRACT $\{(a_i, b_i), (a_{\bar{i}}, b_{\bar{i}})\}$ from first row of the table

EXPAND $a_i, b_i, a_{\bar{i}}, b_{\bar{i}}$ as polynomials in x (3)

REDUCE $\{(a_i, b_i), (a_{\bar{i}}, b_{\bar{i}})\}$ to $\{(a_i, b_i), (a'_{\bar{i}}, b'_{\bar{i}})\}$ (4)

OUTPUT $\{(a_i, b_i), (a'_{\bar{i}}, b'_{\bar{i}})\}$

Finally, the rest of the table is modified to incorporate the changes made to the basis elements (at (2)). The output values of the basis elements may be taken from the first row of the table, but these will need to be expanded in terms of x (at (3)) if the the corresponding value of α is not zero. A reduction step (4) may be required to convert the output Gröbner basis to a reduced Gröbner basis. Details of these computations, including several shortcuts that minimise their complexity, may be found in [5].

4 Parametrisation of rational interpolations

In order to give a complete parametrisation of the rational interpolations of the given sequence of points for which $\mathbf{A}_r$ holds (with $\mathbf{A}_r$ now interpreted as a minimality condition *among rational interpolations*) we need to ensure that the zeros of the denominators avoid the components α_J and also that the numerators and corresponding denominators have no common factors. The following theorem gives the details.

Theorem 5. *Let m denote the minimum value of m_r among rational interpolations a/b and let $\mathcal{B} = \{(a_1, b_1), (a_2, b_2)\}$ be the reduced Gröbner basis of the module M of weak interpolations. Then the rational interpolations a/b satisfying $m_r(a, b) = m$ are as follows.*

- I *(unique solution) If $m_r(a_1, b_1) < m_r(a_2, b_2)$ and $b_1(\tau_I) \neq 0$ for all I then $a_1 \neq 0$ and a_1/b_1 is the unique interpolation.*
- II *(parametrised solution) If $m_r(a_1, b_1) < m_r(a_2, b_2)$ and $b_1(\tau_I) = 0$ for some I, or if $m_r(a_1, b_1) = m_r(a_2, b_2)$, then there is a family of interpolations defined by $\dfrac{fa_1 + \gamma a_2}{fb_1 + \gamma b_2}$ for $f \in A, \gamma \in F$, where f is an arbitrary polynomial of degree $m_r(a_2, b_2) - m_r(a_1, b_1)$, with the restriction $(fb_1 + \gamma b_2)|_{\tau_I} \neq 0$ for all I. The number of independent parameters is $\delta f + 1 = m_r(a_2, b_2) - m_r(a_1, b_1) + 1$.*

5 Finite precision effects

In this section we indicate how the algorithm may be used in the presence of round-off error to determine rational approximations. Numerical experiments are presented for (a) the partial realisation problem, (b) the classical rational interpolation problem, and (c) rational approximation of e^x.

(a) Partial realisation. A set of 1000 trials was carried out, in each of which a sequence (1.1) was constructed of 20 points of the form $(0, \beta)$ with β randomly chosen in $[-1, 1]$. The corresponding degree 19 polynomial h in (1.6) was used as input to the algorithm with computation to d decimal places for $d = 9, 10, \ldots$ The output (a, b) was used to form the expansion

$$bh - a \equiv s_0 + s_1 x + \cdots + s_{19} x^{19} \bmod x^{20}$$

and the weak interpolation (a, b) was accepted if

$$\Delta = \left(\sum_{k=0}^{19} s_k \right)^{\frac{1}{2}} < 10^{-6}. \tag{5.1}$$

In each trial the value of d was increased until this condition was satisfied. The results are given in Table 2.

(b) Rational interpolation. A further 1000 trials were carried out, in each of which a rational function a/b was constructed with $\delta a = 2, \delta b = 3$ and with random coefficients in the interval $[-1, 1]$, apart from $b_3 = 1$. For each function, samples $\beta_i, i = 1, \ldots, 6$ were taken for values of α_i chosen at random in the interval $[-4, 4]$, and these were used in the algorithm to construct estimates $\hat{a}, \hat{b}$ using the minimal element of the reduced Gröbner basis. The differences $a_j - \hat{a}_j, b_j - \hat{b}_j$ were compared and the estimate was accepted if

$$\Delta = \left(\sum_{i=0}^{2} (a_j - \hat{a}_j)^2 + \sum_{i=0}^{3} (b_j - \hat{b}_j)^2 \right)^{\frac{1}{2}} < 10^{-6}. \tag{5.2}$$

Again, computations were carried out to $d = 9, 10, \ldots$ decimal places until acceptance was obtained. In approximately 3% of the trials this did not succeed with $d \leq 15$. However, it became apparent on considering the graph of a/b for these cases, that the interval in which the samples α_i were taken was too wide, since it contained one or more poles of the function. These cases were recomputed with the α_i chosen randomly in the interval $[-2, 2]$ or $[-1, 1]$ and all then met the acceptability criterion with $d \leq 14$. Including these cases the results of the experiment are shown in Table 3.

Table 2. Partial realisation

d	*Number of trials passing* (5.1)
9	667
10	265
11	54
12	14
Total	1000

Table 3. Rational interpolation

d	*Number of trials passing* (5.2)
9	304
10	324
11	240
12	116
13	13
14	3
Total	1000

(c) Rational approximation of e^x. The approximation of e^x is often used as a test of interpolation or approximation algorithms (see [3, 6], for example). Taking 5 equally spaced interpolation points in the interval $[-1, 1]$ and applying our algorithm, we obtain the following interesting approximation

$$f(x) = \frac{12.2531 + 6.12448x + x^2}{12.2531 - 6.12448x + x^2}.$$

The divergence between this and e^x is very small as indicated by the maximum error

$$|e^x - f(x)| \leq 0.00037$$

over the interval. This should be compared with the corresponding $(2, 2)$ Padé approximant at $x = 0$ for e^x which is given by

$$f(x) = \frac{12 + 6x + x^2}{12 - 6x + x^2}$$

and for which the maximum error in the interval $[-1, 1]$ is 0.004.

6 Conclusion

We have presented a new approach to the problem of representing a function by rational approximations, defined by specifying the values of the function and its derivatives at multiple points in the domain. Our technique involves the construction of a reduced Gröbner basis of a certain polynomial module and therefore requires exact computations in a finite field or the field of rational numbers. However, experimental evidence suggests that the algorithm may also be used successfully in a finite precision environment over the real or complex numbers.

Bibliography

1. A.C. Antoulas and B.D.O. Anderson, On the scalar rational interpolation problem, *IMA J. Math. Control and Inform.*, 3 (1986), 61–88.

2. S.R. Blackburn, A generalised rational interpolation problem and the solution of the Welch–Berlekamp key equation, *Designs, Codes, and Cryptography*, 11, No. 3, (1997), 223-234.

3. D. Braess, *Nonlinear Approximation Theory*, Springer–Verlag, Berlin, Heidelberg, 1986.

4. P. Fitzpatrick, On the key equation, *IEEE Trans.* IT–41, No. 5 (1995), 1290–1302.

5. P. Fitzpatrick, On the scalar rational interpolation problem, to appear in *Mathematics of Control, Signals, and Systems*, 9 (1996), 352-369.

6. A. Ralston, P. Rabinowitz, *A First Course in Numerical Analysis*, McGraw–Hill Kogakusha, Tokyo, 1978.

Gröbner Basis Design of Incomplete Chebyshev Polynomials

Ivan W. Selesnick

Department of Electrical Engineering, Polytechnic University, Brooklyn, USA

Abstract

This paper describes a novel technique using Gröbner bases for constructing monic polynomials of the form $x^k P(x)$ that best approximate 0 in the Chebyshev sense, an approximation problem discussed by Souto in his 1970 dissertation [5] on filter design. In this paper, a new differential equation in two polynomials is given and it is suggested that a Gröbner basis be used to obtain the sought coefficients. The resulting polynomials can be used to design analogue and digital IIR filters the properties of which are between those of the classical Butterworth and Chebyshev filters of types I and II.

1 Introduction

This paper describes a novel technique using Gröbner bases for constructing monic polynomials of the form $x^k P(x)$ that best approximate 0 in the Chebyshev sense. To our knowledge, of the techniques for constructing such polynomials, this is the first that does not employ an iterative numerical algorithm (for example, Remez-like algorithms and the differential correction algorithm). The problem of constructing such polynomials is a classical one in approximation theory.

From the set of monic n^{th} degree polynomials, the polynomial x^n is an optimal approximation to 0 according to a Taylor series (derivative) criterion. The Chebyshev polynomial is an optimal approximation to 0 according to the the L_∞ or Chebyshev criterion.

In his 1970 dissertation, "A Mixed Flat and Equal-ripple Criterion for Filter Design" [5], Souto was interested in polynomials that mix these two approximation criteria. Specifically, he sought a monic polynomial with a specified number of roots at $x = 0$ and a specified number of equal-sized ripples over $[0, 1]$. Such polynomials are optimal approximations to zero over $[0, 1]$ in the Chebyshev sense subject to the constraint on the number of roots at $x = 0$. See Figure 1.

Both x^n and $T_n(x)$ are used in filter design, and it is of interest to expand the available methods for filter design by using these mixed-type polynomials, as Souto did in [5]. One reason for considering such polynomials in filter design is the flexibility they provide. The number of derivative constraints directly

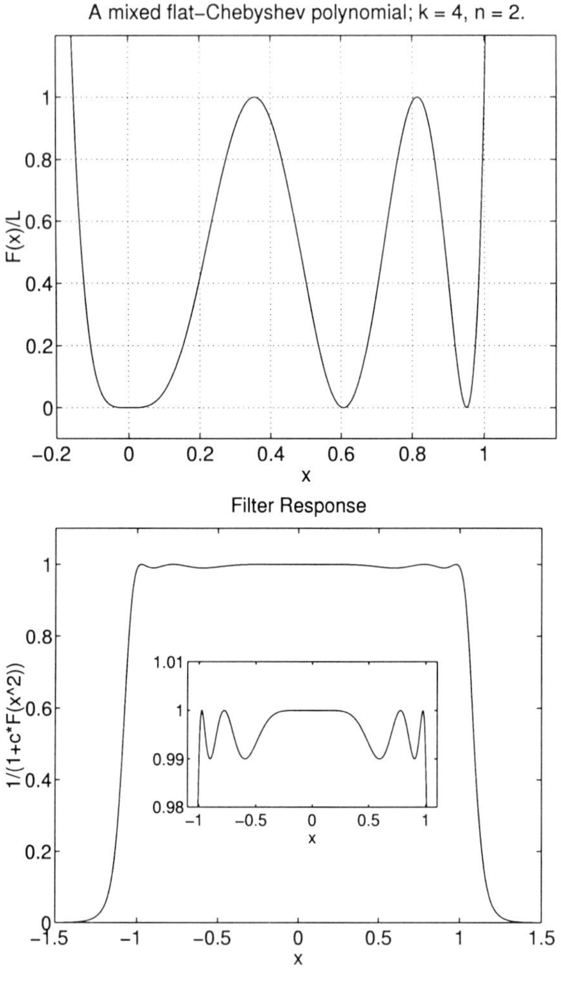

Figure 1. Above, $F_{4,2}(x) = x^4\left(0.57750 - 1.55857x + x^2\right)^2$, $L = 3.5849E-04$. Below, filter response corresponding to $F_{4,2}(x)$

influences the Chebyshev norm of the monic polynomial of fixed degree that best approximates 0 over $[0,1]$ in the Chebyshev sense. For a fixed degree, the minimum achievable Chebyshev norm varies with the number of derivative constraints at $x = 0$.

Properties of mixed flat-Chebyshev polynomials have been studied by Borwein and Erdélyi in [1], in which these polynomials are called *incomplete* Chebyshev polynomials (for some of their coefficients are zero). However, we have not seen the approach for constructing these polynomials, given in this paper, elsewhere.

2 Gröbner bases

The fact that the Gröbner basis gives a solution to this problem is very interesting. Roughly, given a system of multi-variable polynomials, the Gröbner basis is a new system of multi-variable polynomials having the same solution set [2]. The (lexical) Gröbner basis is one in which back-substitution is possible, as in Gaussian elimination in linear algebra. The (lexical) Gröbner basis is similar to the row-reduced echelon form in linear algebra in that the last polynomial in the basis is a polynomial in a single variable. When the roots of this polynomial are known, they can be substituted into the other polynomials of the basis, etc. If there are a finite number of solutions to the original system of equations, this use of the Gröbner basis make it possible to compute them by computing the roots of a set of polynomials.

3 Chebyshev polynomials

In this section the Chebyshev polynomials are derived in a way that can be (roughly) generalized to the desired polynomials. Consider the polynomials $T_n(x)-1$ and $T_n(x)+1$ with n even. The roots of these polynomials in the interval (0,1) are double roots, and the roots of $T_n(x) - 1$ at $x = 1$ and $x = -1$ are simple. Because the derivative of $T_n(x)$ is zero exactly at the roots of $T_n(x) + 1$ and $T_n(x) - 1$, $T_n'(x)$ shares its roots (which are simple) with $(T_n(x) - 1)(T_n(x) + 1)$. By counting roots and degrees, and by making leading coefficients equal, the well known relationship

$$n^2(T_n^2(x) - 1) = (T_n'(x))^2(x^2 - 1) \tag{3.1}$$

is obtained [4]. The monic polynomial $M(x) = \frac{1}{2^{n-1}}T_n(x)$ satisfies $n^2(M^2(x) - \frac{1}{2^{2(n-1)}}) = (M'(x))^2(x^2 - 1)$.

While most presentations of the Chebyshev polynomials solve this differential equation to obtain $\arccos(T_n) = n\arccos(x)$ at this point, here we take another approach. By substituting a generic polynomial with unknown coefficients into (3.1) a set of nonlinear equations for the coefficients of T_n is obtained. For example, if $n = 4$, then the equations obtained by substituting

$M(x) = x^n + \sum_{i=0}^{i=n-1} c_i x^i$ into the differential equation above require that the following eight expressions be 0:

$$-8\,c_3 \tag{3.2}$$
$$-16 - 16\,c_2 - 7\,c_3^2 \tag{3.3}$$
$$-20\,c_2 c_3 - 24\,c_1 - 24\,c_3 \tag{3.4}$$
$$-26\,c_1 c_3 - 32\,c_0 - 16\,c_2 - 9\,c_3^2 - 12\,c_2^2 \tag{3.5}$$
$$-12\,c_2 c_3 - 8\,c_1 - 28\,c_1 c_2 - 32\,c_0 c_3 \tag{3.6}$$
$$-32\,c_0 c_2 - 6\,c_1 c_3 - 4\,c_2^2 - 15\,c_1^2 \tag{3.7}$$
$$-4\,c_1 c_2 - 32\,c_0 c_1 \tag{3.8}$$
$$\frac{1}{4} - c_1^2 - 16\,c_0^2. \tag{3.9}$$

Although these are, at first glance, nonlinear equations in the coefficients c_i, notice that they can be solved simply by making a sequence of substitutions. The first equation in this list requires that c_3 is zero (this is as expected since $T_n(x)$ is an even polynomial for even n). Having ascertained the value of c_3, the second equation in this list gives $c_2 = -1$. When the equations are used in this order, the coefficients of $T_4(x)$ are easily obtained. The same is true for all $T_n(x)$.

4 Derivation of a Gröbner basis solution

To obtain polynomials approximating 0 by a mixed flat-Chebyshev criterion, the following form will be used:

$$F_{k,n}(x) = x^k (A(x))^2 \tag{4.1}$$

where $A(x)$ is a polynomial of degree n. The degree of $F_{k,n}(x)$ is $2n + k$. This form incorporates the flatness requirement at $x = 0$. A polynomial $A(x)$ is sought, so that $F(x)$ has equal-sized ripples over $[0, 1]$.

In order to carry out a procedure to find $A(x)$ that is similar to the procedure used above to obtain the Chebyshev polynomials, it is first necessary to find an equation analogous to (3.1). The polynomial $F_{k,n}(x)$ we seek has n double roots in the interval (0,1), equal-sized ripples of size L in (0,1), and takes on the value L at $x = 1$. See Figure 1. As above, we consider the roots of the polynomials $F(x)$, $F(x) - L$, and the roots of the derivative of $F(x)$. It follows that for the polynomial $F(x)$ we seek, $F(x)(F(x) - L)$ shares a number of roots with $F'(x)$. The roots of $F(x)(F(x) - L)$ in (0,1) are double roots and the roots of $F'(x)$ are simple.

Using (4.1) leads to a relationship between $x^k A(x)^2 - 1$ and $(2xA'(x) + kA(x))(x - 1)$. By matching degrees and leading coefficients, this leads to the requirement, that for the nth degree polynomial $A(x)$ we seek, there must exist a positive number L and a monic polynomial $U(x)$ of degree $k - 1$ such that

$$(2n + k)^2 (x^k A^2(x) - L) = (2xA'(x) + kA(x))^2 (x - 1) U(x). \tag{4.2}$$

After substituting generic monic polynomials with unknown coefficients into (4.2) and equating like powers of x, the resulting equations are not essentially linear as they are in the traditional Chebyshev case. For example, the equations associated with $F_{4,2}(x)$ require that each of the following eight expressions be zero (where $A(x) = a_0 + a_1 x + x^2$ and $U(x) = u_0 + u_1 x + u_2 x^2 + x^3$):

$$-64\,u_2 + 32\,a_1 + 64 \tag{4.3}$$

$$-96\,u_2 a_1 + 28\,a_1^2 - 64\,u_1 + 64\,u_2 + 96\,a_1 + 64\,a_0 \tag{4.4}$$

$$-36\,u_2 a_1^2 - 96\,u_1 a_1 + 96\,u_2 a_1 - 64\,u_2 a_0 + 36\,a_1^2 + 80\,a_1 a_0 - 64\,u_0 + 64\,u_1 + 64\,a_0 \tag{4.5}$$

$$-36\,u_1 a_1^2 + 36\,u_2 a_1^2 - 48\,u_2 a_1 a_0 - 96\,u_0 a_1 + 96\,u_1 a_1 - 64\,u_1 a_0 + 64\,u_2 a_0 + 48\,a_1 a_0 + 48\,a_0^2 + 64\,u_0 \tag{4.6}$$

$$-36\,u_0 a_1^2 + 36\,u_1 a_1^2 - 48\,u_1 a_1 a_0 + 48\,u_2 a_1 a_0 - 16\,u_2 a_0^2 + 96\,u_0 a_1 - 64\,u_0 a_0 + 64\,u_1 a_0 + 16\,a_0^2 \tag{4.7}$$

$$36\,u_0 a_1^2 - 48\,u_0 a_1 a_0 + 48\,u_1 a_1 a_0 - 16\,u_1 a_0^2 + 16\,u_2 a_0^2 + 64\,u_0 a_0 \tag{4.8}$$

$$16\,(3\,u_0 a_1 - u_0 a_0 + u_1 a_0)\,a_0 \tag{4.9}$$

$$16\,u_0 a_0^2 - 64\,L. \tag{4.10}$$

Note that $U(x)$ does appear linearly in (4.2), and so the coefficients of $U(x)$ can be found in terms of the coefficients of $A(x)$ by linear algebra. However, after $U(x)$ is eliminated from the equations, the resulting equations are not inviting. Again, using $F_{4,2}(x)$ for example, $u_2 = \frac{a_1}{2} + 1$, $u_1 = -\frac{5\,a_1^2}{16} + \frac{a_1}{2} + 1 + a_0$ and $u_0 = \frac{3\,a_1^3}{16} - \frac{5\,a_1^2}{16} + \frac{a_1}{2} - \frac{3\,a_1 a_0}{4} + a_0 + 1$. The Equations (4.3–4.10) above, after eliminating $U(x)$ become:

$$-\frac{27\,a_1^4}{4} + 12\,a_1^3 + 32\,a_1^2 a_0 - 20\,a_1^2 - 48\,a_1 a_0 - 16\,a_0^2 + 32\,a_1 + 64\,a_0 + 64 \tag{4.11}$$

$$-\frac{a_1}{4}(27\,a_1^4 - 72\,a_1^3 - 120\,a_1^2 a_0 + 120\,a_1^2 + 288\,a_1 a_0 + 32\,a_0^2 - 192\,a_1 - 384\,a_0 - 384) \tag{4.12}$$

$$\frac{27\,a_1^5}{4} - 9\,a_1^4 a_0 - \frac{45\,a_1^4}{4} - 15\,a_1^3 a_0 + 41\,a_1^2 a_0^2 + 18\,a_1^3 + 16\,a_1^2 a_0 - 48\,a_1 a_0^2 - 16\,a_0^3 + 36\,a_1^2 + 32\,a_1 a_0 + 64\,a_0^2 + 64\,a_0 \tag{4.13}$$

$$3\,(3\,a_1^3 - a_1^2 a_0 - 5\,a_1^2 - 12\,a_1 a_0 + 4\,a_0^2 + 8\,a_1 + 16\,a_0 + 16) a_1 a_0 \tag{4.14}$$

$$3\,a_1^3 a_0^2 - 5\,a_1^2 a_0^2 - 12\,a_1 a_0^3 + 8\,a_1 a_0^2 + 16\,a_0^3 + 16\,a_0^2 - 64\,L. \tag{4.15}$$

Using the purely lexical ordering of monomials with $a_1 < a_0$, the Gröbner basis [2] of the first two of these equations are given by the following polynomials:

$$31050\,a_1^8 - 427491\,a_1^7 + 3623544\,a_1^6 - 24731232\,a_1^5 - 23761024\,a_1^4 + 63315456\,a_1^3 + 40826880\,a_1^2 + 62595072\,a_0^2 - 234078208\,a_1 - 250380288\,a_0 - 250380288, \tag{4.16}$$

$$-2808\,a_1^8 + 56241\,a_1^7 - 548712\,a_1^6 + 4023840\,a_1^5 - 10057216\,a_1^4 - 46133760\,a_1^3 + 6377472\,a_1^2 + 62595072\,a_1 a_0 + 18120704\,a_1, \tag{4.17}$$

and

$$27\,a_1^9 - 432\,a_1^8 + 4032\,a_1^7 - 29184\,a_1^6 + 35840\,a_1^5 + 57344\,a_1^4 - 196608\,a_1^3 - 262144\,a_1^2 - 262144\,a_1. \tag{4.18}$$

Notice that (4.18) is a polynomial in a single variable. Therefore, the coefficient a_1 of $A(x)$ can be found by computing the appropriate root of (4.18). The coefficient a_0 of $A(x)$ can then be computed by (4.17). The roots of (4.18) are

$$\{0,\ -1.55857,\ -0.50503 \pm 0.82455I,\ 1.91389 \pm 9.02319I,\ 2.19563 \pm 1.65705I,\ 10.34958\}. \tag{4.19}$$

Discarding the roots off the real line, the appropriate root can be chosen by graphing the resulting polynomial $F(x)$ over $(0,1)$ for each real root. It turns out that the appropriate root to use for the value of a_1 is -1.55857. The resulting value for a_0 from (4.17) is 0.577508. The polynomial we seek is thus

$$F_{4,2}(x) = x^4\left(0.577508 - 1.55857x + x^2\right)^2.$$

The size of the ripple, L, is given by (4.10), which evaluates to $3.5849E-04$. The polynomial $U(x)$ corresponding to this solution is $0.0042998 + 0.0391110x + 0.2207130x^2 + x^3$. The frequency response magnitude squared of an analogue filter constructed using $F_{4,2}(x)$ is also illustrated in Figure 1.

5 Remarks

Chebyshev polynomials have many special properties. Are any of these properties also satisfied by polynomials $F_{k,n}(x)$? For example, the Chebyshev polynomials obey a second order linear differential equation as well as the first order nonlinear differential Equation (3.1). There also exist recurrence relations for the Chebyshev polynomials. Most important for us, however, are the locations of the roots of the polynomials. The roots of the Chebyshev polynomials are well known to be samples of the cosine function, but, there is no similarly simple expression for the roots of $F_{n,k}$. Erdélyi, however, has found estimates of the locations of the zeros of these polynomials [3].

6 Conclusion

This paper has described a novel technique using Gröbner bases for constructing monic polynomials of the form $x^k P(x)$ that best approximate 0 in the Chebyshev sense. In practice, the use of Gröbner bases as described here is suitable only for relatively small degree problems. However, using an appropriate change of variables and using a different ordering of monomials, may make the use of Gröbner bases more practical for greater degrees.

Acknowledgment

This work has been supported by Nortel (previously known as BNR) and by NSF grant MIP-9316588.

Bibliography

1. P. Borwein and T. Erdélyi. *Polynomials and Polynomial Inequalities.* Springer-Verlag, 1995.
2. D. Cox, J. Little, and D. O'Shea. *Ideals, varieties, and algorithms : an introduction to computational algebraic geometry and commutative algebra.* Springer-Verlag, 1991.
3. T. Erdélyi, 1995. Personal communication.
4. T. W. Körner. *Fourier Analysis.* Cambridge University Press, 1988.
5. F. S. Souto. *A Mixed Flat and Equal-Ripple Criterion for Filter Design.* PhD thesis, Rice University, 1970.

Fractals, Linear Channels and Delay Methods

D.S. Broomhead, J.P. Huke and M.R. Muldoon

Department of Mathematics, University of Manchester Institute of Science and Technology

Abstract

We model the response of a linear channel to random digital input, obtaining a stochastic system of the sort Michael Barnsley calls an Iterated Function System (IFS). We show that modern, geometric methods for time series analysis, originally developed for signals from deterministic sources, may be applied in this genuinely stochastic setting.

1 Introduction

The last few years have seen a growth of interest in two particular developments of signal processing theory: techniques which exploit non-gaussian properties of noise through the use of higher order moments (there is a huge literature see for example [5]); and the use of the so-called "method of delays" together with its rigorous justification based on the theory of dynamical systems [6, 7]. These approaches represent extremes. On the one hand, the method of delays is based on the theory of finite-dimensional, *deterministic*, nonlinear systems. This leads to a theory of nonlinear signal processing which is based on a natural and powerful treatment of nonlinearity but which lacks a traditional concept of noise. On the other hand the use of higher order moments is an inherently statistical treatment which naturally characterises deviations from simple central limit statistics. Nonlinearity arises here more as a pertubation, and much of the organisation and structure which is attendant on nonlinear mechanisms is lost. An essential goal of any theory of signal processing that pretends to be complete is the unification of these extremes. Such a theory should contain linear signal processing as a limiting case.

Many of the obstacles to achieving this goal are a consequence of a lack of common language or even a set of common examples. In this paper we make a move towards the goal by introducing a class of simple examples. These show how the method of delays can be extended to incorporate non-deterministic dynamics; they illustrate how complex dynamics and non-gaussian statistics arise from an apparantly trivial, benign nonlinearity, and—by the same token—demonstrate the fragility of the linearity assumption on which most of our signal processing is based.

1.1 The basic example

Consider a linear channel described by its internal state u which, in the absence of external influences, evolves as a damped harmonic oscillator:

$$\ddot{u} + \gamma\dot{u} + \omega_0^2 u = 0.$$

We chose this model for simplicity rather than from necessity; any other evolution equation having a globally attracting, hyperbolic fixed point would do as well.

Now imagine that this channel is forced by a digital transmission, here modelled as a random binary signal, so that the dynamics become

$$\ddot{u} + \gamma\dot{u} + \omega_0^2 u = \sigma(t), \tag{1.1}$$

where the forcing process $\sigma(t)$ switches randomly between ± 1 every τ seconds[1]. Figure 1 shows a typical realization of $\sigma(t)$ and the corresponding evolution for $u(t)$.

The first step toward a geometric analysis of this model is to simplify the dynamics. The second order system (1.1) is equivalent to a two dimensional, first-order one which, during one of the intervals when $\sigma(t)$ is constant, takes the form:

$$\frac{d}{dt}\begin{pmatrix} u \\ \dot{u} \end{pmatrix} = \begin{bmatrix} 0 & 1 \\ -\omega_0^2 & -\gamma \end{bmatrix}\begin{pmatrix} u \\ \dot{u} \end{pmatrix} + \begin{pmatrix} 0 \\ \sigma_n \end{pmatrix}, \tag{1.2}$$

where σ_n is the value of $\sigma(t)$ when $n\tau \le t < (n+1)\tau$. During this interval the channel evolves toward an equilibrium point,

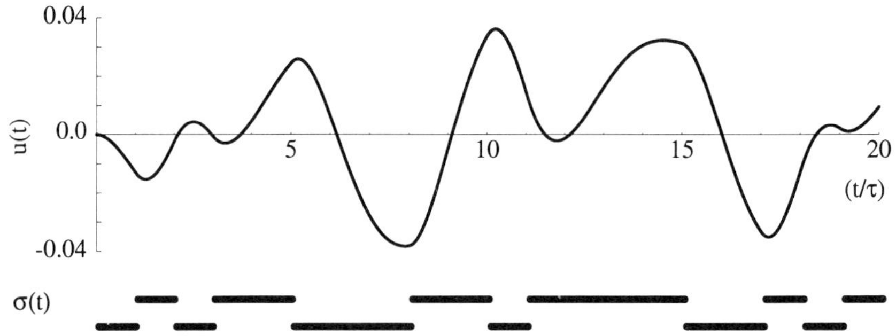

Figure 1. A realization of the digital forcing function and the corresponding evolution of the channel's state. Here $\omega_0 = 2\pi$; $\gamma = 0.8\,\omega_0$ (underdamped) and $\tau = 0.2$

[1]The linearity of the channel implies that the amplitude of the forcing term can be rescaled arbitrarily.

$$\begin{pmatrix} u_\star \\ \dot{u}_\star \end{pmatrix} = \begin{bmatrix} -\gamma/\omega_0^2 & -1/\omega_0^2 \\ 1 & 0 \end{bmatrix} \begin{pmatrix} 0 \\ -\sigma_n \end{pmatrix} = \begin{pmatrix} \sigma_n/\omega_0^2 \\ 0 \end{pmatrix} \equiv \begin{pmatrix} \xi_n \\ 0 \end{pmatrix},$$

that depends on the value σ_n of the forcing. If we consider motion relative to the equilibrium, changing variables to $v = u - \xi_n$, the dynamics become

$$\frac{d}{dt}\begin{pmatrix} v \\ \dot{v} \end{pmatrix} = \begin{bmatrix} 0 & 1 \\ -\omega_0^2 & -\gamma \end{bmatrix} \begin{pmatrix} v \\ \dot{v} \end{pmatrix} \equiv \Omega \begin{pmatrix} v \\ \dot{v} \end{pmatrix}$$

with solutions

$$\begin{pmatrix} v(t) \\ \dot{v}(t) \end{pmatrix} = e^{\Omega t} \begin{pmatrix} v(0) \\ \dot{v}(0) \end{pmatrix}.$$

That is, the channel relaxes toward equilibrium in a way that is independent of the fixed point's location.

1.2 Reduction to discrete random dynamics

Suppose that at time $n\tau$ the channel is in the state

$$\begin{pmatrix} u(n\tau) \\ \dot{u}(n\tau) \end{pmatrix} \equiv \begin{pmatrix} u_n \\ \dot{u}_n \end{pmatrix}.$$

Over the next τ seconds the forcing will be constant and equal to σ_n, so the system will tend toward the equilibrium point at $(\xi_n, 0)$, reaching

$$\begin{aligned} \begin{pmatrix} u_{n+1} \\ \dot{u}_{n+1} \end{pmatrix} &= e^{\Omega\tau} \begin{pmatrix} u_n - \xi_n \\ \dot{u}_n \end{pmatrix} + \begin{pmatrix} \xi_n \\ 0 \end{pmatrix} \\ &= e^{\Omega\tau} \begin{pmatrix} u_n \\ \dot{u}_n \end{pmatrix} + (I - e^{\Omega\tau}) \begin{pmatrix} \xi_n \\ 0 \end{pmatrix} \end{aligned} \tag{1.3}$$

after τ seconds. Here I is the identity matrix.

As the forcing function assumes just two values, the time-τ evolution is given by one of two maps, each of the form (1.3). That is, the dynamics reduce to the random composition of a pair of maps. Further, each of these maps is affine and provided there is some dissipation ($\gamma < 0$), each is a strict contraction. This model is an example of a very rich class of stochastic systems that Michael Barnsley has called Iterated Function Systems (IFS), [1, 2]. In the next section we will review some of the results he and his collaborators have proved, then show how to use them to analyse the stochastically driven channel.

2 Iterated function systems

A *(hyperbolic) IFS with probability* consists of three things: a complete metric space (X, d), a finite collection of contraction mappings $W_k : X \to X$ with

$k \in \{1, \ldots, K\}$ and a collection of probabilities $\{p_k\}_{k=1,K}$, one for each map. A single step of the dynamics involves acting with one of the W_k, chosen at random according to the probabilites $\{p_j\}$. The study of these systems falls into two parts: an examination of their (often beautiful) attracting sets and an analysis of how the points of a typical trajectory are distributed over the attractor. For the former, the study of the attractor as a set, the $\{p_k\}$ are not important and for the moment we will ignore them. The main theorem establishes the existence of global attractors:

Theorem 1. (Barnsley) *Given a hyperbolic IFS, the transformation W that maps compact subsets of X to other compact subsets of X,*

$$W(B) = \bigcup_{k=1}^{K} W_k(B), \tag{2.1}$$

has a unique, globally attracting fixed point $\mathcal{A}$ satisfying

$$\mathcal{A} = \bigcup_{k=1}^{K} W_k(\mathcal{A}).$$

Note that there is nothing probabilistic about the set-to-set mapping in Equation (2.1). In this theorem the construction of $\mathcal{A}$ proceeds by following all of the branches of the IFS at once.

These attractors are often fractal and indeed, there is a two-map IFS acting on the real line whose attractor is that oldest of fractals, the famous middle-thirds Cantor set:

$$\begin{aligned} W_1(x) &= \tfrac{1}{3}x \qquad & p_1 &= \tfrac{1}{2} \\ W_2(x) &= \tfrac{1}{3}x + \tfrac{2}{3} \qquad & p_2 &= \tfrac{1}{2}. \end{aligned} \tag{2.2}$$

Figure 2 shows this set, while Figure 3 shows two attractors of our channel's IFS, one for an underdamped channel and the other from an overdamped one.

The probabilities p_k become important when one is interested in statistical properties of the IFS, such things as time averages along an oribit or averages over many trajectories. The main theorem here is:

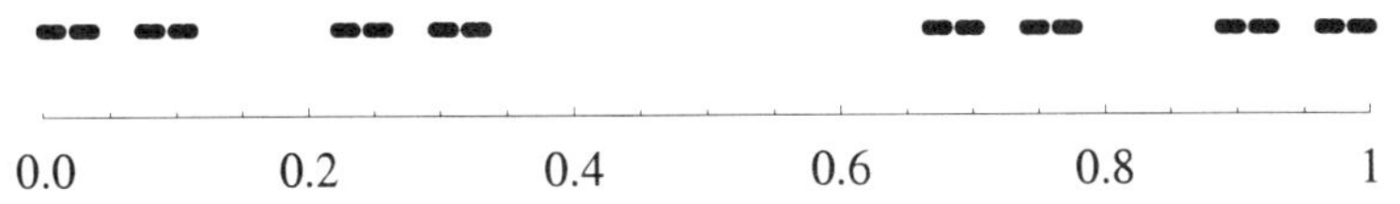

Figure 2. The middle-thirds Cantor set, generated by applying the set-to-set map (2.1) 10 times to the closed interval [0, 1]

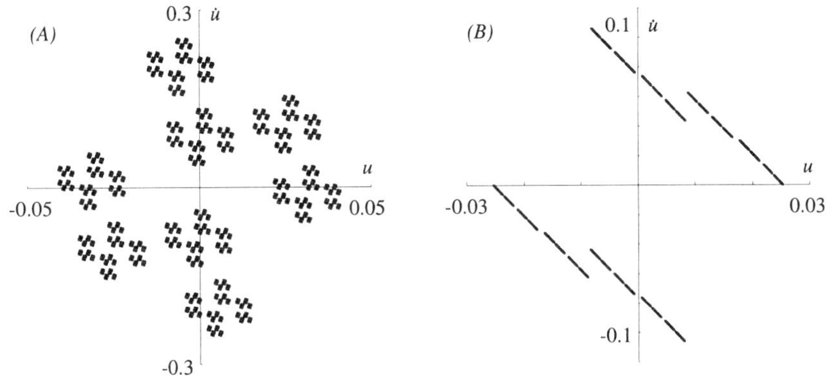

Figure 3. Attractors for the model (1.3). In both cases, $\omega_0 = 2\pi$ and $\tau = 0.2$, but in panel (A) $\gamma = 0.8\,\omega_0$ producing an underdamped system, while in panel (B) $\gamma = 2.2\,\omega_0$, an overdamped regime. Both figures show the image of the set $\{(-1/\omega_0^2,\, 0),\ (1/\omega_0^2,\, 0)\}$ after 10 iterations of the set-to-set map (2.1)

Theorem 2. (Elton) *There is a unique invariant measure with support $\mathcal{A}$ and this measure is ergodic for almost all trajectories of the IFS: that is, for almost all sequences of the W_k.*

This makes explicit the relationship between the attracting set of Theorem 1 and the invariant set generated by random composition of the W_k. Theorem 2 says that there is a single probability measure on $\mathcal{A}$ that describes the long-time statistics gathered along orbits of almost all the points in $\mathcal{A}$ under almost all the sequences of maps consistent with the p_k. This measure is an extra structure that rests on top of the attracting set. If, for example, we chose different values for the probabilities in (2.2), say $p_1 = 0.8$ and $p_2 = 0.2$, then a typical orbit would still wander over a set that looked like Figure 2, but it would spend much more of its time in the left half of the interval.

The fractal structures in Figures 2 and 3 are a direct consequence of the randomness of IFSs: considered individually, each of the maps W_k gives rise to trivial dynamics characterized by a single attracting fixed point. It is only the random choice of map, which drives the orbit first toward the fixed point of one W_j, then, typically, toward one of the others, that gives rise to the complicated attractors pictured above.

2.1 Labelling points in the attractor

In each of the particular examples shown in Figure 3 the attractor $\mathcal{A}$ divides into two disjoint pieces, each being the image of $\mathcal{A}$ under one of the maps (1.3)

$$\mathcal{A} = W_1(\mathcal{A}) \bigcup W_2(\mathcal{A}).$$

Each of these can be subdivided similarly:

$$\begin{aligned}\mathcal{A} &= W_1\left(W_1(\mathcal{A}) \cup W_2(\mathcal{A})\right) \bigcup W_2\left(W_1(\mathcal{A}) \cup W_2(\mathcal{A})\right) \\ &= W_1 \circ W_1(\mathcal{A}) \bigcup W_1 \circ W_2(\mathcal{A}) \bigcup W_2 \circ W_1(\mathcal{A}) \bigcup W_2 \circ W_2(\mathcal{A}).\end{aligned}$$

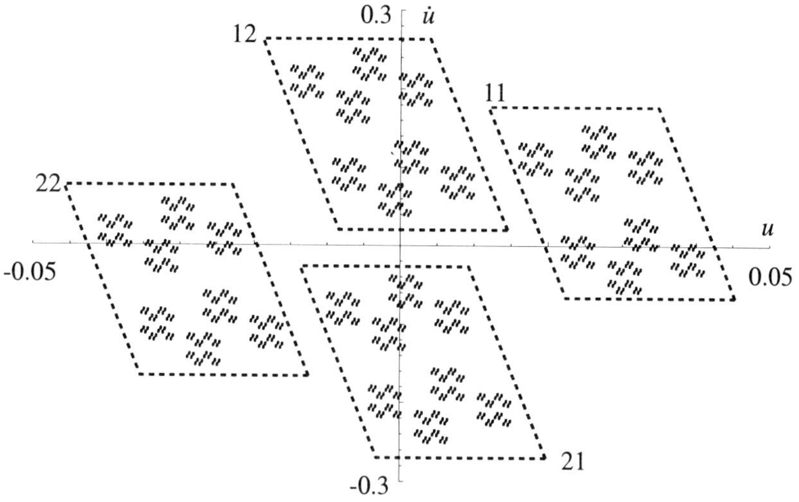

Figure 4. The attractor of the underdamped system with images of $\mathcal{A}$ labelled by the sequences of maps that produce them. The labels read right-to-left, so "21" means "act first with W_1, then with W_2"

As Figure 4 shows, we can address the pieces of $\mathcal{A}$ with the sequence of maps used to generate them. As the labels grow longer, the pieces they address become smaller, suggesting that we might ultimately be able to label individual points. There are two theorems bearing on this idea.

Theorem 3. *Every point in the attractor of a hyperbolic IFS can be associated with its own (infinitely long) label.*

Theorem 4. *Given a hyperbolic IFS and its attractor $\mathcal{A}$, each point $x \in \mathcal{A}$ has a* unique *label if and only if*

$$W_j(\mathcal{A}) \bigcap W_k(\mathcal{A}) = \emptyset \tag{2.3}$$

for all pairs of maps W_j and W_k in the IFS.

When both these theorems apply each point in $\mathcal{A}$ is uniquely labelled by its own history (the IFS is said to be *totally disconnected*). For the IFS channel this means that in principle one could infer the entire history of the transmission from an observation of the channel's current state, $(u_n, \dot{u}_n)$. We explore this idea in the next section, where we will also show that one can infer $(u_n, \dot{u}_n)$ from a few successive values of u_j; that is, one can in principle infer the channel's entire history from an observation of its recent past.

3 Delay-embedding and decision boundaries

The attractors pictured in Figure 3 lie in the channel's natural phase space where the coordinates are u and $\dot{u}$, but one would like to analyse only the channel's state, $u(t)$, or better still, the time series $\{u_0, u_1, \cdots\}$ of states measured at the baud rate of the transmission. A natural approach is to use the method of delays, that is, to consider vectors $(u_n, u_{n+1}, \cdots, u_{n+d-1})$ of successive measurements. In the problem at hand, one can work out the corresponding changes of variables explicitly: we begin with the case $d = 2$. Starting with Equation (1.3), one can obtain a recursion relation connecting successive values of u,

$$u_{n+1} = e_{11}^{\Omega\tau} u_n + e_{12}^{\Omega\tau} \dot{u}_n + \left(1 - e_{11}^{\Omega\tau}\right) \xi_n,$$

where $e_{11}^{\Omega\tau}$ is the 11 entry of the matrix exponential $e^{\Omega\tau}$. This leads to the change of coordinates:

$$\begin{pmatrix} u_n \\ u_{n+1} \end{pmatrix} = A_1 \begin{pmatrix} u_n \\ \dot{u}_n \end{pmatrix} + T_1(\xi_n), \tag{3.1}$$

where

$$A_1 \equiv \begin{pmatrix} 1 & 0 \\ e_{11}^{\Omega\tau} & e_{12}^{\Omega\tau} \end{pmatrix} \quad \text{and} \quad T_1(\xi_n) \equiv \begin{pmatrix} 0 \\ \left(1 - e_{11}^{\Omega\tau}\right) \xi_n \end{pmatrix}.$$

For fixed ξ_n this is an inevertible transformation, and as $\xi_n = \pm 1/\omega_0^2$, there are actually two such transformations, so the $d = 2$ delay-embedding should

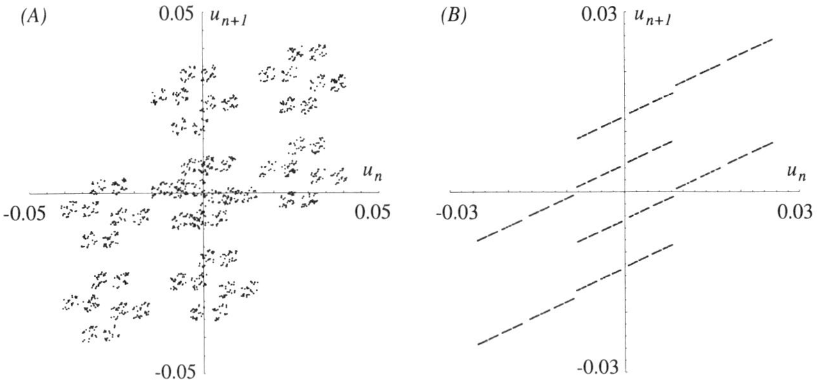

Figure 5. The images of the attractors from Figures 3 in the $(u_n,\, u_{n+1})$ space

produce two copies of the fractal we saw in $(u,\, \dot{u})$ space. Figure 5 bears this out. It also leads to the question, does a totally disconnected fractal in $(u,\, \dot{u})$ space give rise to a totally disconnected fractal in (u_n, u_{n+1}) space?

Whether it does or not, the mapping that produced Figure 5 is impractical; even if we knew that the blur around the origin in the underdamped case consisted of two disjoint clouds with distinct, two-symbol labels, they are too closely mingled to allow us clearly to distinguish the correct label for a given point. Adding one more observation—looking at $(u_n,\, u_{n+1},\, u_{n+2})$—removes the ambiguity. The dynamics become:

$$\begin{pmatrix} u_{n+2} \\ \dot{u}_{n+2} \end{pmatrix} = e^{2\Omega\tau} \begin{pmatrix} u_n \\ \dot{u}_n \end{pmatrix} + e^{\Omega\tau} \left(I - e^{\Omega\tau}\right) \begin{pmatrix} \xi_n \\ 0 \end{pmatrix} + \left(I - e^{\Omega\tau}\right) \begin{pmatrix} \xi_{n+1} \\ 0 \end{pmatrix}$$

which leads to the recursion relation

$$u_{n+2} = e_{11}^{2\Omega\tau} u_n + e_{12}^{2\Omega\tau} \dot{u}_n + \left(e_{11}^{\Omega\tau} - e_{11}^{2\Omega\tau}\right) \xi_n + \left(1 - e_{11}^{\Omega\tau}\right) \xi_{n+1}.$$

The corresponding change of coordinates is:

$$\begin{pmatrix} u_n \\ u_{n+1} \\ u_{n+2} \end{pmatrix} = A_2 \begin{pmatrix} u_n \\ \dot{u}_n \end{pmatrix} + T_2(\xi_n,\, \xi_{n+1}), \tag{3.2}$$

where

$$A_2 \equiv \begin{pmatrix} 1 & 0 \\ e_{11}^{\Omega\tau} & e_{12}^{\Omega\tau} \\ e_{11}^{2\Omega\tau} & e_{12}^{2\Omega\tau} \end{pmatrix}$$

and

$$T_2(\xi_n,\, \xi_{n+1}) \equiv \begin{pmatrix} 0 \\ \left(1 - e_{11}^{\Omega\tau}\right) \xi_n \\ \left(1 - e_{11}^{\Omega\tau}\right) \xi_{n+1} + \left(e_{11}^{\Omega\tau} - e_{11}^{2\Omega\tau}\right) \xi_n \end{pmatrix}.$$

The transformation (3.2) depends on both ξ_n and ξ_{n+1}, each of which can assume two values, for a total of four distinct changes of coordinates, each of which has the same form; all four have the same linear part, the matrix A_2, but are distinguished by the translations $T_2(\xi_n, \xi_{n+1})$. This structure is enough to ensure that the four images of the attractor are distinct (see Figure 6).

To see why, first notice that the matrix A_2 determines the linear subspace to which the fractal is mapped; the columns of A_2, viewed as vectors in $\mathbf{R}^3$, span this subspace. The vector product of these columns is thus a vector normal to the linear subspace:

$$\vec{n} \equiv \begin{pmatrix} e_{11}^{\Omega\tau} e_{12}^{2\Omega\tau} - e_{11}^{2\Omega\tau} e_{12}^{\Omega\tau} \\ -e_{12}^{2\Omega\tau} \\ e_{12}^{\Omega\tau} \end{pmatrix}.$$

The projection of the translations $T_2(\xi_n, \xi_{n+1})$ onto this normal,

$$\begin{aligned} \vec{n} \cdot T_2(\xi_n, \xi_{n+1}) \quad = \quad & \left[e_{11}^{\Omega\tau} \left(e_{12}^{\Omega\tau} + e_{12}^{2\Omega\tau}\right) - e_{12}^{\Omega\tau} e_{11}^{2\Omega\tau} - e_{12}^{2\Omega\tau}\right] \xi_n + \\ & e_{12}^{\Omega\tau} \left(1 - e_{11}^{\Omega\tau}\right) \xi_{n+1}, \end{aligned}$$

is generally nonzero for all four choices of (ξ_n, ξ_{n+1}), so the four images of the fractal lie in distinct parallel planes.

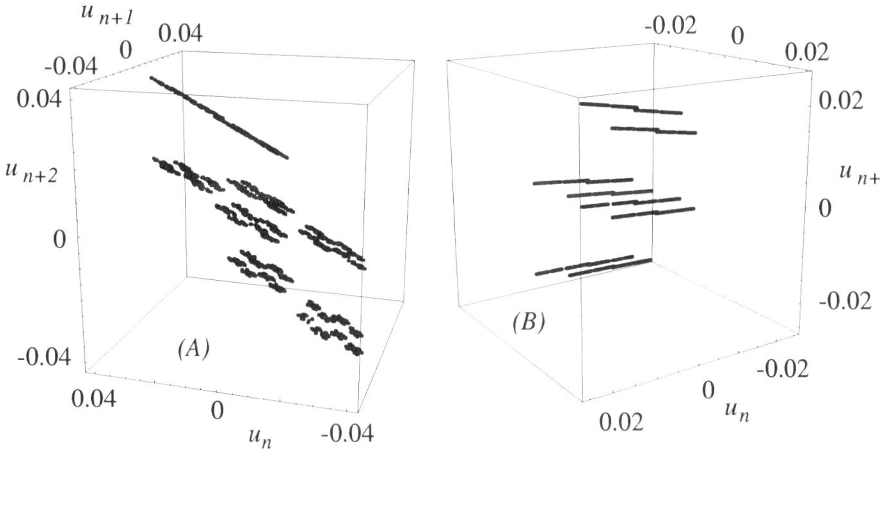

Figure 6. The images of the attractors from Figures 3 in the (u_n, u_{n+1}, u_{n+2}) space

So, where have we got to? We have demonstrated that the dynamics of a recursive channel driven with a random binary input are equivalent to an IFS. We have seen that the channel dissipation means that the IFS has a fractal attractor in its natural phase space. And finally, we have established that the use of delay coordinates recovers finitely many distinct images of the attractor. Further, these images are distiguished by the translations $T_2(\xi_n, \xi_{n+1})$ which depend on the recent history of the input.

The final observation can be exploited for channel equalisation. The normal vector $\vec{n}$ may be thought of as an FIR filter which when applied to the time series u_n will yield one of 4 values. That is to say, by applying $\vec{n}$ one can recover the two most recent symbols from input sequence σ_n. Indeed, as we recover two symbols at a time, but only one of them is new, we can demand self-consistency of our estimated symbol sequence and thereby detect single-bit errors. This approach amounts to the construction of linear decision boundaries—leaves that lie between the sheets that contain images of the IFS's attractor.

Although a similar approach would work for more general channel models, the resulting algorithms would not be quite so simple. Higher order linear channels could still be equalised with an FIR-filter analogous to $\vec{n}$, but would require higher dimensional embeddings—in general a $(k+1)$-dimensional embedding for a k-th order system—and so would have to distinguish a larger number of images of the channel's attractor. A similar framework can be applied to nonlinear channel models. In the treatment above, we needed the existence of attractors and the possibility of using time delays to embed them. Both these ingredients are available for general hyperbolic IFSs [2, 4]. Implementation would require higher dimensional embeddings—perhaps as many as $(2k+1)$ dimensions for a k-th order system—as well as non-planar decision boundaries, which could be implemented with, say, radial-basis function networks or neural nets.

4 Afterword

This example is a small step on the way to a larger prize. We have shown for a particular example that the method of delays can be applied to a stochastically forced linear channel. The generalisation to higher order linear channels is obvious. The generalisation to nonlinear channels, although less obvious, is nevertheless straightforward [4]. The novelty of this approach is that it begins to establish a connection between classical statistical time series analysis and the sort of modern geometric analysis that has come out of the experimental study of dynamical systems.

This is a connection that could prove fruitful. For example, our stochastic forcing process $\sigma(t)$ could be interpreted differently—as cross-talk from an adjacent digital channel. As such, it would be a credible model for an important, non-gaussian noise source in digital and mixed analogue/digital systems. In the same way that signal processing algorithms have been developed around the assumption of an ARMA model of the noise, we envisage a circle of techniques which exploit the structure of these IFS digital noise models. As well as the channel equalisation scheme sketched above there is an immediate application to signal separation, in which one could exploit the embedding properties of the IFS noise to separate it from a band-limited analogue signal. The essential idea, which involves the construction of a nonlinear inverse for a linear filter, appears in [3]. The construction is possible because of nonlinearity—embedding methods provide a way to exploit the nonlinear couplings relating different frequency channels—a design principle with larger scope.

Acknowledgements

We would like to thank Jaroslav Stark and Mike Davies for stimulating discussions and DRA (Malvern) for supporting MRM.

Bibliography

1. M.F. Barnsley and S. Demko, "Iterated Function Systems and the Global Construction of Fractals", *Proc. Roy. Soc. of London A* **339**, pp. 243-375, (1985).

2. M.F. Barnsley, *Fractals Everywhere*, Academic Press, London & San Diego, (1988).

3. D.S. Broomhead, J.P. Huke and M.A.S. Potts, "Canceling deterministic noise by constructing nonlinear inverses to linear filters", *Physica D* **89**, pp. 439-458, (1996).

4. J. Stark, D.S. Broomhead, M.E. Davies and J.P. Huke, "Takens Embedding Theorems for Forced and Stochastic Systems", submitted to the *Proceedings of the 2nd World Congress of Nonlinear Analysts, Athens, Greece, July 1996.*

5. T. Subba Rao and K.C. Indukumar, " Spectral and Wavelet Methods for the Analysis of Nonlinear and Nonstationary Time Series", *J. Franklin Inst.* **333B**, pp. 425-452, (1996)

6. F. Takens, Detecting strange attractors in turbulence, in *Dynamical Systems and Turbulence*, (Springer Lecture Notes in Mathematics, vol. 898) edited by D. A. Rand and L.-S. Young, Springer-Verlag, Berlin, 1981.

7. T. Sauer, J. A. Yorke and M. Casdagli, "Embedology", *J. Stat. Phys.* **65**, pp. 579-616, (1991).

Blind Binary Signal Separation using Joint Diagonalization Techniques

Alle-Jan van der Veen

Department of Electrical Engineering, Delft University of Technology, The Netherlands

Abstract

The blind separation of multiple co-channel binary digital signals using an antenna array involves finding a factorization of a data matrix X into $X = AS$, where all entries of S are $+1$ or -1. It is shown that this problem can be solved exactly and non-iteratively, via a certain generalized eigenvalue decomposition. As indicated by simulations, the algorithm is robust in the presence of noise. An interesting implication is that certain cluster segmentation problems can be solved using eigenvalue techniques.

1 Introduction

A core problem in the area of blind signal separation/equalization is the following. Consider d independent sources, transmitting binary symbols $\{+1, -1\}$ at equal rates in a wireless scenario. The signals are received by a central antenna array, consisting of M antennas. Assuming synchronized sources, equal transmission delays, negligible delay spread, and sampling at the bit rate, each antenna receives a linear combination of the transmitted symbol sequences $\mathbf{s}_i = [s_i(T), s_i(2T), \cdots, s_i(NT)]$ $(i = 1, \cdots, d)$, leading to the well-known data model

$$X = AS = \mathbf{a}_1\mathbf{s}_1 + \cdots + \mathbf{a}_d\mathbf{s}_d\,, \qquad S_{ij} \in \{+1, -1\}\,. \tag{1.1}$$

Here, $X = [x_{ik}]$ with $x_{ik} = x_i(kT)$ $(i = 1, \cdots, M;\ k = 1, \cdots, N)$ is a complex matrix containing the received data during N symbol periods. $A \in \mathbb{C}^{M\times d}$ is the array response matrix, $S \in \{\pm 1\}^{d\times N}$ contains the transmitted bits. In the blind signal separation scenario, both A and S are unknown, and the objective is, given X, to find the factorization $X = AS$ such that S belongs to the binary alphabet. Alternatively, we try to find a weight matrix W of full row rank d such that $S = W^{\mathrm{H}}X$. Uniqueness of this factorization is important, and was established in [1]: if A is full rank and the columns of S exhaust all 2^{d-1} distinct (up to a sign) possibilities, then this is sufficient for the factorization to be unique up to trivial permutations and scalings by ± 1 of the rows of S and columns of A. Hence, once any such factorization of X is found, S contains the binary signals that were originally transmitted, or their negative, but not some ghost signal.

This scenario by itself is perhaps naive, but it is the core problem in more realistic blind (FIR-MIMO) scenarios [2], where long delay multipath is allowed, and sources are not synchronized and are modulated by arbitrary pulse shape functions. This problem is separable into a blind multi-user equalization stage and a separation problem which is precisely of the form (1.1). Several other binary modulation schemes such as MSK or biphase (Manchester) codes are easily converted to fit the model as well [3].

One of the first papers to consider this problem appeared in full as [1]. In that paper, arbitrary finite alphabets are considered although only BPSK was tested extensively. The problem was cast into an optimization problem, $\min \| X - AS \|_F$, subject to $S_{ij} \in \{\pm 1\}$. Two fixed-point iteration algorithms were proposed, one called ILSE which is based on column-wise enumeration of candidate matrices S, and a second called ILSP which uses alternating projections:

ILSE
for $k = 1, 2, \cdots$
a. $\mathbf{s}_i^{(k)} = \arg\min_{\mathbf{s}_i \in \Omega} \|\mathbf{x}_i - A^{(k)}\mathbf{s}_i\|, \quad i = 1, \cdots, N$
b. $A^{(k+1)} = XS^{(k)\dagger}$

ILSP
for $k = 1, 2, \cdots$
a. $S^{(k)} = \text{Proj}_\Omega[A^{(k)\dagger}X]$
b. $A^{(k+1)} = XS^{(k)\dagger}$

The operator Proj_Ω denotes element-wise projection (rounding) on to the alphabet; in our case we have $\Omega = \{\pm 1\}$.

The main concern with these algorithms is their initialization and lack of global convergence. Depending on the initialization, the algorithms can converge to a local minimum, and restarts are needed if not all independent signals are found. If successful, ILSE is a conditional maximum likelihood estimator. ILSP is suboptimal but much cheaper to compute, and can be used to initialize ILSE. Later, an unconditional maximum likelihood technique for the estimation of A was derived [4], here called the UML. The algorithm consists of a fixed-point iteration as well, and also requires an accurate initialization. Its performance is similar to ILSE.

Several people noted that the problem (1.1) is essentially a clustering problem, as illustrated in Figure 1 for the case of $d = 2$, $M = 2$. In the absence of noise, X can contain only 2^d distinct vectors. To estimate A, it suffices to determine a suitable assignment of these vectors (or cluster centers) to constellation vectors, *i.e.*, the columns of S, taking symmetry into account. A non-iterative combinatorial algorithm based on such ideas, called SD, was presented in [5]. With noise, however, the segmentation and hence the estimation of the cluster centers is difficult and limits the performance of the algorithm.

The main point of this contribution is the observation that there exists a non-iterative algorithm that finds the factorization (1.1) exactly and algebraically, by reducing it to a joint diagonalization problem, which is a (generalized) eigen-

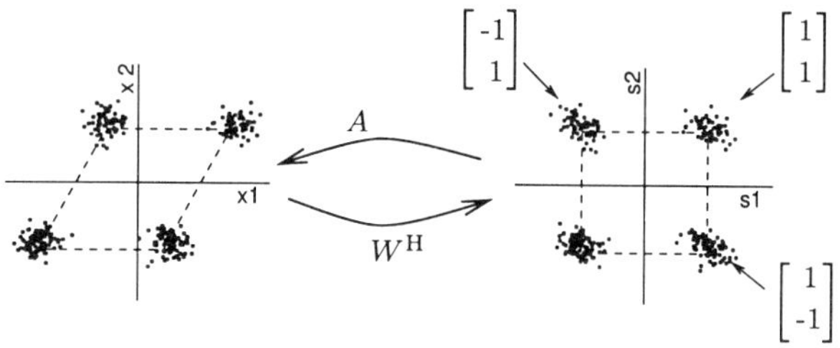

Figure 1. A maps a source constellation onto a transformed constellation

value problem. The algorithm is robust in the presence of noise, as demonstrated by simulations. Apart from certain details, it is in fact an almost trivial specialization of a recently developed "analytical constant modulus algorithm" (ACMA) [6], which solves the factorization $X = AS, |S_{ij}| = 1$ for complex matrices A and S, as it is straightforward to restrict S to be real as well. Nonetheless, the problem is sufficiently important to warrant a separate mentioning, especially since it implies the interesting observation that some clustering problems can be solved by eigenvalue techniques.

It should be noted that the blind binary source separation problem fits into the more general class of source separation based on observed linear instantaneous mixtures. Of particular interest here are algorithms that use the statistical independence of the souces, which has led to "independent component analysis" and related high-order statistics techniques, viz. among others [7–9]. Rather intriguingly, these methods are also based on joint diagonalizations, in this case of cumulant matrices. Related applications of joint diagonalizations can be found in the chapter by Demoor and Delathauwer, elsewhere in this volume.

2 Algorithm

The blind binary souce separation problem is to find a factorization $X = AS$ where $S_{ij} \in \{\pm 1\}$. Since S is real-valued, it is advantageous to write

$$X = AS \quad \Leftrightarrow \quad \left[\begin{array}{c} \mathrm{Re}(X) \\ \mathrm{Im}(X) \end{array} \right] = \left[\begin{array}{c} \mathrm{Re}(A) \\ \mathrm{Im}(A) \end{array} \right] S \quad \Leftrightarrow \quad X_R = A_R S\,,$$

with obvious definitions of $X_R \in \mathbb{R}^{2M \times N}$ and $A_R \in \mathbb{R}^{2M \times d}$. This forces S to be real, and at the same time, A_R is usually much better conditioned than A. Equivalently, the problem is to find all independent vectors $\mathbf{w}_R \in \mathbb{R}^{2M}$ such that $\mathbf{w}_R^T X_R = \mathbf{s}$ has entries $(\mathbf{s})_k \in \{\pm 1\}$.

Without noise, X_R is rank-deficient, which leads to ambiguities in $\mathbf{w}_R$. To avoid this, the first step in the algorithm is to reduce the dimension of $\mathbf{w}_R$ from $2M$ to d. Thus let $X_R = \hat{U}\hat{\Sigma}\hat{V}$ be an "economy-size" SVD for X_R, where

$\hat{U}: 2M \times d$ has orthogonal columns, $\hat{\Sigma}: d \times d$ is a diagonal matrix containing the non-zero singular values of X_R, and $\hat{V}: d \times N$ has orthogonal rows, which form a basis for the row span of X_R. Thus, the problem is equivalent to finding all independent vectors $\mathbf{w} \in \mathbb{R}^d$ such that

$$\mathbf{w}^T \hat{V} = \mathbf{s}, \qquad (\mathbf{s})_k \in \{\pm 1\}. \tag{2.1}$$

The alphabet condition is written as

$$s \in \{\pm 1\} \quad \Leftrightarrow \quad (s-1)(s+1) = 0 \quad \Leftrightarrow \quad s^2 = 1 \tag{2.2}$$

(with possible extensions to other constellations). Denoting the k-th column of $\hat{V}$ by $\mathbf{v}_k$, substitution of (2.1) into (2.2) leads to

$$\mathbf{w}^T \mathbf{v}_k \mathbf{v}_k^T \mathbf{w} = 1, \qquad k = 1, \cdots, N. \tag{2.3}$$

Similar equations arose in the solution of the constant modulus problem [6], where we had $|s|^2 = 1$ rather than $s^2 = 1$. As in [6], the conditions can be rewritten in a linear form by using Kronecker products: $[\mathbf{v}_k \otimes \mathbf{v}_k]^T [\mathbf{w} \otimes \mathbf{w}] = 1$, but in the present case the Kronecker product vectors have duplicate entries which can (and have to) be removed. Thus define, for a $d \times d$ real symmetric matrix $Y = [y_{ij}]$, a scaled stacking of the lower triangular part of the columns:

$$\text{rvec}(Y) := [y_{11} \quad y_{21}\sqrt{2} \ \cdots \ y_{d1}\sqrt{2} \quad y_{22} \quad y_{32}\sqrt{2} \ \cdots \ y_{d,d-1}\sqrt{2} \quad y_{dd}]^T$$

where $\text{rvec}(Y) \in \mathbb{R}^{d(d+1)/2}$. This allows to write (2.3) as

$$[\text{rvec}(\mathbf{v}_k \mathbf{v}_k^T)]^T \text{rvec}(\mathbf{w}\mathbf{w}^T) = 1, \qquad k = 1, \cdots, N. \tag{2.4}$$

After collecting all rows $\text{rvec}(\mathbf{v}_k \mathbf{v}_k^T)^T$ into a matrix P, the problem reduces to finding all independent vectors $\mathbf{y}$ satisfying

$$P\mathbf{y} = \mathbf{1}, \qquad \mathbf{y} = \text{rvec}(\mathbf{w}\mathbf{w}^T),$$

where $\mathbf{1} = [1 \cdots 1]^T$. Hence, we have replaced the quadratic Equations (2.3) by a linear system $P\mathbf{y} = \mathbf{1}$, subject to a quadratic constraint which imposes a certain structure on $\mathbf{y}$.

The remaining steps are identical to the procedure in [6], and only summarized here. First transform the linear system to an equivalent system $\hat{P}\mathbf{y} = \mathbf{0}$. Let Q be an orthogonal (Householder) transformation such that $Q\mathbf{1} = [\sqrt{N} \quad \mathbf{0}_{N-1}]^T$, and let $\hat{P}$ be the last $N-1$ rows of QP (*i.e.*, the first row is removed), then, up to a scaling, solving $P\mathbf{y} = \mathbf{1}$ is equivalent to solving

$$\hat{P}\mathbf{y} = \mathbf{0}, \qquad \mathbf{y} \neq \mathbf{0}. \tag{2.5}$$

The general solution of (2.5) has the form

$$\mathbf{y} = \alpha_1 \mathbf{y}_1 + \cdots + \alpha_\delta \mathbf{y}_\delta, \qquad (\alpha_i \in \mathbb{R}, \sum |\alpha_i| \neq 0),$$

where $\{\mathbf{y}_i\}$ is a basis of the null space of $\hat{P}$, and δ is defined to be the dimension of this space. In the presence of noise, both the basis and its dimension are estimated by an SVD of $\hat{P}$. Since we know that there are d linearly independent solutions $\mathbf{w}$, and since linearly independent vectors $\mathbf{w}$ lead to linearly independent vectors $\mathbf{y} = \text{rvec}(\mathbf{w}\mathbf{w}^T)$, there are at least d independent solutions to (2.5): $\delta \geq d$. On the other hand, if sufficient conditions are imposed by $\hat{P}$, then the dimension of the null space of $\hat{P}$ will not be larger than d. In particular, if $N \gg 2^{d-1}$ then we expect $\delta = d$ with high probability, as is argued later in this section. For this property to hold, it is essential to have used rvec in (2.4), or else $\hat{P}$ has duplicate columns and the dimension of the kernel will be too large. (This is precisely the reason why BPSK and MSK signals were noted exceptions in the ACMA algorithm [6].)

At this point, we have obtained a basis of solutions $\{\mathbf{y}_i\}$, but since the basis is arbitrary, each $\mathbf{y}_i$ is probably not of the form $\text{rvec}(\mathbf{w}\mathbf{w}^T)$. To force the structural property $\mathbf{y} = \text{rvec}(\mathbf{w}\mathbf{w}^T)$, write $Y_i = \text{rvec}^{-1}(\mathbf{y}_i)$, which gives

$$\mathbf{w}\mathbf{w}^T = \alpha_1 Y_1 + \cdots + \alpha_\delta Y_\delta \,. \tag{2.6}$$

We have to find all d parameter vectors $[\alpha_1 \cdots \alpha_\delta]$ such that the resulting linear combination of the matrices $\{Y_i\}$ is of rank 1 and symmetric, in which case it can be factored as $\mathbf{w}\mathbf{w}^T$. As discussed in [6], this is essentially a generalized eigenvalue problem. For $d = 2$, it is a 2×2 matrix pencil problem with a closed-form solution, for $d > 2$ and with noise, there is in general no exact such decomposition, but we can try to make the linear combination as close to rank 1 as possible. The symmetry property is automatic because for real-valued α-vectors, $Y_1, \cdots, Y_\delta$ are symmetric by construction.

A technique for computing all α-vectors is detailed in [6], for the general complex case, but a specialization to the present real case is immediate. For each α-vector, the corresponding $\mathbf{w}$ follows from (2.6).

Since d is typically small, the computational effort required by the diagonalization step is negligible in comparison with computing the SVD of $\hat{P}$. This brings the overall computational complexity of the algorithm to around $\mathcal{O}((M^2 + d^4)N)$. There are interesting possibilities for updating the null space estimate of $\hat{P}$ using subspace trackers.

Rank of P

From the above, it is clear that the dimension δ of the null space of $\hat{P}$ plays an important role. Using similar arguments as in [6], one can show that this dimension is independent of A as long as A has full rank. Furthermore, $\delta = d$ iff $\hat{P}$ has rank $\frac{1}{2}d(d+1) - d = \frac{1}{2}d(d-1)$, *i.e.*, iff P has rank $\frac{1}{2}d(d-1) + 1$. For this, it is *sufficient* that S contains all 2^{d-1} essentially distinct constellation vectors ("essentially" meaning beyond a factor ± 1). A conjecture that any subset of $\frac{1}{2}d(d-1) + 1$ constellation vectors out of these 2^{d-1} would already be sufficient turns out not to be true because of linear dependencies. For example, for $d = 5$ it was found that some subsets of 11 constellation vectors only give rank(P) = 10 rather than 11, and that at least 13 constellation vectors are needed to guarantee rank(P) = 11. An experiment for $d = 10$ shows that 380 vectors out of 512 is still not sufficient in all cases.

For $N \gg 2^{d-1}$ and assuming equal probabilities on the occurrence of any constellation vector, a lower bound on the probability p that S contains all 2^{d-1} constellation vectors is given in [1, Section A.2] as $p \geq 1 - 2^{d-1}(1 - 2^{-(d-1)})^N$. This also gives a lower bound on the probability that $\delta = d$, be it rather pessimistic because there are many subsets that are sufficient as well.

3 Simulations

To test the algorithm, the following scenario is considered. We have $d = 4$ equipowered sources, with directions-of-arrival $-3°, 0°, 4°, 8°$ with respect to the array broadside. The sensor array is a uniform linear array consisting of $M = 6$ omnidirectional antennas spaced at $\lambda/2$. An arbitrary initial phase of each signal is incorporated in A. The condition number of the complex-valued A is about 300, so that the problem looks quite ill-conditioned. However, since we try to recover *real* signals, the true conditioning of the problem is determined by $A_R = [\text{Re}(A)^T \;\; \text{Im}(A)^T]^T$. Unlike the complex case, the conditioning of A_R is very much dependent on the initial (random) phases of the signals: it can be as low as 3, or as high as 200. The median of the distribution was found to be 9.5, with a standard deviation of 8.4, so that the problem is medium-conditioned in the majority of cases.

The signal-to-noise ratio (SNR) is defined with respect to signal 1. We took $N = 100$ snapshots and a total of 2000–8000 Monte Carlo runs. The bit-error rate (BER) is the total BER over all d signals. The percentage of cases where not all d signals are recovered is defined as the recovery failure rate (RFR) and is listed separately. These cases are omitted from the BER statistics.

RACMA is compared to ILSP [1], ILSE [1], UML [4] and SD [5]. Of the latter algorithms, only SD does not require an explicit initial guess for A. ILSP is initialized with $A_0 = I_{M \times d}$. ILSE and UML require a more accurate initialization, and we use the result of the ILSP algorithm for that. It is also possible to use the result of RACMA to initialize ILSP, ILSE and UML, which can improve results because RACMA is not statistically optimal (in fact it is biased). UML requires an estimate of the noise power.

Figures 2(a) and 2(b) shows the performance for the case where the initial phases of the signals are selected randomly for every Monte-Carlo run. Figures 2(c) and 2(d) shows the corresponding RFR. It is seen that SD is effective at high SNRs, but as a non-iterative combinatorial method, it is easily confused at low SNRs where it fails to recover all sources in a majority of cases. The performance of UML is virtually the same as that of ILSE, except that its capture performance is slightly better at low SNRs. RACMA has a BER performance similar to ILSP(A), although for low SNR it is less successful in recovering all d signals. By itself, it is suboptimal, but provides a good initial point for ILSE

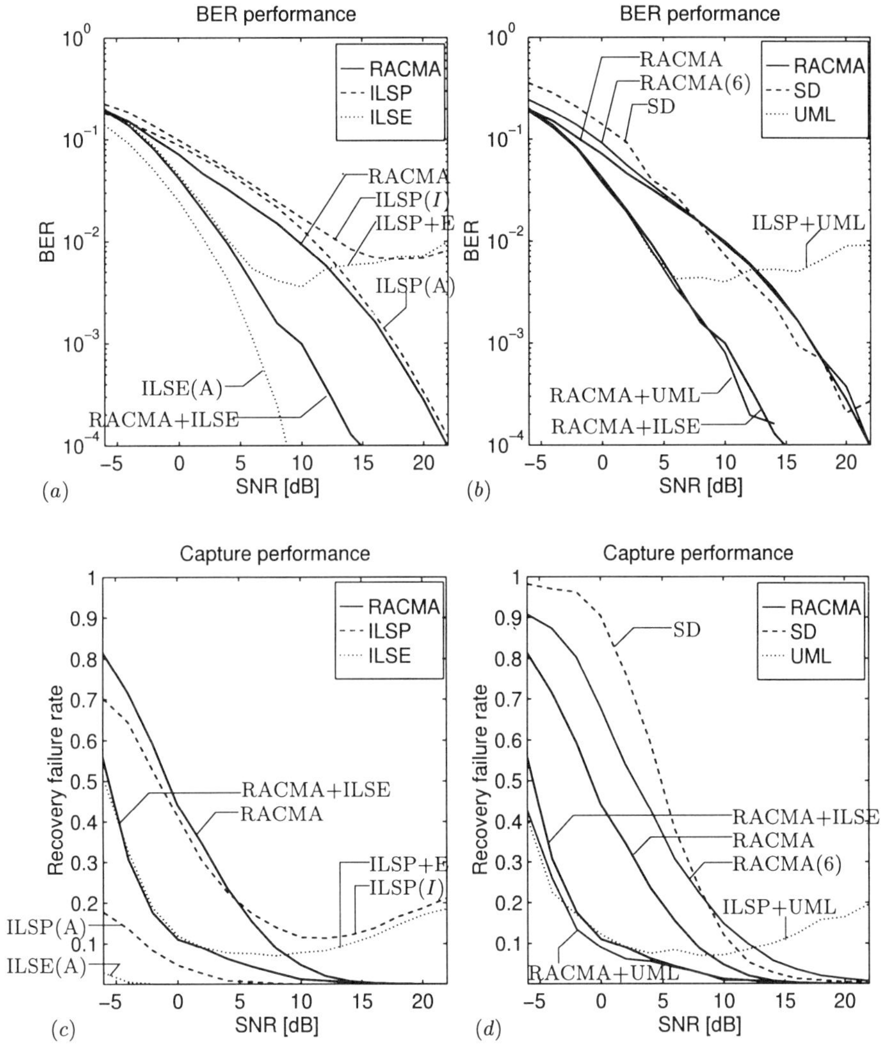

Figure 2. (a) and (b) BER performance for A with random signal phase; and (c) and (d) corresponding failure rate (cases where not all signals are recovered)

or UML. It does not reach the performance of ILSE(A), because of the inherent squaring in the algorithm, which somewhat amplifies the noise.

Figure 2(b) also shows the effect of overestimating d in RACMA: "RACMA(6)" lists the case where $d = 6$ is used in the SVD of X and the construction of P. The BER performance is almost the same as RACMA, but it becomes 3 dB less effective in capturing all signals.

4 Application to clustering

As remarked in the introduction, for discrete signals the $X = AS$ factorization problem is essentially a cluster segmentation problem. This implies that certain cluster segmentation problems can be solved using eigenvalue techniques, which might provide an interesting alternative to the usual iterative algorithms. Although the present algorithm expects the cluster centers to lie on the vertices of a parallelepiped, some generalizations to other configurations are possible.

To illustrate this, consider Figure 3, which shows two clusters arbitrarily located in a two-dimensional space. This is a special case of our data model: we have $M = 2$ real-valued sensors, and the received data can be written as

$$X = [\mathbf{a}_1 \quad \mathbf{a}_2] \begin{bmatrix} 1 & 1 & \cdots \\ \pm 1 & \pm 1 & \cdots \end{bmatrix} + E \tag{4.1}$$

where E signifies the additive noise. Hence we can set $d = 2$ "sources", although the first source is in fact constant (+1). Since we receive 2^{d-1} essentially different constellation vectors, this is sufficient for the factorization to be unique. Figure 3(b) shows the singular values of $\hat{P}$, which is an $N-1$ by 3 matrix. Clearly, there are $\delta = 2$ small singular values (they would be zero in the noise-free case). With more noise, the small singular values of $\hat{P}$ are increased (Figure 3(d)), and it becomes hard to detect from the singular values that there are really two clusters rather than one. Nonetheless, if it is decided that $\delta = 2$, then the resulting cluster centers are still close to the true centers (indicated by a "×"). The singular values could be used for a hypothesis test to distinguish between the presence of one versus two clusters.

A similar example is a case were we have two clusters but only one sensor ($M = 1$), for example if we receive a single binary source, distorted by an arbitrary offset c, scaling k, and additive zero mean white noise:

$$x_i = k \cdot s_i + c + e_i\,, \qquad i = 1, \cdots, N\,.$$

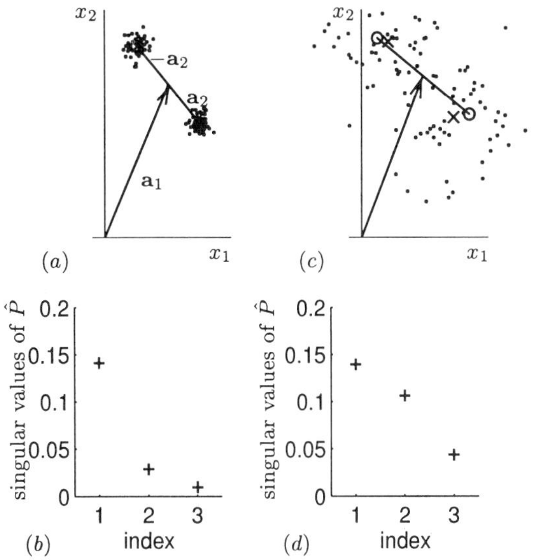

Figure 3. Example of cluster segmentation using RACMA. (a) and (b) low noise; and (c) and (d) high noise

To fit this to an $X = AS$ model where $M \geq d$, we can consider an augmented data matrix,

$$X_e = \begin{bmatrix} 1 & 1 & \cdots & 1 \\ x_1 & x_2 & \cdots & x_N \end{bmatrix} = \begin{bmatrix} 1 & 0 \\ c & k \end{bmatrix} \begin{bmatrix} 1 & 1 & \cdots & 1 \\ s_1 & s_2 & \cdots & s_N \end{bmatrix} + \begin{bmatrix} 0 & 0 & \cdots & 0 \\ e_1 & e_2 & \cdots & e_N \end{bmatrix}.$$

It is instructive to partially work out the expressions for the estimation of $\mathbf{w}$ (and hence of k and c, since $\mathbf{w} = [-ck^{-1} \;\; k^{-1}]^T$). In the model $P\mathbf{y} = \mathbf{1}$, we here define for convenience $\mathbf{y} = [w_1^2 \;\; w_1 w_2 \;\; w_2^2]^T$ and the rows of P by $[1 \;\; 2x_k \;\; x_k^2]$. The transformation by Q maps the condition $P\mathbf{y} = \mathbf{1}$ to $\big(\mathbf{1}^T P\mathbf{y} = n\,,\; \hat{P}\mathbf{y} = \mathbf{0}\big)$. Using the simple structure of Q, we then find

$$\begin{aligned} R \;:=\; & \hat{P}^T\hat{P} \;=\; P^T(I - \tfrac{1}{N}\mathbf{1}\cdot\mathbf{1}^T)P \\ \sim\; & \begin{bmatrix} 0 & 0 & 0 \\ 0 & 4(\Sigma_2 - \frac{1}{N}\Sigma_1^2) & 2(\Sigma_3 - \frac{1}{N}\Sigma_1\Sigma_2) \\ 0 & 2(\Sigma_3 - \frac{1}{N}\Sigma_1\Sigma_2) & \Sigma_4 - \frac{1}{N}\Sigma_2^2 \end{bmatrix} \\ & + \begin{bmatrix} 0 & 0 & 0 \\ 0 & 4E_2 & \frac{4}{N}\Sigma_1 E_2 \\ 0 & \frac{4}{N}\Sigma_1 E_2 & E_4 - \frac{5}{N}E_2^2 + \frac{4}{N}\Sigma_2 E_2 \end{bmatrix} \end{aligned}$$

where $\Sigma_n := \sum_1^N x_i^n$, $E_n := \sum_1^N e_i^n$, and "$\sim$" denotes equality in mean. In the absence of noise, R has two zero eigenvalues, with eigenvectors $\mathbf{y}_1 = [1 \;\; 0 \;\; 0]^T$

and $\mathbf{y}_2 = [0 \;\; a \;\; b]^T$, say. The joint diagonalization step collapses: it directly follows that $\mathbf{w} = \alpha[a \;\; b]^T$, for some scaling α which can be estimated from the condition $\mathbf{1}^T P \mathbf{y} = n$. This example shows that the algorithm is in fact a square-root method based on 4-th order moments of the data.

With noise, it is clear that the nonzero block $\underline{R}$ of R is biased, an effect which so far has not been taken into account. A correction is possible if the moments of the noise are known. For example, for Gaussian noise with variance σ^2, we have $E_2 \sim N\sigma^2$, $E_4 \sim 3N\sigma^4$, so that the the bias term is asymptotically given by

$$\underline{R}_E = \sigma^2 \begin{bmatrix} 4N & 4\Sigma_1 \\ 4\Sigma_1 & 4\Sigma_2 - 2N\sigma^2 \end{bmatrix}.$$

If we neglect the term $-2N\sigma^4$, then the noise variance can be readily estimated as the (smallest) eigenvalue of the pencil $(\underline{R}, \underline{R}_E)$, since this is the value that will make $\underline{R} - \lambda \underline{R}_E$ singular again.

Some simulation results are given in Figure 4. In this simulation, $k = 0.3$, $c = 1.2$, $N = 15$, and the results are averaged over 5000 Monte-Carlo runs. It

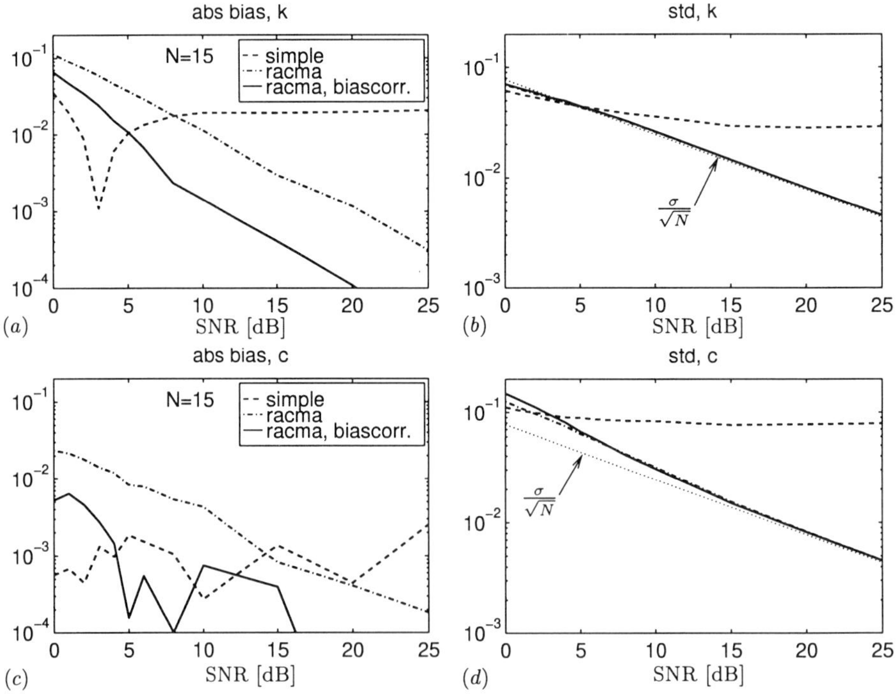

Figure 4. Parameter estimation of a model $x_i = ks_i + c + e_i$

is observed that the bias correction has little influence on the variance of the estimates, which quickly converge to $\frac{1}{N}\sigma^2$: the variance of the ML estimators of k and c for known $\mathbf{s}$ and sufficiently small $\sum_1^N s_i$. The "simple estimator" is $\hat{c} = \frac{1}{N}\sum_1^N x_i$, $\hat{k} = \frac{1}{N}\sum |x_i - \hat{c}|$, which assumes that there is an equal number of $+1$ and -1 in the data batch. For small N or otherwise asymmetric sources, this estimator is not very good.

Bibliography

1. S. Talwar, M. Viberg, and A. Paulraj, "Blind estimation of synchronous co-channel digital signals using an antenna array. Part I: Algorithms," *IEEE Trans. Signal Processing*, vol. 44, pp. 1184–1197, May 1996.

2. A.J. van der Veen, S. Talwar, and A. Paulraj, "Blind estimation of multiple digital signals transmitted over FIR channels," *IEEE Signal Processing Letters*, vol. 2, pp. 99–102, May 1995.

3. A.J. van der Veen and A. Paulraj, "Singular value analysis of space-time equalization in the GSM mobile system," in *Proc. IEEE ICASSP*, (Atlanta (GA)), pp. 1073–1076, IEEE, May 1996.

4. B. Halder, B.C. Ng, A. Paulraj, and T. Kailath, "Unconditional maximum likelihood approach for blind estimation of digital signals," in *Proc. IEEE ICASSP*, vol. 2, (Atlanta (GA)), pp. 1081–1084, May 1996.

5. K. Anand, G. Mathew, and V.U. Reddy, "Blind separation of multiple co-channel BPSK signals arriving at an antenna array," *IEEE Signal Proc. Letters*, vol. 2, pp. 176–178, Sept. 1995.

6. A.J. van der Veen and A. Paulraj, "An analytical constant modulus algorithm," *IEEE Trans. Signal Processing*, vol. 44, pp. 1136–1155, May 1996.

7. L. Tong, Y. Inouye, and R.-W. Liu, "Waveform-preserving blind estimation of multiple independent sources," *IEEE Trans. Signal Processing*, vol. 41, pp. 2461–2470, July 1993.

8. J.F. Cardoso and A. Souloumiac, "Blind beamforming for non-Gaussian signals," *IEE Proceedings F (Radar and Signal Processing)*, vol. 140, pp. 362–370, Dec. 1993.

9. P. Comon, "Independent component analysis, a new concept?," *Signal Processing*, vol. 36, pp. 287–314, Apr. 1994.

Adaptive Algorithms for Single User Blind Channel Equalization

Piet Vandaele and Marc Moonen[1]

ESAT - Katholieke Universiteit Leuven, Heverlee, Belgium

Abstract

In this report we present two adaptive algorithms for blind channel equalization in a single user environment. The algorithms form a cheaper alternative to an earlier presented adaptive SVD + Viterbi algorithm. The first step in the algorithms, consists of an adaptive matrix singular value decomposition (SVD) for a (virtual) channel identification type operation. In the first algorithm, the subsequent symbol detection is done by means of a recursive total least squares algorithm while the second algorithm uses a recursive least squares (Kalman filter) approach. It is shown that the performance of the SVD + Viterbi algorithm is closely approximated, especially for the SVD + total least squares algorithm, while the computational complexity is dramatically reduced.

1 Introduction

The problem of blind channel identification/equalization using second-order statistics or equivalent deterministic properties of the oversampled channel output has drawn considerable attention recently. Most of the algorithms developed up till now, however, are based on block processing and have a high computational complexity, which forms an impediment for real-time implementation. In [1] we presented a fully adaptive SVD + Viterbi algorithm based on the block algorithm of [2].

Here we present two cheaper alternatives (also based upon [2]), which are also fully adaptive but have a much lower computational complexity. The algorithms estimate the transmitted symbol sequence without explicitly computing the channel. This is useful for very short data sequences, where the number of

[1]Piet Vandaele is a Research Assistant supported by the Flemish I.W.T. (the Flemish Institute for the Advancement of Scientific-Technological Research in Industry). Marc Moonen is a Research Associate with the Fund for Scientific Research - Flanders (Belgium) (F.W.O.). This research work was carried out at the ESAT laboratory of the Katholieke Universiteit Leuven, in the framework of a Concerted Action Project of the Flemish Community, entitled *Model-based Information Processing Systems* (GOA/MIPS/95/99/3) as well as the IT-program of the I.W.T., *Integrating Signal Processing Systems* (ITA/GBO/T23) and was partially sponsored by IMEC (Flemish Interuniversity Microelectronics Center). The scientific responsibility is assumed by its authors.

channel parameters is of the same order of magnitude as the number of symbols to be estimated. This particularly makes sense in the context of mobile telecommunications where averaging over long data sequences is not meaningful due to the highly time varying nature of the environment. The first algorithm is based on an adaptive SVD + recursive total least squares (RTLS) solution. The second algorithm is based on an adaptive SVD + Kalman filter. We will show that the performance of the algorithms comes close to that of the SVD + Viterbi algorithm while the complexity is dramatically reduced. Further we will show that the algorithms allow to track a time varying channel.

In Section 2 the data model is presented. Section 3 reviews the block processing approach of [2]. In Section 4 two new adaptive algorithms are derived. In Section 5, the performance of the new algorithms is investigated. Finally, in Section 6 some conclusions are drawn.

2 Data model

The received signal for linear digital modulation over a linear channel with additive noise, is

$$y(t) = \sum_k h(t - kT) \cdot x[k] + n(t) \tag{2.1}$$

where the $x[\cdot]$ are the transmitted symbols, T is the symbol period, $h(t)$ is the composite channel impulse response (which includes transmitter and channel filters), and $n(t)$ is additive noise. The channel is assumed to be FIR with duration of approximately LT.

With an oversampling factor M, the sampling instants for the received signal are $t_o + (k + \frac{i-1}{M}) \cdot T$ for integer k and $i = 1, 2, \ldots, M$. It is common to use a so-called polyphase description

$$\begin{cases} y_i[k] &= y(t_o + (k + \frac{i-1}{M}) \cdot T) \\ n_i[k] &= n(t_o + (k + \frac{i-1}{M}) \cdot T) \\ h_i[k] &= h(t_o + (k + \frac{i-1}{M}) \cdot T) \end{cases} \tag{2.2}$$

and to view the oversampled received signal as an M-channel output signal at the symbol rate [3, 4]. Define the output vector $\mathbf{y}[k] = [y_1[k] \ldots y_M[k]]^T$, the input vector $\mathbf{x}[k] = [x[k-L] \ldots x[k]]^T$ and the noise vector $\mathbf{n}[k] = [n_1[k] \ldots n_M[k]]^T$ then

$$\mathbf{y}[k] = \underbrace{\begin{bmatrix} h_1[L] & \ldots & h_1[1] & h_1[0] \\ \vdots & & & \vdots \\ h_M[L] & \ldots & h_M[1] & h_M[0] \end{bmatrix}}_{\boxed{H}} \mathbf{x}[k] + \mathbf{n}[k]. \tag{2.3}$$

Spatial oversampling, *i.e.* using multiple antennas at the receiver, fits into the same framework by considering $y_1[k], \ldots, y_M[k]$ as the outputs of M receiving antennas. A combination of spatial and temporal oversampling may also be used. From here on we may therefore consider M to be the product of the spatial and temporal oversampling factor.

For the sake of short notation, it is assumed that $\mathbf{n}[k] = 0$. The computational scheme to be presented, involves an SVD which is assumed to be robust against such additive noise [5].
With the above input/output-formula, a data model can be put up as follows (which has been the starting point for many algorithmic developments already). Define

$$Y_{k|k+i-1}^{(j)} = \begin{bmatrix} \mathbf{y}[k] & \mathbf{y}[k+1] & \dots & \mathbf{y}[k+j-1] \\ \mathbf{y}[k+1] & \mathbf{y}[k+2] & \dots & \mathbf{y}[k+j] \\ \mathbf{y}[k+2] & \mathbf{y}[k+3] & \dots & \mathbf{y}[k+j+1] \\ \vdots & \vdots & & \vdots \\ \mathbf{y}[k+i-1] & \mathbf{y}[k+i] & \dots & \mathbf{y}[k+i+j-2] \end{bmatrix} \tag{2.4}$$

(the superscript refers to the number of columns, the subscript refers to the time indices in the first column), and with a similar notation

$$X_{k-L|k+i-1}^{(j)} = \begin{bmatrix} x[k-L] & \dots & x[k-L+j-1] \\ x[k-L+1] & \dots & x[k-L+j] \\ \vdots & & \vdots \\ x[k+i-1] & \dots & x[k+i+j-2] \end{bmatrix} \tag{2.5}$$

then,

$$Y_{k|k+i-1}^{(j)} = \underbrace{\begin{bmatrix} \boxed{H} & 0 \; 0 & \dots \\ 0 \; \boxed{H} & 0 & \dots \\ 0 \; 0 & \boxed{H} & \dots \\ & \vdots & \\ 0 \; 0 & \dots & \boxed{H} \end{bmatrix}}_{\mathcal{H}} X_{k-L|k+i-1}^{(j)}. \tag{2.6}$$

Here, $Y_{k|k+i-1}^{(j)}$ is a known matrix. The aim is to compute the symbol sequence $X_{k-L|k+i-1}^{(j)}$ (or equivalently the vector $X_{k-L|k-L}^{(i+j+L-1)}$) from $Y_{k|k+i-1}^{(j)}$, with or without computing $\mathcal{H}$ explicitly.

3 Block processing algorithm

The adaptive algorithms derived here are based on the algorithm of Liu & Xu [2]. Therefore we first review their original approach. The algorithm [2] is based on a singular value decomposition of $Y^{(j)}_{k|k+i-1}$ (which is a "short-fat" matrix, *i.e.* with many more columns than rows)

$$Y^{(j)}_{k|k+i-1} = U \cdot \Sigma \cdot V^H. \tag{3.1}$$

It is assumed that $\mathcal{H}$ is of full column rank and hence $M \cdot i \geq (L+i)$. Then a symbol sequence is sought that best matches the row space of $Y^{(j)}_{k|k+i-1}$, *i.e.*

$$\min_{X^{(j)}_{k-L|k+i-1}} \|X^{(j)}_{k-L|k+i-1} \cdot V^{\perp}\|_F^2 \tag{3.2}$$

where $\|\cdot\|_F$ denotes the Frobenius norm and $V^{\perp}$ is the (approximate) null space of $Y^{(j)}_{k|k+i-1}$ ($Y^{(j)}_{k|k+i-1} \cdot V^{\perp} \approx 0$), which is extracted from the V-matrix of the SVD (columns of $V^{\perp}$ correspond to singular values below the "noise threshold"). This is equivalent to

$$\min_{X^{(i+j+L-1)}_{k-L|k-L}} \|X^{(i+j+L-1)}_{k-L|k-L} \cdot \begin{bmatrix} V^{\perp} & 0 & \cdots & 0 \\ & V^{\perp} & & \\ 0 & & \ddots & \\ & & & V^{\perp} \end{bmatrix} \|_F^2 \tag{3.3}$$

(subject to some constraint to avoid the trivial solution) which may be solved by means of standard least squares techniques. One may for example estimate the transmitted symbol sequence as a scaled version of the left singular vector corresponding the smallest singular value of the above matrix.

4 Adaptive algorithms

In order to make the above algorithm adaptive, a new SVD is computed in each time step (*i.e.* in each symbol period),

$$Y^{(j)}_{k|k+i-1} = U_k \cdot \Sigma_k \cdot V_k^H. \tag{4.1}$$

Note that j, *i.e.* the number of columns of $Y^{(j)}_{k|k+i-1}$, will be much smaller than in the block processing approach of [2]. It is again assumed that $\mathcal{H}$ is of full column rank. A crucial observation is now that the columns of $V_k^{\perp}$ may be viewed as a number of ("virtual") FIR channels that produce a zero-output when fed with a specific segment of the symbol sequence:

$$X^{(j)}_{k-L|k+i-1} \cdot V_k^{\perp} = 0. \tag{4.2}$$

After rearranging the indices, one obtains an equivalent equation:

$$X^{(j)}_{k|k} \cdot \underbrace{[V^{\perp}_{k-i+1} \cdots V^{\perp}_{k} \cdots V^{\perp}_{k+L}]}_{\mathbf{V}^{\perp}_{\mathbf{k}}} = 0 \tag{4.3}$$

where $X^{(j)}_{k|k} = \begin{bmatrix} x[k] & \cdots & x[k+j-1] \end{bmatrix}$. In [1] it is shown that the above equation uniquely determines the transmitted sequence $X^{(j)}_{k|k}$ if the matrices $X^{(j)}_{k-L-i+1|k+1}, \ldots, X^{(j)}_{k-L|k+i}, \ldots, X^{(j)}_{k-1|k+i+L-1}$ are of full row rank, which implies $j > (i+L)$. $\mathbf{V}^{\perp}_{\mathbf{k}}$ is a $j \times N$ matrix with $N = (i+L) \cdot (j-(i+L))$.

Increasing j provides a better noise averaging, but on the other hand, for highly time-varying channels, averaging over long data sequences may not be meaningful and so the usage of smaller matrices may be imposed by practical considerations anyway.

At this point we have derived a relationship between the (desired) inputs and our "virtual" channel. It can now be used in several ways to restore the original symbol sequence from the knowledge of the "virtual" FIR channels $\mathbf{V}^{\perp}_{\mathbf{k}}$ and their zero outputs.

For the ease of notation, we use a BPSK modulation format further on.

4.1 Viterbi algorithm [1]

One straightforward way to exploit the above relationship is to apply it in a Viterbi algorithm [6]. State transitions in the Viterbi trellis are then governed by the cost function $\|X^{(j)}_{k|k} \cdot \mathbf{V}^{\perp}_{\mathbf{k}}\|^2_F$. The non-triviality constraint comes from the finite alphabet property of the digital input symbols.

4.2 Recursive total least squares algorithm

Note that Equation (4.3) can be rewritten as:

$$\mathbf{V}^{\perp\mathbf{T}}_{\mathbf{k}} \cdot X^{(1)}_{k|k+j-1} = 0. \tag{4.4}$$

The problem can now be viewed as a total least squares problem [7], with a constant norm non-triviality constraint. This then has to be solved in a recursive way.

First we will present an algorithm with growing matrix dimensions (which computes an estimate $\hat{X}^{(1)}_{1|k+j-1}$ at time k), from which we will then derive an approximate algorithm with computational requirements that do not grow in time (which computes an estimate $\hat{X}^{(1)}_{k|k+j-1}$ at time k).

4.2.1 RTLS algorithm: version 1

At time $k-1$, a new $\mathbf{V}^{\perp\mathbf{T}}_{\mathbf{k-1}}$ is computed which provides the above homogeneous equations in a corresponding data segment $X^{(1)}_{k-1|k+j-2}$. At that moment, all the previous equations in $\mathbf{V_1^{\perp T}}$, $\mathbf{V_2^{\perp T}}$,..., $\mathbf{V_{k-2}^{\perp T}}$ together with the new equations in $\mathbf{V_{k-1}^{\perp T}}$ may be assembled in one matrix to form one large overdetermined set of homogeneous equations in $X^{(1)}_{1|k+j-2}$. The total least squares solution $X^{(1)}_{1|k+j-2}$ then equals the right singular vector (up to a scaling factor) associated with the smallest singular value of:

$$\underbrace{\begin{bmatrix} \boxed{\mathbf{V_1^{\perp T}}} & 0\ 0\ \ldots & \\ 0 & \boxed{\mathbf{V_2^{\perp T}}} & 0\ \ldots \\ & \ddots & \\ 0\ 0\ \ldots & & \boxed{\mathbf{V_{k-1}^{\perp T}}} \end{bmatrix}}_{\mathcal{V}_{k-1}} . \tag{4.5}$$

This right singular vector may now be computed in an iterative way, *i.e.* for the computation of the right singular vector at time k one can use the results obtained at time step $k-1$. The singular vector is updated in two steps.

The first step is a QR-updating operation in which the triangular R-factor of the QR-decomposition of the matrix $\mathcal{V}_{k-1}$ is tracked [8].

In the second step the singular vector is computed through inverse iteration [8] using the R-factor computed in the first step.

- **QR updating**

  Denote the QR-decomposition of $\mathcal{V}_{k-1}$ as:

$$\mathcal{V}_{k-1} = Q_{k-1} \cdot R_{k-1}. \tag{4.6}$$

  Assume that R_{k-1} is known at time step $k-1$. In a first step we will now show how to update R_{k-1} to R_k.

  To go from time step $k-1$ to k, two operations have to be performed on the $\mathcal{V}_{k-1}$ matrix. First we have to add a column of zeros to $\mathcal{V}_{k-1}$, then we can append the rows of

$$\Upsilon_k = \begin{bmatrix} 0_{N\times(k-1)} & \boxed{\mathbf{V_k^{\perp T}}} \end{bmatrix} \tag{4.7}$$

  to the updated $\mathcal{V}_{k-1}$ (where $0_{x\times y}$ denotes an $x \times y$-matrix filled with zero entries). The first operation implies that R_{k-1} is extended with an extra row and column of zeros. Then this modified R_{k-1} can be updated to R_k using QR-updating:

* $E = \begin{bmatrix} R_{k-1} & 0 \\ 0 & 0 \end{bmatrix}$
* for $r = 1, \ldots, N$

$$\left[\frac{E}{0}\right] \Leftarrow Q_r^H \cdot \left[\frac{E}{\Upsilon_k(r,:)}\right] \tag{4.8}$$

  end
* $R_k = E$

where N is the number of rows in Υ_k. The Q_r^H matrices are accumulations of orthogonal matrix transformations which cancel out the bottom part in the right hand side matrices.

- **Inverse iteration**

The singular vector v_k, corresponding to the smallest singular value of R_k, is now computed using R_k in an inverse iteration procedure [8]. Inverse iteration is based upon the following principle. In each iteration the component of v_k (as defined in the loop below) in the direction of the eigenvector associated with the smallest eigenvalue of $(R_k^H \cdot R_k)$ is amplified.

* initialize v_k:

$$v_k = \begin{bmatrix} v_{k-1}^H & 0 \end{bmatrix}^H \tag{4.9}$$

* iteration:
  for $r = 1, \ldots, P$

$$R_k^H \cdot R_k \cdot s_k = v_k \tag{4.10}$$

$$v_k = \frac{s_k}{||s_k||_F} \cdot \sqrt{k+j-1} \tag{4.11}$$

  end

Since the right singular vectors of R_k equal the eigenvectors of $(R_k^H \cdot R_k)$, the desired right singular vector can be computed (up to a scaling factor) using the above procedure. The first equation in the iteration loop has to be solved for s_k. The last equation is a constant norm constraint. In the noiseless case, the entries in v_k should ideally have an absolute value 1. So they equal the BPSK constellation symbols up to a rotation. Equation (4.10) can be refined as:

$$R_k^H \cdot w_k = v_k \tag{4.12}$$

$$R_k \cdot s_k = w_k. \tag{4.13}$$

The first equation is solved for w_k (using forward substitution) and the second one for s_k (using back substitution). Clearly the operations required in this procedure grow in time since the dimensions of R_k and R_k^H grow. The next section explains how to transform the above algorithm into an algorithm with constant computational requirements.

4.2.2 RTLS algorithm: version 2

The algorithm with constant computational requirements is based on the following approximation. At time step k, we assume that the first $k-1$ components of the vector v_k have been accurately estimated, so that they can be replaced by a fixed value in all subsequent starting vectors for the inverse iteration steps, namely:

$$v_k(1:k-1) \approx \hat{x}(1:k-1). \tag{4.14}$$

This means that the inverse iteration step will iterate only over the last j components of the vector v_k, *i.e.* $v_k(k:k+j-1)$. Equations (4.10) and (4.11) can now be rewritten as (using Equations (4.12) and (4.13)):

1. compute $w_k(1:k-1)$ (forward substitution):

$$w_k(1) = \frac{\hat{x}(1)}{R_k^H(1,1)} \tag{4.15}$$

for $z = 2 : k-1$

$$w_k(z) = \frac{\hat{x}(z) - \sum_{q=1}^{z-1} R_k^H(z,q) \cdot w_k(q)}{R_k^H(z,z)} \tag{4.16}$$

end

2. compute $w_k(k:k+j-1)$ (forward substitution):
for $z = k : k+j-1$

$$w_k(z) = \frac{v_k(z) - \sum_{q=1}^{z-1} R_k^H(z,q) \cdot w_k(q)}{R_k^H(z,z)} \tag{4.17}$$

end

3. compute $s_k(k:k+j-1)$ (back substitution):

$$s_k(k+j-1) = \frac{w_k(k+j-1)}{R_k(k+j-1,k+j-1)} \tag{4.18}$$

for $z = k+j-2 : -1 : k$

$$s_k(z) = \frac{w_k(z) - \sum_{q=z+1}^{k+j-1} R_k(z,q) \cdot s_k(q)}{R_k(z,z)} \tag{4.19}$$

end

4. finally, the new values for the last j entries in v_k are obtained as:

$$v_k(k:k+j-1) = \frac{s_k(k:k+j-1)}{||s_k(k:k+j-1)||_F} \cdot \sqrt{j}. \tag{4.20}$$

Steps 2-3-4 are executed P times. At this point, the algorithm with constant computational requirements is only one step away. It will now be shown how one can avoid the computation of $w_k(1:k-1)$ in step 1. To this aim, the first two steps are redefined as:

1. define the $j \times 1$ vector f_k as:

$$f_k = \sum_{q=1}^{k-1} R_k^H(k:k+j-1,q) \cdot w_k(q) \tag{4.21}$$

   for $k > 1$ and $f_1 = 0_{j\times 1}$.

2. compute $w_k(k:k+j-1)$

$$w_k(k) = \frac{v_k(k) - f_k(1)}{R_k^H(k,k)}$$

   for $z = k+1 : k+j-1$

$$w_k(z) = \frac{v_k(z) - f_k(z-k+1) - \sum_{q=k}^{z-1} R_k^H(z,q) \cdot w_k(q)}{R_k^H(z,z)} \tag{4.22}$$

   end

Now f_k can be computed recursively, *i.e.* using f_{k-1}, which will then allow to set up a fully adaptive algorithm. The crucial observation is that the first $k-1$ rows of R_{k-1} remain unchanged in the QR-updating step after iteration $k-1$, see Equation (4.7). This can be stated more formely as:

$$R_l(1:k-1,:) = \begin{bmatrix} R_{k-1}(1:k-1,:) & 0_{(k-1)\times(l-k+1)} \end{bmatrix} \tag{4.23}$$

for $l \geq k$. Using the Equations (4.15) and (4.16) this leads to (for $l \geq k$):

$$w_l(1:k-2) = w_{k-1}(1:k-2). \tag{4.24}$$

Using the above two equations it can be shown that f_k can be computed recursively as:

$$f_k = \begin{bmatrix} f_{k-1}(2:j) \\ 0 \end{bmatrix} + \frac{\hat{x}(k-1) - f_{k-1}(1)}{R_{k-1}^H(k-1,k-1)} \cdot \begin{bmatrix} R_{k-1}^H(k:k+j-2,k-1) \\ 0 \end{bmatrix} \tag{4.25}$$

for $k > 1$ and $f_1 = 0_{j\times 1}$. Note that this recursive formulation avoids the explicit computation of $w_k(1:k-1)$. The proof is omitted here due to space limitations.

The above recursive formula for f_k allows us to set up the following algorithm (at this point, time indices are omitted since matrices are overwritten):

1. initialization:

$$\begin{aligned} v &= \begin{bmatrix} 1 & 0_{1\times(j-1)} \end{bmatrix}^T \\ R &= 0_{j\times j} \\ f &= 0_{j\times 1}. \end{aligned}$$

2. iteration equations:
   for $k = 1, \ldots$

   - update R
     for $r = 1, \ldots, N$

     $$\left[\frac{R}{0}\right] \Leftarrow Q_r^H \cdot \left[\frac{R}{\mathbf{V_k^T}(r,:)}\right]$$

     end
   - update the v-vector,
     for $r = 1, \ldots, P$

     $$w(1) = \frac{v(1) - f(1)}{R^H(1,1)}$$

     for $z = 2 : j$

     $$w(z) = \frac{v(z) - f(z) - \sum_{q=1}^{z-1} R^H(z,q) \cdot w(q)}{R^H(z,z)}$$

     end

     $$v(j) = \frac{w(j)}{R(j,j)}$$

     for $z = j-1 : -1 : 1$

     $$v(z) = \frac{w(z) - \sum_{q=z+1}^{j} R(z,q) \cdot v(q)}{R(z,z)}$$

     end

     $$v = \frac{v}{||v||_F} \cdot \sqrt{j}$$

     end
   - make a decision:

     $$\hat{x}[k] = \frac{\Re(v(1))}{|\Re(v(1))|}$$

- append a column of zeros:

$$\begin{aligned} f &= \begin{bmatrix} f(2:j) \\ 0 \end{bmatrix} + \frac{\hat{x}[k] - f(1)}{R^H(1,1)} \cdot \begin{bmatrix} R^H(2:j,1) \\ 0 \end{bmatrix} \\ R &= \begin{bmatrix} R(2:j,2:j) & 0_{(j-1)\times 1} \\ 0_{1\times(j-1)} & 0 \end{bmatrix} \\ v &= \begin{bmatrix} v(2:j) \\ 0 \end{bmatrix} \end{aligned}$$

end

4.3 Kalman filter algorithm

The second algorithm presented here is based on an RLS (Kalman filter) algorithm. First we shall embed Equation (4.4) (which is repeated here),

$$\mathbf{V_k^{\perp T}} \cdot X^{(1)}_{k|k+j-1} = 0 \tag{4.26}$$

in a state space model.

Select the number of states as $j-1$, then the state equation at time step k reads:

$$\underbrace{X^{(1)}_{k+1|k+j-1}}_{S_{k+1}} = \underbrace{\begin{bmatrix} 0 & 1 & 0 & \cdots & 0 \\ 0 & 0 & 1 & \cdots & 0 \\ \vdots & & & & \vdots \\ 0 & & & & 1 \\ 0 & \cdots & & & 0 \end{bmatrix}}_{A} \cdot \underbrace{X^{(1)}_{k|k+j-2}}_{S_k} + \underbrace{\begin{bmatrix} 0 \\ 0 \\ \vdots \\ 1 \end{bmatrix} \cdot x[k+j-1]}_{q_k}. \tag{4.27}$$

Since the input to the system is unknown, it is considered as a noise term q_k with zero mean. Its variance Q_k equals (for a BPSK constellation):

$$Q_k = \begin{bmatrix} 0 & 0 & \cdots & 0 \\ 0 & \cdots & & 0 \\ \vdots & & & \vdots \\ 0 & 0 & \cdots & 1 \end{bmatrix}. \tag{4.28}$$

To derive the output equation, we start with Equation (4.26), at time step $k-1$:

$$\mathbf{V_{k-1}^{\perp T}} \cdot X^{(1)}_{k-1|k+j-2} = 0. \tag{4.29}$$

In order to avoid a trivial solution, it is assumed that a reliable estimate of $x[k-1]$, *i.e.* $\hat{x}[k-1]$ is available at time k. This allows us to write down the following output equation

$$\underbrace{-\mathbf{V_{k-1}^{\perp T}}(:,1) \cdot \hat{x}[k-1]}_{Z_k} = \underbrace{\mathbf{V_{k-1}^{\perp T}}(:,2:j)}_{C_k} \cdot S_k + w_k \tag{4.30}$$

where w_k represents measurement noise. Its covariance matrix W_k is assumed to be a multiple of the identity matrix, *i.e.* $W_k = \sigma^2 I_{N\times N}$. Further it is assumed that the noise processes q_k and w_k are uncorrelated.

At this point the above equations could straightforwardly be applied in a conventional Kalman filter. We propose however to use the numerically more stable square root formulation of the Kalman filter, which can be derived using the approach of [9]. The idea is to rewrite the Kalman filter equations, such that the state vector S can be computed by solving a least squares problem of the form $A \cdot S = b$. This problem can be solved through QR-factorization:

$$Q^H \left[\begin{array}{c|c} A & b \end{array}\right] = \left[\begin{array}{c|c} R & d \\ \hline 0 & * \end{array}\right] \tag{4.31}$$

with R upper triangular. The least squares solution S is then obtained through back substitution:

$$S \stackrel{LS}{=} R^{-1} \cdot d. \tag{4.32}$$

Due to space limitations the derivation of the equations is omitted, we only give the final algorithm:

1. initialization:

$$\begin{aligned} d &= 0_{(j-1)\times 1} \\ R &= I_{(j-1)\times(j-1)}. \end{aligned}$$

2. iteration equations:
   for $k = 2, \ldots$

   - update R
     for $r = 1, \ldots, N$

$$\left[\begin{array}{c|c} R & d \\ \hline 0 & * \end{array}\right] \Leftarrow Q_r^H \cdot \left[\begin{array}{c|c} R & d \\ \hline C_k(r,:) & Z_k(r) \end{array}\right]$$

     end

   - $$v(j-1) = \frac{d(j-1)}{R(j-1, j-1)}$$

     for $z = j-2 : -1 : 1$

$$v(z) = \frac{d(z) - \sum_{q=z+1}^{j-1} R(z,q)\cdot v(q)}{R(z,z)}$$

     end

- make a decision:
$$\hat{x}[k] = \frac{\Re(v(1))}{|\Re(v(1))|}.$$
-
$$\begin{aligned} R &= \begin{bmatrix} R(2:j-1, 2:j-1) & 0_{(j-2)\times 1} \\ 0_{1\times(j-2)} & -1 \end{bmatrix} \\ d &= \begin{bmatrix} d(2:j-1) \\ 0 \end{bmatrix} \end{aligned}$$

end

5 Simulation results

Here we compare the performance of the algorithms presented above with the algorithm [1] based on an adaptive SVD and the Viterbi algorithm. The modulation format is BPSK. We use 3 antennas which are 2 times oversampled. The channel impulse response coefficients are given in Table 1. The SNR is defined as
$$SNR = 10 \cdot \log_{10} \frac{E\{||H \cdot \mathbf{x}[k]||^2\}}{E\{||\mathbf{n}[k]||^2\}}$$
and the constellation variance as:
$$var(k) = E\{|v(k) - x[k]|^2\}.$$
The parameters i and j were set to 1 and 8. First we observed the convergence behaviour of the RTLS and the Kalan solution over bursts of 100 symbol periods. The variance is plotted in Figure 1 for SNR=0 dB and SNR=10 dB, averaged over 50 runs.

Both the RTLS and the Kalman algorithm need approximately j time steps to converge. The "steady state" variance of the Kalman filter, however, is much higher then that of the RTLS solution. Notice that there is no big difference in convergence behaviour between the RTLS algorithm with 1 and 5 iterations.

Next, we show some simulation results for a time varying channel. In Figure 3, the typical time variation of a tap over 100 symbol periods is shown. The full line represents the real part, the dotted line the imaginary part. The convergence behaviour for SNR = 0 dB and 10 dB is shown in Figure 2. For

Table 1. Channel impulse response

$$\mathrm{H} = \begin{bmatrix} -0.0084 + 0.0005i & 0.3048 - 0.1433i & 0.3780 - 0.4281i & 0.1836 - 0.0976i & 0.0363 - 0.0000i \\ 0.0952 - 0.0139i & 0.4206 - 0.3499i & 0.2543 - 0.3002i & 0.1273 - 0.0042i & 0 \\ -0.0056 - 0.0002i & -0.1598 + 0.5564i & -0.2432 + 0.7285i & 0.0808 + 0.0576i & 0.0263 + 0.0020i \\ -0.0494 + 0.1666i & -0.2751 + 0.8429i & -0.0650 + 0.3288i & 0.0903 + 0.0021i & 0 \\ -0.0037 - 0.0004i & -0.5749 + 0.0941i & -0.6107 + 0.2087i & 0.0758 + 0.0104i & 0.0193 - 0.0015i \\ -0.1871 + 0.0112i & -0.8016 + 0.2017i & -0.1652 + 0.1113i & 0.0664 - 0.0144i & 0 \end{bmatrix}$$

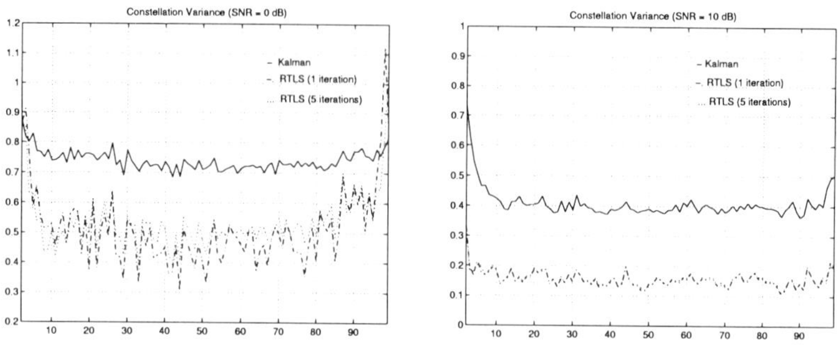

Figure 1. Variance estimation

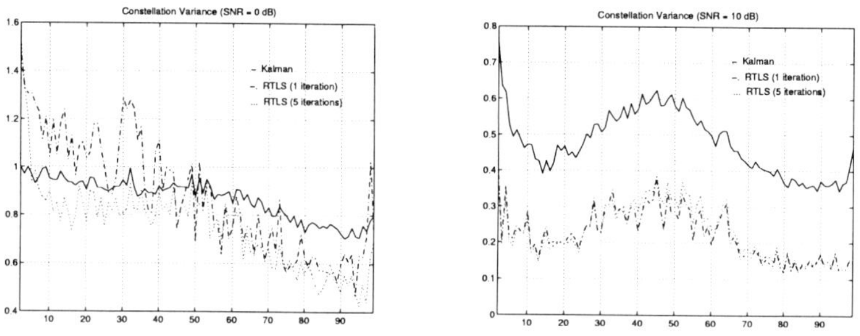

Figure 2. *Variance estimation*

SNR=0 dB there is a clear difference in convergence speed between the RTLS algorithm with 1 and 5 iterations. For SNR=10 dB the results are similar to the non-timevarying case, except for the fact that the power dip in the middle of the burst (which is present in almost all channels) leads to a temporary increase in variance.

Finally we did some BER simulations with the time varying channel model. Bursts consisted of 100 symbols, we averaged over 50 runs, the results are shown in Table 2. The performance of the RTLS algorithm comes fairly close to that of the Viterbi algorithm while its complexity is no longer exponential in j. Finally, note that for a sufficiently high SNR, all algorithms allow to track the time-varying channel.

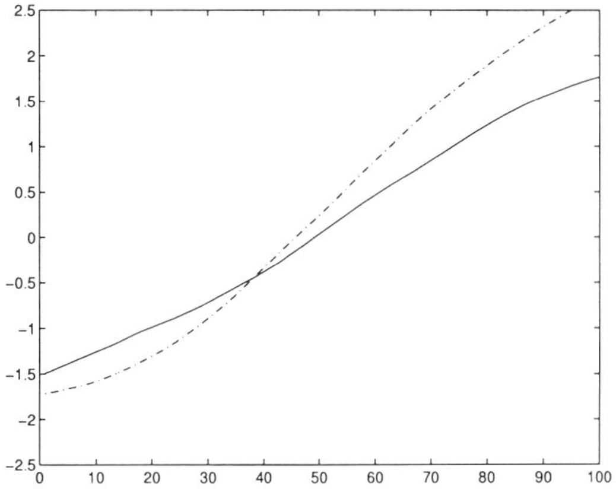

Figure 3. Typical time variation of one tap during a burst

Table 2. BER estimates (in %)

alg./SNR	0 dB	2 dB	4 dB	6 dB	8dB	10dB
Viterbi	16.32	8.56	3.32	0.24	0.40	0.04
RTLS (5 it.)	16.28	8.72	3.60	1.50	0.16	0
RTLS (1 it.)	17.36	8.62	3.74	1.64	0.24	0
Kalman	23.90	9.70	3.76	1.94	0.66	0.12

6 Conclusions

In this report we have developed two adaptive algorithms for blind channel equalization of a single user. The first step in both algorithms consists of an adaptive SVD for a (virtual) channel identification type operation. The second step in the first algorithm is based on a recursive total least squares (RTLS) scheme while the second algorithm uses a Kalman filter (RLS) approach. Both algorithms form a cheaper alternative to an earlier presented adaptive SVD + Viterbi algorithm. The simulation results show that the performance of the SVD + Viterbi algorithm is closely approximated while the computational complexity is strongly reduced.

Bibliography

1. P. Vandaele and M. Moonen. An 'SVD + Viterbi' Algorithm for Adaptive Blind Equalization of Mobile Radio Channels. Technical Report 96-79, K.U.Leuven ESAT-SISTA, 1996.

2. H. Liu and G. Xu. A Deterministic Approach to Blind Symbol Estimation. *IEEE Signal Processing Letters*, pages 205–207, December 1994.

3. D.T.M. Slock and C.B. Papadias. Blind Fractionally-Spaced Equalization Based on Cyclostationarity. In *Proc. Vehicular Technology Conf., Stockholm, Sweden*, June 1994.

4. L. Tong, G. Xu, and T. Kailath. A New Approach to Blind Identification and Equalization of Multipath Channels. In *Proc. of the 25th Asilomar Conference on Signals, Systems & Computers, Pacific Grove, CA*, pages 856–860, November 1991.

5. B. De Moor. The Singular Value Decomposition and Long and Short Spaces of Noisy Matrices. *IEEE Trans. Signal Processing*, pages 2826–2838, September 1993.

6. G.D. Jr. Forney. The Viterbi Algorithm. *Proceedings of the IEEE*, (3):268–278, 1973.

7. S. Van Huffel and J. Vandewalle. *The Total Least Squares Problem, Computational Aspects and Analysis.* SIAM, 1991.

8. G. H. Golub and C. F. Van Loan. *Matrix Computations.* The Johns Hopkins University Press, 1989.

9. C.C. Paige and M. Saunders. Least Squares Estimation of Discrete Linear Dynamic Systems Using Orthogonal Transformations. *SIAM J. Numer. Anal.*, (2):180–193, 1977.

Assessing Stability for Adaptive Filtering Algorithms in Signal Processing

Richard C. Le Borne[1]

Department of Mathematics, University of Tennessee at Chattanooga, USA

Abstract

The techniques employed for analyzing algorithms in numerical linear algebra have evolved significantly since the 1940s. However, the structure of algorithms in signal processing prevent the direct application of these techniques when assessing their numerical properties. This paper discusses the main numerical analysis techniques for signal processing algorithms that are currently in the literature. In addition, to provide a comparative framework, we review the concepts of stability, conditioning, and perturbation sensitivity that form the foundation to the generalized approach for the analysis of algorithms in numerical linear algebra. Finally, for the lattice update structure, we demonstrate that some applications do exist where the conditioning of the problem and the uncertainty of the update parameters can be integrated together through the use of the general perturbation theorem from numerical linear algebra.

1 Introduction

For signal processing applications that can be linearly modeled and thus written in the form $Ax = b$ the field of numerical linear algebra offers many tools to assist with algorithm design as well as algorithm analysis. For example, matrix decomposition methods, such as the QR-decomposition, have led to a number of recursive least-squares algorithms for signal processing problems. Additionally, numerical linear algebra provides a framework for analyzing an equation's sensitivity to perturbations as well as analyzing an algorithm's behavior under the effects of imprecise arithmetic operations (finite precision). Other disciplines outside signal processing such as system theory and statistics also provide tools which may be used for the analysis of signal processing algorithms.

The analysis of the numerical properties of many signal processing algorithms (such as the recursive least-squares adaptive filtering algorithms) has, however, been hindered by problems not usually found in numerical linear algebra algorithms; nonlinear equations for parameter updates as well as recursive-in-time algorithms (and the associated long term effects of finite precision arithmetic)

[1]The work of this author was supported by NSF-NATO grant GER-9552624 and NSF grant MIP-9625482.

are examples. Furthermore, the meaning of "stability" can differ greatly between disciplines and not all definitions, when satisfied, provide a guarantee that the computed solution is accurate enough to be meaningful to the user. In numerical linear algebra, an algorithm's analysis must address the impact of the problem's conditioning as well as the expected accuracy of the computed solution before a stability analysis is said to have been completed. For signal processing algorithms, this goal has proven, thus far, to be an elusive quest. One cause, for recursive algorithms at least, is the fact that the "problem" changes with time. Furthermore, the diverse interpretations given for algorithmic stability in signal processing has had a correspondingly broadening effect on stability analyses; a stability analysis can be said to have been performed if any one of a number of definitions for stability can be proven (or disproven).

It has become evident, therefore, that extracting the true benefit from an analysis of a signal processing algorithm is not as straight forward as for an algorithm in numerical linear algebra. Nonetheless, these analyses do provide important insights into the performance capabilities of a signal processing algorithm and may, perhaps, form the basis of a much needed, coherent framework for the numerical analysis of signal processing algorithms.

We provide in this paper a review of the various ideas of stability for algorithms in signal processing. We highlight in Section 2 definitions for stability that are commonly used in numerical linear algebra as well as perturbation theorems that serve as the prerequisite to a stability analysis. Attention will be given in Section 3 to definitions for stability that have historically been intended for theoretical applications (infinite precision) but recently have been applied and interpreted for computer-based applications (finite precision). Section 4 demonstrates the potential effects that an ill-conditioned problem can have on the numerical performance of the a posteriori recursive least squares lattice (RLSL) algorithm, which is reported to have good numerical properties. This will illustrate how perturbation theorems from numerical linear algebra may be used to include conditioning with the numerical effects of finite precision arithmetic for analyzing signal processing algorithms. Finally, Section 5 provides a summary. To facilitate this discussion we find it useful to treat the terms *method* and *algorithm* distinctly. For clarity, we will discuss each definition in the context of solving $Ax = b$.

Definition 1. *A* method *is defined to mean the general steps, unique and non-unique, that are necessary to solve a problem.*

There are many ways to solve $Ax = b$. Examples include the method that requires the explicit use of the inverse of A, assuming it exists, the method of Gaussian elimination, or the method that performs a QR decomposition of A and then uses it to find x via a matrix-vector multiplication and a backsolve. A decision at this level, with no further detail, will define the approach that will be used to solve the problem ($Ax = b$). Sufficient information is provided so that an analysis can be performed to assess the method's sensitivity to perturbations.

Definition 2. *An* algorithm *is defined to mean the detailed and unique manner in which the finite sequence of a method's steps are performed.*

Once we have chosen the method to solve our problem, we need to further clarify how the method will be employed. For example, matrix inversion, Gaussian Elimination, and QR-decomposition are all methods that will solve this problem. If we wish to use the method of Gaussian Elimination, will we use pivoting, and if so, will we choose partial pivoting or complete pivoting? These decisions provide the additional detail necessary to define the algorithm. At this level, sufficient detail is provided so that bounds for the perturbed quantities can be found (using models for the computer arithmetic). Typically, one would refer to this as an analysis of the finite precision effects.

We point out that when deciding what method to use it is possible to provide the same detail for the method as given for the algorithm. In this case the algorithm would also be the method. We will try to maintain a distinction, however.

2 Stability of linear systems

Suppose we have a problem $P \in \mathcal{P}$ in nature that we wish to study. Because of limitations, our formulation will be inexact, so instead we must study the (hopefully) nearby problem $\hat{P} \in \mathcal{P}$. If x is the solution to P, and $\hat{x}$ is the solution to $\hat{P}$, it would be useful to know under what circumstances the error ρ,

$$\text{dist}(\hat{x}, x) = \rho \tag{2.1}$$

will be small (in some sense). For example, $\text{dist}(\hat{x}, x) = \hat{x} - x$ is one possibility. The formulation (2.1) is referred to as the *absolute error* in the solution. By replacing the right hand side of (2.1) with the quantity $\epsilon \cdot x$, where ϵ is small (in some sense), the equality becomes known as the *relative error*. Typically, it is also of interest to understand the relationship between a perturbation in the problem and the resulting perturbation in the solution. If $\hat{P} \approx P$, under what circumstances will (2.1) be true (in either the absolute or the relative sense)?

The perturbation $\hat{P}$ of the problem P in nature makes no assumption regarding the manner in which the arithmetic is to be performed. In fact, even if $\hat{P}$ was solved using exact arithmetic we must still expect the solution $\hat{x}$ to differ from x since we are solving different (but related) problems. Studying the relationship between a perturbation in the problem and the resultant solution will be independent of the method, the algorithm, or indeed the implementation. It was eluded to in Section 1 that numerical linear algebra enjoys the benefits from an understanding of the condition number and its involvement with a generalized perturbation theorem. This was not always the case.

2.1 Wilkinson and Gaussian elimination

As a pioneer in numerical linear algebra, J.H. Wilkinson became one of the most celebrated researchers in the field. In [1], one of his last publications, he discusses some of the early problems hindering the research of algorithmic numerical stability in linear algebra: during the 1940s, there lacked a systematic analysis of rounding errors and use of norms in matrix algebra. In this time period, it was widely accepted that Gaussian Elimination was unstable. In fact, it was very stable given the manner in which it was being implemented at the time (desk machines). The paper of von Neumann and Goldstine [2] arrested most of the fears regarding the stability of Gaussian elimination, but to quote Wilkinson in [1], "... The reader is faced with page after page of detailed rounding error analysis and quite soon his senses are numbed. He may well emerge nursing the illusion that he has understood it in that he has followed the derivation of each line from the previous one, but there is more to 'understanding' than that". Wilkinson was not criticizing the important work of von Neumann and Goldstine, he was underscoring the importance of a systematic analysis of rounding errors as well as the use of norms in matrix algebra.

In the subsequent years that followed, the notion of problem conditioning became a standard tool for analyzing algorithms in numerical linear algebra. The term algorithmic stability has branched into no less than four classifications: forward, backward, strong, and weak stability. In short, it is now well-recognized that a stable algorithm in linear algebra can produce a solution having no usable accuracy if the problem was ill-conditioned from the start. There are concise theorems that bound the absolute or relative change in the solution given bounds on the absolute or relative perturbations in the problem. In the next section we will formally review a general perturbation theorem due to Wilkinson.

2.2 Conditioning

Returning to the general problem of Section 2, let us assume that it can be formulated as $Ax = b$. Due to restrictions, however, we can only work with a nearby problem, $\hat{A}\hat{x} = \hat{b}$, $\hat{A} \cong A$, $\hat{b} \cong b$. For an appropriate norm, what can be asserted regarding $\| x - \hat{x} \|$? The following theorem due to Wilkinson [3] provides an upper bound:

Theorem 3. *For A nonsingular, $\mathcal{K}(A) \equiv \| A \| \cdot \| A^{-1} \|$, and $\frac{\|A-\hat{A}\|}{\|A\|} < \frac{1}{\mathcal{K}(A)}$, then $\hat{A}$ is nonsingular and*

$$\frac{\| x - \hat{x} \|}{\| x \|} \leq \frac{\mathcal{K}(A)}{1 - \mathcal{K}(A)\frac{\|A-\hat{A}\|}{\|A\|}} \left[\frac{\| A - \hat{A} \|}{\| A \|} + \frac{\| b - \hat{b} \|}{\| b \|} \right]. \tag{2.2}$$

Theorem 3 states that the relative error in the solution can not be expected to have more accuracy than that of the problem. Furthermore, for a large condition-number, $\mathcal{K}(A) \geq 1$, the problem is said to be ill-conditioned. In this case the theorem tells us that the solution could be much less accurate when compared to the uncertainty in the problem's representation.

Note, however, that Theorem 3 says nothing about the numerical performance of a method or an algorithm! Regardless of the method or algorithm used to find the solution x, the bound (2.2) is valid. This is the key point that manifested the confusion regarding the stability of Gaussian elimination; sometimes it gave a good result when the calculations were performed on the desk machines of the 1940s, other times it gave a very bad result. The explanation was in the variation of the condition number of the underlying problem from one case to the next. In the next section we review some of the definitions of stability that are useful in numerical linear algebra.

2.3 Stability

The term stability pertains to the effects of finite precision on the algorithm that is used to find the solution of some problem of interest. As given in Definition 2, our use of the term algorithm refers to a series of instructions that uniquely defines the steps in the method. From this, one can see that it is possible to have unusable computed results because of the algorithm rather than the method. One well-known example is the Gram-Schmidt (GS) algorithm. In its original form (referred to as the Classical GS algorithm) it suffers from potential stability problems. However, a subtle change to the sequencing of steps defines a new algorithm (referred to as the Modified GS algorithm) that is stable.

What guarantees can be given to the user of an algorithm that has been deemed stable? This depends mostly on the definition of stability. Since there can be, and are, many different meanings to the word, this is an important issue. However, as long as the use of the word "stable" is well-defined, there usually is not a problem with its correct interpretation. We will now give four definitions of stability that are useful in numerical linear algebra. The following will define stability in terms of matrices A belonging to some class of matrices $\mathcal{A}$. The matrix A given in each of these definitions may or may not represent the problem we wish to solve; this issue is the focus of the *perturbation analysis*, here we are only interested in the impact the algorithm has on the solution (using computer arithmetic). We begin with the more intuitive definition that provides the greatest guarantees that an algorithm will provide a good solution.

Definition 4. *An algorithm for solving a nonsingular system of linear equations $Ax = b$ in finite precision arithmetic is* forward stable *if, for every $A \in \mathcal{A}$ and for every b, the computed solution $\hat{x}$ will be near the exact solution $x = A^{-1}b$.*

If a forward stable algorithm is used then the computed solution $\hat{x}$ will be close to the exact solution x (i.e., we will have a good solution to the problem we asked the computer to solve). Definition 4 is quite powerful but often difficult to prove. The following definition shifts the interest from the computed solution, $\hat{x}$, to the perturbed problem that $\hat{x}$ solves exactly.

Definition 5. *An algorithm for solving a nonsingular system of linear equations* $Ax = b$ *in finite precision arithmetic is* backward stable *if, for every* $A \in \mathcal{A}$ *and for every* b*, the computed solution* $\hat{x}$ *is the exact solution to a nearby system* $\hat{A}\hat{x} = \hat{b}$*, where* $\hat{A} \cong A$*, and* $\hat{b} \cong b$*.*

Backward stable algorithms do not necessarily produce computed solutions $\hat{x}$ that are close to the exact solution x. This form of stability only guarantees that the computed solution is the exact solution to a nearby problem. For example, if the original matrix A is nearly singular then a nearby problem can produce solutions very different from x. However, Wilkinson's perturbation bound given in Theorem 3 can be used with Definition 5 to assure that any well-conditioned matrix A will produce a solution $\hat{x}$ that is close to x. This is an example how the perturbation theorem can be used in two ways; for assessing the sensitivity of the solution to small changes in the problem, and to assess the effect of finite precision arithmetic on the computed solution when we know (from a backward error analysis) how much the problem was changed during the execution of the algorithm. The following two definitions are due to Bunch [10].

Definition 6. *An algorithm for solving linear equations* $Ax = b$ *in finite precision arithmetic is* strongly stable *if, for every* $A \in \mathcal{A}$ *and for every* b*, the computed solution* $\hat{x}$ *is the exact solution to a nearby system* $\hat{A}\hat{x} = \hat{b}$*, where* $\hat{A} \in \mathcal{A}$*,* $\hat{A} \cong A$*, and* $\hat{b} \cong b$*.*

An algorithm that is strongly stable has the added feature of ensuring the computed solution be the exact solution to a problem of the same class. For example, suppose a certain circuit design yields equations that form a matrix A lying in the class $\mathcal{A}$ of symmetric matrices. If we solve the system using an algorithm that has been shown to be strongly stable for this class $\mathcal{A}$, then we are guaranteed that there exists some $\hat{A} \cong A$ also in class $\mathcal{A}$, and some $\hat{b} \cong b$, such that $\hat{A}\hat{x} = \hat{b}$.

Definition 7. *An algorithm for solving linear equations* $Ax = b$ *in finite precision arithmetic is* weakly stable *if, for every well-conditioned* $A \in \mathcal{A}$ *and for every* b*, the relative error* $\frac{\|x - \hat{x}\|}{\|x\|}$ *is small.*

Weak stability combines the algorithm's performance under machine arithmetic with the problem's conditioning. It assures us that an algorithm will produce a good answer as long as we provide it with a well-conditioned problem. In most cases, this is all that one can expect.

The definitions of this section and Wilkinson's perturbation theorem of the previous section demonstrate the subtle differences between the classes of stability and the connection that stable algorithms have regarding the problems that they are asked to solve. In our distinction between a method and an algorithm we associate a perturbation analysis with the method and the stability analysis with the algorithm. The next section will address these same issues for the field of signal processing.

3 Stability and adaptive filtering in signal processing

Although the notion of a problem's conditioning and an algorithm's stability is included in the body of literature that addresses the numerical aspects of signal processing algorithms [4, 5, 6, 7, 8], what is lacking is a systematic approach. Referring to the numerical stability analysis of the Recursive Least Squares (RLS) algorithm, Haykin in [4] (page 768), states "... Having said this, however, we lack a unified treatment of finite-precision effects in the RLS family of adaptive filtering algorithms that goes beyond a single error propagation". It is interesting that the comment of Haykin assimilates well the comments and afterthoughts of Wilkinson regarding the numerical analysis of algorithms in linear algebra in the 1940s. For the field of signal processing, however, and amidst the many noteworthy papers on the subject of stability, it is still quite uncertain if, after having been told that an algorithm is stable, a good solution will always be given by that algorithm.

The main difficulty with the analysis of many of the algorithms used in signal processing is that the problem being solved is inherently recursive; at each time instant there is a new, but related problem to be solved. With few exceptions this renders the standard analysis techniques from numerical linear analysis virtually useless. One example of an exception can be found for the stationary input case. The Levinson-Durbin algorithm for Toeplitz systems of equations was shown by Cybenko [9] to be weakly stable in the sense of Definition 7.

3.1 Conditioning

It was brought out in Section 2.3 that a stable algorithm can give a bad solution. In fact, for a backward stable algorithm this is exactly the case. In signal processing one should not be surprised if the same situation was found to be true. The difference, however, is the definition of a condition number. Since the algorithms in signal processing are often time-recursive, their solution applies to many problems. Given this, if an algorithm gives an inaccurate solution, how will one know if it is because it is unstable or because the underlying problem is ill-conditioned? Unfortunately there lacks a general perturbation bound similar in purpose to Theorem 3. It is possible, however, to use (2.2) in a limited manner to link the underlying conditioning of the problem with the effects of finite precision. This will be discussed in greater detail in Section 4.

3.2 Stability

The numerous articles on algorithmic stability give evidence to the interest this subject has maintained in signal processing. Examples covering the past three decades include [6, 7, 9, 11]. To the author's knowledge, Regalia in [7], and Slock in [8], were first to address, in a general manner, the subject of stability and conditioning in the field of signal processing. The fields of system science and numerical analysis were used to introduce the notion of backward consistency and minimality to signal processing.

It remains true, however, that an algorithm can be deemed stable in a variety of ways. It is therefore important to understand the relationship that stability has between the fields in which it is used. Outside the field of signal processing, analysis techniques for algorithmic stability are also found in fields such as numerical analysis, system theory, and statistics. It can therefore become a challenge to interpret the context in which stability is applied. We will discuss stability as it pertains to each of these subject fields. The backward consistency concept and minimality will be discussed separately in Section 3.2.4.

3.2.1 System theory

We present next a few of the meanings that have been intended when implying algorithmic stability in the context of system theory. This involves the classical Lyapunov stability problem for nonlinear difference equations of the form

$$\hat{\xi}(t) = f(\hat{\xi}(t-1), z(t)), \tag{3.1}$$

$$\hat{\xi}(t_0) = \xi(t_0) + \delta\xi(t_0). \tag{3.2}$$

Here the state vector at time t is denoted by $\xi(t)$, and the state perturbed by the quantity $\delta\xi(t_0)$ at the initial time t_0 is denoted by $\hat{\xi}(t)$. Many recursive least squares algorithms can be represented by (3.1) and (3.2). An important paper by Ljung and Ljung [12] discussed the stability properties for various recursive least squares methods and algorithms (our notation in this subsection was chosen to be consistent with theirs). Stability, as discussed in [12], involves the behavior of a single perturbation in the initial conditions and its effect on the output, $\eta(t) = h(\hat{\xi}(t), z(t))$, where $z(t)$ is the input. If the effect of the perturbation decays in time, the algorithm is said to be stable. The rate at which it decays defines the particular type of stability.

When a recursive least squares algorithm can be represented by (3.1) and (3.2), one can conceptually regard $\xi(t)$ as an output and analyze the effects of perturbations to it. Ljung and Ljung [12] performed a perturbation analysis for various methods for solving the recursive least squares problem: the recursive least squares method, the fast least squares method, and the fast lattice least squares method. Using Definitions 1 and 2, the so-called *basic RLS*, *conventional LS*, and the *UD-factorization LS* algorithms fall into the method of recursive least squares since each algorithm is defined by the manner in which it works with the correlation matrix. In our context, the fast least-squares algorithm is also a method since it offers the same detailed information at both levels. Finally, the fast lattice algorithm discussed in [12] is one of a number of fast lattice algorithms. The fast lattice method is defined separately since it has a unique structure for updating its forward and backward residuals.

Definition 8. *The solution $\xi(t)$ (corresponding to $\delta\xi(t_0) = 0$) is said to be* <u>stable</u> *if* $\forall \epsilon \geq 0, \exists\, \delta$ *such that* $\|\delta\xi(t_0)\| < \delta$ *implies that* $\|\hat{\xi}(t) - \xi(t)\| < \epsilon, \forall\, t \geq t_0$.

Stability in this sense provides the assertion that any solitary perturbation (of small enough modulus) will have a bounded effect on the state vector for all subsequent t. If one wishes to apply this definition to finite precision effects, the $\delta\xi(t_0)$ will be determined (i.e., a finite precision analysis) and it is then the task to find ϵ as small as possible but still satisfy the requirements of the definition. Ljung and Ljung did not attempt to use this definition in this manner, however. They focused their analysis on the propagation of $\delta z(t)$ and $\delta\xi(t)$ to future time instants by studying the propagation of a single error in the state vector occurring at time t_0.

Definition 9. *The solution $\xi(t)$ (corresponding to $\delta\xi(t_0) = 0$) is said to be* asymptotically stable *if it is stable and there exists a $\delta > 0$ such that $||\delta\xi(t_0)|| < \delta$ implies that $||\hat{\xi}(t) - \xi(t)|| \to 0$ as $t \to \infty$.*

The idea is that if $\hat{\xi}(t) \to \xi(t)$, then $\hat{\eta}(t) \to \eta(t)$ (given a single perturbation and a large enough value for t). This definition, however, falls short from providing a guarantee that if $\hat{\xi}(t) \approx \xi(t)$, then $\hat{\eta}(t) \approx \eta(t)$.

Definition 10. *The solution $\xi(t)$ (corresponding to $\delta\xi(t_0) = 0$) is said to be* exponentially stable *if it is asymptotically stable and $\exists\ C > 0$ and $0 < \lambda < 1$ such that $||\hat{\xi}(t) - \xi(t)|| \leq C\lambda^{t-t_0}$, $t \geq t_0$ if $||\delta\xi(t_0)|| \leq \delta$.*

Knowledge of the minimal rate in which $\hat{\xi}(t) \to \xi(t)$ is the added benefit from having the condition for exponential stability satisfied. However, even for exponentially stable algorithms there is no guarantee that $\hat{\eta}(t) \approx \eta(t)$ when $\hat{\xi}(t) \approx \xi(t)$.

These definitions, however, apply more to the method than to the algorithm since different algorithms can not be expected to have the same stability properties (using the definitions from Section 2.3). In their analysis, Ljung and Ljung considered the three least squares algorithms to have the same stability properties since they were all found to be algebraically equivalent through a change of basis to the state vector. Interpreting this in terms of Definitions 1 and 2, the three RLS algorithms belong to the same method. Their work must then be considered as an analysis of a method's sensitivity to perturbations rather than an algorithm's stability under finite precision arithmetic (in the sense of Section 2.3). In this regard, stability, asymptotic stability, and exponential stability pertain to the sensitivity, asymptotic sensitivity, and the exponential sensitivity to independent and isolated perturbations of a method.

3.2.2 Statistics

Algorithmic stability analyses in the past have also employed the use of statistics. This technique seems reasonable since the processes involved are assumed to be stochastic. Its use has mainly been in the analysis of arithmetic effects of update parameters other than the filter output. The technique generally employed, under some broad assumptions regarding the input processes (gaussian, first and second moments, for example), is to construct a model for arithmetic error growth for one or more of the update recursions to the algorithm of interest. If a definition for stability is not explicitly given in this type of an analysis, it usually is implicitly defined to mean non-divergence of the filter output. Examples include [13], [14], [15], and [16].

In [13], Ardalan and Alexander discuss the tradeoff between the effects of additive noise and the numerical roundoff error when choosing a value for the user-defined forgetting factor for the exponentially windowed RLS algorithm. Their analysis relied on the expectation operator to make meaningful interpretations of the algorithm and its relationship to the forgetting factor. Interesting is their reference to [12] when claiming that the algorithm under study was proven to be stable in the sense of single error perturbations. As pointed out in Section 3.2.1, the definitions for stability that were used in [12] pertained more to perturbation theorems than for guaranteeing that the algorithm's solution would be meaningful.

The equations for predicting the steady-state mean-squared values of the accumulated numerical errors in the computation of the variables of a QR-decomposition-based lattice predictor algorithm were given by Syed and Mathews in [14]. Their analysis provided an intuitive framework that could assist a potential user of the algorithm with decisions regarding issues of numerical error propagation. Although their analysis was not intended to prove or disprove the stability of the algorithm under study, the various interpretations of an algorithm's stability were eluded to when previous related efforts were discussed.

3.2.3 Signal processing

Stability with respect to the field of signal processing addresses bounded input, bounded output stability (BIBO) [5] or, equivalently, the stable behavior of the transfer function [6]. Leung and Haykin [5] proved that the recursive QRD-LS algorithm (when implemented using a finite-precision systolic array) is BIBO-stable. That is, for all bounded input signals the output of the algorithm will always be bounded. This form of stability does not, however, guarantee that the output of the algorithm will be meaningful (as pointed out in [14]).

For the case of the linear prediction of the output of an all-pole filter, Makhoul presented a class of stable recursive lattice methods in [6]. Stability in this case referred to the stability of the transfer function $H(z)$ where

$$H(z) = \frac{G}{A(z)}, \quad A(z) = \sum_{k=0}^{p} a_k z^{-k}, \quad a_0 = 1. \tag{3.3}$$

Here G denotes the gain factor, a_k the predictor coefficients, and p the number of poles (predictor coefficients) in the model. This requirement simplifies to the condition that the reflection coefficients are strictly less than one in modulus.

This stability definition, as in the previous example, does not assert a meaningful result from the algorithm. For this case it will be demonstrated in Section 4 that the condition number of the 2×2 matrix defining the lattice structure has a great effect on the accuracy of the forward and backward residuals, and hence, the output of the algorithm.

3.2.4 Stability and conditioning

Stability and problem conditioning were first addressed as a general technique for assessing the effects of computer arithmetic concurrently and independently by Regalia and Slock in [7] and [8], respectively. These works relied on the concepts of backward consistency from numerical analysis as well as minimality and reachability from system theory. Due to the similarities in both works we will only focus on one, namely [7].

Given that the prediction portion of a fast least-squares algorithm can be written in the form of (3.1) and (3.2), that is, a nonlinear function, the input-output relationship of the prediction part of the filter can be interpreted as a nonlinear mapping from a domain (input) to a range (output). The domain $\mathcal{D}$ consists of all possible input processes and the range, classified in terms of the arithmetic precision, results from the mapping defined by f, the nonlinear function: $\mathcal{S}_i$, if infinite precision is assumed (i.e., $\mathcal{D} \stackrel{f}{\to} \mathcal{S}_i$), and $\mathcal{S}_f$ for finite precision (i.e., $\mathcal{D} \stackrel{f}{\to} \mathcal{S}_f$). When the algorithm is run using finite precision arithmetic, the definition for backward stability presented as Definition 5 can easily be applied; we attempt to find an input process near the original such that the filter's (mapping) computed output is the exact realization using infinite precision. The algorithm will be backward consistent as long as $\mathcal{S}_f \subset \mathcal{S}_i$. It will be backward stable as long as there exists at least one input process near the original input process.

The set $\mathcal{S}_i$ was termed the stability domain. Unstable error propagation implies that the prediction part of the adaptive filter algorithm has evolved, under the effects of finite precision, to a state that, under infinite precision arithmetic, cannot be reached by *any* input process (i.e. it is not backward consistent). The state, given infinite precision and a persistently exciting input process (i.e. the underlying correlation matrix is nonsingular), will be restricted to the interior of $\mathcal{S}_i$ as long as it is minimal. It will evolve on the boundary of $\mathcal{S}_i$, denoted $\partial\mathcal{S}_i$, if and only if the input process is perfectly predictable (i.e. the underlying correlation matrix is singular). For the pth-order forward/backward prediction problem ξ is minimal if it has dimension $2p+1$. Furthermore, the conditioning of the input sequence (the condition number of the underlying correlation matrix) is inversely related to the distance between ξ and $\partial\mathcal{S}_i$. Measuring the distance, therefore, is a way to assess the conditioning of the underlying problem. In the next section we will introduce this concept in terms of the lattice formulation for updating the forward and backward residuals of the one-step forward and backward prediction problem.

4 Conditioning and stability for the a posteriori direct RLSL algorithm

In this section we will combine the notion for BIBO stability discussed in Section 3.2.3 with conditioning and stability domains discussed in Section 3.2.4. We show that the numerical behavior of a Least-Squares-Lattice (LSL) algorithm relies on the conditioning of the underlying problem. In turn, the conditioning is inversely proportional to the proximity that the regression coefficients have to two sides of a triangular stability boundary.

4.1 Lattice structure

It is well known (for example [4], [7]) that the lattice structure defining the forward and backward residual update recursions can be defined in matrix-vector form as

$$\mathbf{r}_m(n,n) = \mathbf{H}_m(n)\mathbf{r}_{m-1}(n,n-1) \tag{4.1}$$

where we have defined $\mathbf{r}_m(n,p) = (f_m(n), b_m(p))^t$ and

$$\mathbf{H}_m(n) = \begin{pmatrix} 1 & \Gamma^*_{f,m}(n) \\ \Gamma^*_{b,m}(n) & 1 \end{pmatrix}. \tag{4.2}$$

Here, the forward and backward residuals, $f_m(n)$ and $b_m(n)$, at time n, are updated by means of the forward and backward reflection coefficients, $\Gamma_{f,m}(n)$ and $\Gamma_{b,m}(n)$, respectively, as well as the forward and backward residuals from the previous stage.

The perturbations in $\mathbf{H}_m(n)$ has the special off-diagonal structure while the perturbations in $\mathbf{r}_{m-1}(n,n-1)$ will be more general:

$$\delta\mathbf{H}_m(n) = \begin{pmatrix} 0 & \delta\Gamma^*_{f,m}(n) \\ \delta\Gamma^*_{b,m}(n) & 0 \end{pmatrix}, \delta\mathbf{r}_{m-1}(n,n-1) = \begin{pmatrix} \delta f_{m-1}(n) \\ \delta b_{m-1}(n-1) \end{pmatrix}.$$

Although we wish to compute $\mathbf{r}_m(n,n)$ via (4.1) at each given update (time and order), we actually compute the update using quantities which have been perturbed. Mathematically, this can be formulated as

$$\overline{\mathbf{r}_m(n,n)} = (\mathbf{H}_m(n) + \delta\mathbf{H}_m(n))\,(\mathbf{r}_{m-1}(n,n-1) + \delta\mathbf{r}_{m-1}(n,n-1))\,. \tag{4.3}$$

The above representation of (4.3) provides a natural mechanism to study the nonlinear and compounding effects of finite precision arithmetic. It is important to note that the perturbations $\delta\mathbf{H}_m(n)$ and $\delta\mathbf{r}_{m-1}(n,n-1)$ are defined at the algorithmic level, and hence, these terms are algorithm dependent. For example, the direct and indirect RLSL algorithms are distinguished by the manner in which the reflection coefficients are updated. It has been shown in [17] that $\delta\mathbf{H}_m(n)$ will differ between these algorithms. Independent of algorithmic implementation, however, is the sensitivity to perturbations that is governed by the update structure of (4.1). Because of the underlying matrix structure given by the residual update recursions, the perturbation analysis is straight forward once it is realized that we are not, at this juncture, interested in how or where the perturbations occur.

The following results and the experiments given in Section 4.2 are due to Bunch, Le Borne, and Proudler [18]. For the matrix $\mathbf{H}_m(n)$, the following lemma gives the condition number with respect to the 2-norm.

Lemma 11. *Let $\mathbf{H}_m(n)$ be defined by Equation (4.2). The condition number, $\mathcal{K}_2(\mathbf{H}_m(n))$, using the 2-norm for complex matrices is then given by*

$$\mathcal{K}_2(\mathbf{H}_m(n)) = \frac{\max(Y_\pm)}{2\,|q|}, \tag{4.4}$$

where,

$$\begin{aligned} Y_\pm &= k \pm \sqrt{k^2 - 4\,|q|^2}, & (4.5)\\ k &= 2 + |\,\Gamma_{f,m}(n)\,|^2 + |\,\Gamma_{b,m}(n)\,|^2, & (4.6)\\ q &= 1 - \Gamma_{f,m}(n)\,\Gamma_{b,m}(n). & (4.7) \end{aligned}$$

Using the generalized perturbation bounds given in Theorem 3 we have the following result that combines the conditioning of $\mathbf{H}_m(n)$ to perturbations in the forward and backward residuals.

Theorem 12. *Suppose*

$$\mathbf{r}_m(n,n) = \mathbf{H}_m(n)\mathbf{r}_{m-1}(n,n-1)$$

and

$$\overline{\mathbf{r}_m(n,n)} = (\mathbf{H}_m(n) + \delta\mathbf{H}_m(n))\,(\mathbf{r}_{m-1}(n,n-1) + \delta\mathbf{r}_{m-1}(n,n-1))\,.$$

Then the relative error of the updated residual vector is given by

$$\begin{aligned} &\frac{\|\mathbf{r}_m(n,n) - \overline{\mathbf{r}_m(n,n)}\|_2}{\|\mathbf{r}_m(n,n)\|_2} & (4.8)\\ &\leq \frac{\max(Y_\pm)}{2\,|\,1 - \Gamma_{f,m}(n)\,\Gamma_{b,m}(n)\,|} \left\{ \frac{\|\delta\mathbf{H}_m(n)\|_2}{\|\mathbf{H}_m(n)\|_2} + \frac{\|\delta\mathbf{r}_{m-1}(n,n-1)\|_2}{\|\mathbf{r}_{m-1}(n,n-1)\|_2} \right\}. \end{aligned}$$

We see from Theorem 12 that uncertainty in the updated residuals can be magnified by a factor as large as $\mathcal{K}_2(\mathbf{H}_m(n))$ during its time and order updates. In the next section we demonstrate that arbitrarily large condition numbers exist within the set of Bounded-Input-Bounded-Output (BIBO) processes generated from a 2nd order Auto-Regressive (AR) process.

4.2 Conditioning and 2nd order, all-pole AR processes

To maintain a tractable series of experiments we have chosen four case studies in which the direct RLSL algorithm is given the input process, $\mathbf{u}(i)$, $i = 1, 4096$, that is from a 2nd order, all-pole AR process satisfying the difference equation, $v(n) = u(n) + a_1 u(n-1) + a_2 u(n-2)$. Here the process $v(n)$ is a zero mean, normally distributed white noise process with variance σ_v^2. When $a_1, a_2 \in \mathcal{R}$, this process will be ensured to be stable (BIBO stable) as long as the AR coefficients satisfy some well-known inequalities [4] (pp. 120-127).

If we consider the pairing (a_1, a_2), the inequalities defining BIBO stability describe the inner region of the triangle with vertices $(0, -1), (-2, 1)$, and $(2, 1)$ that is given in Figures 1i), 1ii) and 1iii) detail the regions (shaded) for which $\mathcal{K}_2(\mathbf{H}_m(n)) \geq 100$ and 1000, respectively.

The first experiment uses a set of AR coefficients that gives rise to a small condition number for the matrix $\mathbf{H}_m(n)$. The second through fourth experiments use AR coefficients that cause the matrix $\mathbf{H}_m(n)$ to have varying degrees of ill-conditioning.

The direct RLSL algorithm, programmed in C, was run from Matlab[2]. The algorithm used is the same as given in [15]. To compute the normed quantities $\|\delta\mathbf{H}_m(n)\|_2$ and $\|\delta\mathbf{r}_{m-1}(n, n-1)\|_2$, the RLSL algorithm was run in both double and single precision. This effectively gives a normed measurement of the uncertainty in the quantities rather than the norm of the exact errors. The left hand quantity $\|\mathbf{r}_m(n,n) - \overline{\mathbf{r}_m(n,n)}\|_2$ of (4.8) was computed in a like manner.

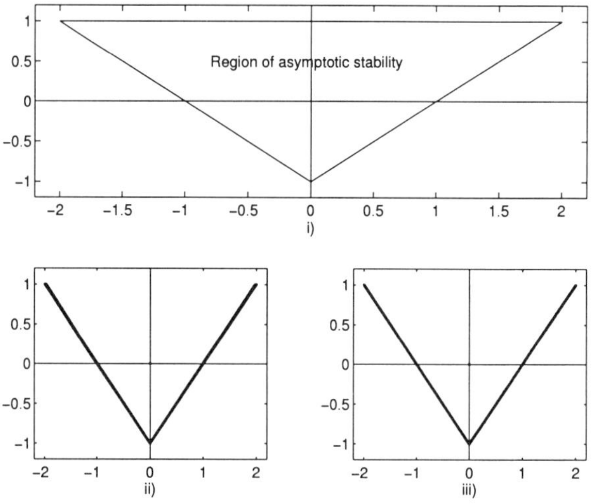

Figure 1. 2nd order AR coefficients (a_1, a_2): i) Triangular region bounding the permissible pairings to ensure asymptotic stability, ii) The thick region depicts all pairings that generate a condition number $\mathcal{K}_2(\mathbf{H}_m(n)) \geq 100$, and iii) $\mathcal{K}_2(\mathbf{H}_m(n)) \geq 1000$

---

[2]The authors thank Richard North for supplying the direct RLSL algorithm.

The details describing each AR process are given in Table 1. The variance of the input process, δ_v^2, was chosen in each case so that the output process, $u(n)$, would have unit variance. It was deemed appropriate to choose at least one experiment from the literature. The first case was selected from [4] (page 123) and the third case studied was selected from [6]. The results from each study are presented in Table 2. Presented in the middle three columns are the mean values for each of the terms on the right-hand side of (4.8). For comparative purposes, the two right-most columns give the computed mean values for each side of (4.8). It is seen that for each case studied, the relative perturbations of $\mathbf{H}_1(n)$ and $\mathbf{r}_0(n, n-1)$ are of order $O(10^{-7})$ and $O(10^{-8})$, respectively. Thus the dramatic differences in performance is due entirely to the conditioning of (4.1). The last two columns of Table 2 for each of these case studies were also graphed at each iteration and are found in Figures 2 and 3. These four cases are graphed with the corresponding computed condition number appearing directly below. In each case the line representing the estimated perturbation bounds (the last column of Table 2) is found above the measured relative uncertainty (the fifth column of Table 2). The computed values of the condition numbers of $\mathbf{H}_1(n)$ also are seen to rapidly converge to their expected values.

Table 1. Four 2nd order AR processes and their associated conditioning

$$v(n) = u(n) + a_1 u(n-1) + a_2 u(n-2)$$

Case	AR coef's		Roots			Expected
#	a_1	a_2	$\lvert p_1 \rvert$	$\lvert p_2 \rvert$	σ_v^2	$\mathcal{K}_2(\mathbf{H}_1(n))$
1	0.100	-0.8	.9	.9	2.70×10^{-1}	3.000×10^0
2	1.970	.990	.99	.99	3.980×10^{-4}	1.980×10^2
3	-1.8	.81	1.0000	.8	3.790×10^{-3}	3.610×10^2
4	1.997	.999	.9995	.9995	3.996×10^{-6}	1.998×10^3

Table 2. Effects from the conditioning of $\mathbf{H}_1(n)$ on the updated residuals

Theorem 12: Computed Mean (2-norm)					
Case #	$\mathcal{K}_2(\mathbf{H}_1(n))$	$\frac{\delta \mathbf{H}_1(n)}{\mathbf{H}_1(n)}$	$\frac{\delta \mathbf{r}_0(n, n-1)}{\mathbf{r}_0(n, n-1)}$	Left Side	Right Side
1	3.0×10^0	1.4×10^{-7}	2.4×10^{-8}	2.0×10^{-7}	4.5×10^{-7}
2	2.0×10^2	1.7×10^{-7}	3.1×10^{-8}	4.2×10^{-6}	4.0×10^{-5}
3	3.5×10^2	3.4×10^{-7}	3.2×10^{-8}	1.2×10^{-5}	1.3×10^{-4}
4	2.0×10^3	1.8×10^{-7}	3.3×10^{-8}	2.0×10^{-5}	4.3×10^{-4}

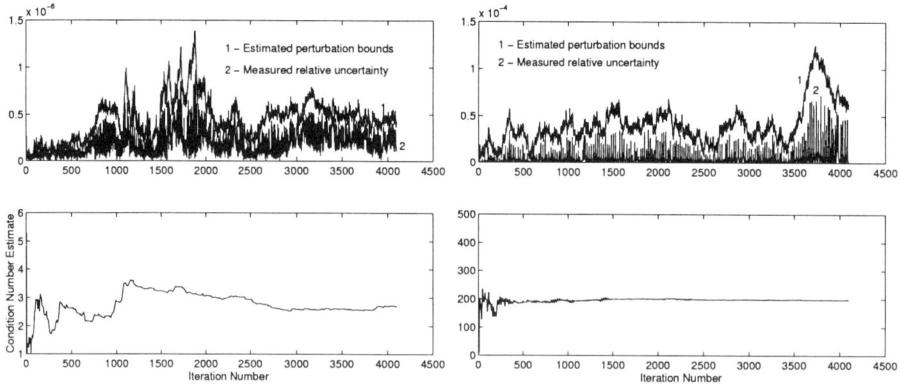

Figure 2. Two AR-models of order 2: Coefficients chosen to yield (left) an $O(1)$ condition number and (right) an $O(10^2)$ condition number

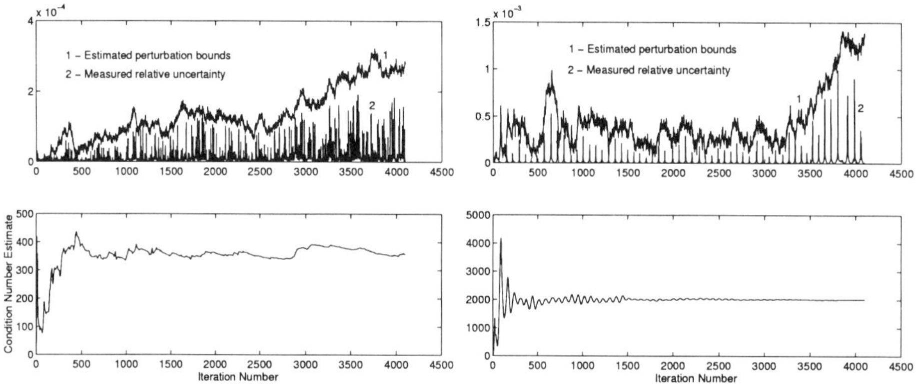

Figure 3. Two AR-models of order 2: Coefficients chosen to yield (left) an $O(10^2)$ condition number and (right) an $O(10^3)$ condition number

5 Conclusion

Analyzing algorithms in numerical linear algebra has evolved significantly since the 1940s. General theorems now exist to measure a method's sensitivity to perturbations and an algorithm's stability in the presence of finite precision. The role of problem conditioning has been thoroughly integrated into standard analysis practices as well. Additionally, experiments are introduced by including a detailed discussion of the condition number of the problems to be presented. However, the structural nature of algorithms in signal processing prevents an easy and direct application of these theorems and concepts.

Although the numerical analysis of algorithms in signal processing lacks a general systematic approach beyond a single error propagation, there are many significant results regarding algorithmic design and performance. To consolidate these and future results into what may eventually be called a systematic approach to assessing the numerical quality of a signal processing algorithm, a first step is to understand what is presently available. The treatment of algorithmic stability for signal processing algorithms, in addition to the use of existing definitions in the field, has looked to other fields such as numerical analysis, system theory, and statistics to analyze an algorithm's performance in finite precision.

The diversity of uses for stability create a challenging atmosphere when tasked with interpreting the results of an analysis. We have consolidated the main numerical analysis techniques for signal processing that are currently in the literature. To provide a comparative framework, we have additionally reviewed the concepts of stability, conditioning, and perturbation sensitivity that form the foundation to the generalized approach for the analysis of algorithms in numerical linear algebra. We have not offered a generalized analysis approach, but we have provided a means to facilitate the comprehension of existing and future analyses of signal processing algorithms. Finally, for the lattice update structure, we have demonstrated that some applications do exist for the use of the general perturbation theorem in numerical linear algebra so that the conditioning of the problem and the uncertainty of the update parameters can be integrated together.

Acknowledgements

The author thanks (alphabetically) James R. Bunch, Marc Moonen, John McWhirter, Ian Proudler, Sanzheng Qiao, and Phillip Regalia for their helpful opinions, suggestions, and comments. The author is also grateful to the Network Theory and Circuit Design Group for the resources that were generously made available while visiting at the Technical University of Munich, Germany.

Bibliography

1. Wilkinson, J.H. (1986). Error analysis revisited. *IMA Bulletin*, *22*, 192–200.

2. von Neumann, J. and Goldstine, H. (1947). Numerical inverting of matrices of high order. *Bull. Amer. Math. Soc.*, *53*, 1021–99.

3. Wilkinson, J.H. (1961). Error analysis of direct methods of matrix inversion. *J. Assoc. Comput. Mach.*, *8*, 281–330.

4. Haykin, S. (1996). *Adaptive Filter Theory* (3rd ed.). Prentice-Hall International, London.

5. Leung, H. and Haykin, S. (1989). Stability of Recursive QRD-LS algorithms using finite-precision systolic array implementation. *IEEE Trans. Acoustics, Speech, and Signal Processing*, *37*, 760–3.

6. Makhoul, J. (1977). Stable and efficient lattice methods for linear predicion. *IEEE Trans. Acoustics, Speech, and Signal Processing*, *ASSP-25*, 423–8.

7. Regalia, P. (1992). Numerical stability issues in fast least-squares adaptation algorithms. *Opt. Engr.*, *31*, 1144–52.

8. Slock, D.T.M. (1992). Backward consistency concept and round-off error propagation dynamics in recursive least-squares algorithms. *Opt. Engr.*, *31*, 1153–69.

9. Cybenko, G. (1980). The numerical stability of the Levinson-Durbin algorithm for Toeplitz systems of equations. *SIAM J. Sci. Stat. Comput.*, *1*, 303–19.

10. Bunch, J. R. (1987). The weak and strong stability of algorithms in numerical linear algebra. *Linear Algebra and its Applications*, *88/89*, 49–66.

11. Luk, F. and Qiao, S. (1989). Analysis of a recursive least-squares signal-processing algorithm. *SIAM J. Sci. Stat. Comput.*, *10*, 407–18.

12. Ljung, S. and Ljung, L. (1985). Error propagation properties of recursive least-squares adaptation algorithms. *Automatica*, *ASSP-21*, 157–67.

13. Ardalan, S. H. and Alexander, S. T. (1987). Fixed-point roundoff error analysis of the exponentially windowed RLS algorithm for time-varying systems. *IEEE Trans. on Acoustics, Speech, and Signal Processing*, *ASSP-35*, 770–83.

14. Syed, M. A. and Mathews, V. J. (1992). Finite precision error-analysis of a QR-decomposition-based lattice predictor. *Optical Engineering*, *31*, 1170–80.

15. North, R., Zeidler, J., Ku, W. and Albert, T. (1993). A floating-point arithmetic error analysis of direct and indirect coefficient updating techniques for adaptive lattice filters. *IEEE Trans. on Signal Processing*, *41*, 1809–23.

16. Ling, F., Manolakis, D. and Proakis, J.G. (1986). Numerically Robust Least-Squares Lattice-Ladder Algorithms with Direct Updating of the Reflection Coefficients. *IEEE Trans. on Acoustics, Speech, and Signal Processing*, *ASSP-34*, 837–45.

17. Le Borne, R. C. (1996). Numerical roundoff effects and the direct and indirect RLSL algorithms. *Proceedings to the 4th International Conference on Mathematics in Signal Processing, Dec.*.

18. Bunch, James R. and Le Borne, Richard C. and Proudler, K. Ian (1997). Tracking ill-conditioning for the RLS-lattice algorithm. *IEE Proceedings Vision, Image and Signal Processing*, to appear.

Constrained Kalman Filter for Least-Squares Estimation of Time-Varying Beamforming Weights

S.D. Hayward

DERA Malvern, Worcestershire

Abstract

In this paper a Kalman filter solution to the problem of adaptive beamforming in a non-stationary signal environment is proposed. We first show how a nonstationary estimation problem can be transformed into a linear stationary one by means of a projection onto a set of time-variant basis functions. The resulting parameter estimation problem is efficiently solved using a Kalman filter. A soft constraint is applied to the Kalman filter solution by introducing "pseudo-measurements". This constraint acts to control the shape of the adapted beampattern, and reduces the amount of data required to achieve a given signal-to-noise ratio at the output of the beamformer.

1 Introduction

There is a wide range of sonar, communications and radar systems for which adaptive beamforming has been proposed as a way of reducing the effects of interference and jamming. Many of these systems are required to operate in a signal environment which changes rapidly with time due to a combination of fluctuating propagation conditions, modulated waveforms and array motion. Surprisingly the ability of adaptive beamforming algorithms to track the time-varying statistics of such signal environments had not been addressed in the literature until it was recently shown that array motion can cause a significant degradation in the performance of the standard least-squares approach [1]. Research on the tracking performance of adaptive filters has been largely confined to the problem of identifying time-varying systems using the Recursive Least Squares (RLS) and Least Mean Squares (LMS) algorithms (see [2] and the references contained therein), where the non-stationary behaviour is universally described by a random-walk model. In contrast it is assumed in this paper that the spatial statistics of the data at the array face vary according to the dynamics of some external process, causing the optimum beamforming weights to vary smoothly with time. A good example of such a process is array rotation which makes signal sources appear to have time-varying directions of arrival.

There are three classes of algorithm described in the literature which may be used for estimating time-varying beamforming weights;

1. adaptive estimation using data windowing;
2. linear time-varying models with time invariant weights;
3. non-linear time-varying models with time invariant weights.

The (RLS) algorithm with exponential fading memory is a popular example from class 1, and is commonly applied to the adaptive beamforming problem. Although it exhibits rapid convergence in a stationary environment its tracking performance has been shown to be poor in general [2], since the choice of exponential weighting factor β involves a trade-off between measurement error and tracking lag error. The RLS algorithm is known to be a special case of the Kalman filter (KF).

Algorithms in class 2 include the parametric methods for estimating the time-varying coefficients of AR and ARMA processes described respectively in [3] and [4]. In [4] Grenier gives the most general solution in terms of transforming the problem of estimating time-varying parameters into one of estimating a larger set of stationary parameters through projection onto a finite-dimensional subspace defined by a known set of basis functions. Both methods are applied to a finite interval of time-series data.

An example of an algorithm from class 3 is the extended Kalman filter (EKF) [5], which is obtained by linearizing the normal KF equations around the state estimate at each iteration. The EKF has been proposed as a way of estimating time-varying signal parameters; see for example [6], in which a comparison is made between the EKF and Grenier's technique as applied to the problem of estimating LFM signal parameters in noise.

In the rest of this paper we develop a hybrid approach to the adaptive beamforming problem which uses Grenier's projection technique, together with the use of an exponentially weighted data window to provide tolerance to modelling errors when a recursive estimation procedure is employed. A recursive solution is desirable in some applications, since it avoids the need to store the large amounts of data required for batch or block processing. An equivalent batch process has been described previously [7]. The following notation will be employed. Vectors are lowercase and underlined, matrices are uppercase and underlined, $\underline{d}^H$ denotes conjugate transpose, $\underline{d}^*$ denotes conjugation, and $E < \underline{d} >$ denotes statistical expectation.

2 Signal model

Using vector notation, the discrete-time data at the sensor outputs is denoted $\underline{x}(k)$ and is assumed to result from the summation of a narrow-band target signal and a noise process at each sensor, according to

$$\underline{x}(k) = s(k)\underline{c} + \underline{\eta}(k). \tag{2.1}$$

At time k the response of the array to the target signal is a "steering vector" $\underline{c}$ scaled by the complex waveform $s(k)$. The nonstationary stochastic noise process $\underline{\eta}(k)$ represents interference, clutter and thermal noise incident on the array, and is assumed to be jointly Gaussian and uncorrelated with the target signal. The output of the beamformer is a linear combination of the sensor outputs at time k with weight vector $\underline{w}(k)$. The adaptive beamforming problem is to choose $\underline{w}(k)$ so that a good estimate of the target signal is obtained.

The signal model of Equation (2.1), with a stationary signal and non-stationary interference, is a reasonable model for the array rotation problem as long as the rotations are small and the desired signal is weak, since the adaptive beamformer is only sensitive to strong sources of interference [1]. If the statistics of the data are known at time k, then the statistically optimum (in the mean square sense) Wiener solution can be computed

$$\underline{w}_W(k) = \frac{\underline{\mathrm{M}}(k)^{-1}\underline{c}}{\underline{c}^H\underline{\mathrm{M}}(k)^{-1}\underline{c}}, \tag{2.2}$$

where an unbiased estimate is obtained by constraining the weight solution to satisy $\underline{w}_W(k)^H\underline{c} = 1$. In practice $\underline{\mathrm{M}}(k) \equiv E < \underline{x}(k)\underline{x}(k)^H >$ is unknown, and in the constrained least-squares solution is replaced by a time average $\hat{\underline{\mathrm{M}}} \equiv \sum_{k=1}^{n} \underline{x}(k)\underline{x}(k)^H$. When the statistics are stationary this can be thought of as approximating the Wiener solution. When the statistics are nonstationary *a priori* knowledge of the dynamics of the generating system may be employed to estimate the time-varying Wiener solution.

3 State-space model of an adaptive beamformer

Since we require a weight solution that evolves with time we start by constructing a linear state-space model of an adaptive beamformer:

$$\underline{w}_a(k+1) = \underline{\Phi}(k+1,k)\underline{w}_a(k) + \underline{\nu}(k); \tag{3.1}$$

$$r(k) = \underline{w}_a{}^H(k)\underline{y}(k) + \epsilon(k). \tag{3.2}$$

Here $\underline{\Phi}(k+1,k)$ is a matrix defining the transition of the weight vector $\underline{w}_a$ from its kth to its $(k+1)$th state, $r(k)$ is a reference signal at time k, and $\underline{y}(k)$ is the data to which the weights are applied. $\underline{\nu}(k)$ is a zero mean Gaussian noise process with covariance matrix $\underline{Q}(k)$ which represents uncertainty in the modelling of $\underline{w}_a$. $\epsilon(k)$ is the difference between the reference and the beamformer output, and so constitutes the error. It is straightforward [8] to transform the constrained LS problem into the canonical form of Equation (3.2).

It can be shown [9] that the RLS algorithm with an exponentially weighted data window assumes a state-space model with $\underline{\Phi}(k+1,k) = \underline{I}$ and $\underline{\nu}(k) = (\beta^{-1}-1)\underline{P}(k)$, where $\underline{P}(k)$ is, in Kalman filter terminology, the state-error covariance matrix. This observation explains why the RLS algorithm with β close to one is well matched to the random-walk model only when $\underline{Q}(k) = \underline{I}$.

To model the dynamic behaviour of $\underline{w}_a(k)$ we follow Grenier and define the weight solution in terms of a set of q basis functions

$$\underline{w}_a(k) \equiv \sum_{i=1}^{q} \underline{\alpha}_i(k) f_i(k), \tag{3.3}$$

where the $\{\underline{\alpha}_i(k)\}$ are unknown vector projection coefficients. If the model in Equation (3.3) is exact then we have transformed the problem of estimating a single nonstationary vector parameter into one of estimating q stationary vector parameters. To allow for modelling errors we assume, in practice, that the $\{\underline{\alpha}_i(k)\}$ are quasi-stationary, that is they vary slowly with time, and are adequately estimated using an algorithm from class 1.

4 General solution

Rewriting the state-space equations in terms of a new parameter vector $\underline{z}(k) \equiv [\ \underline{\alpha}_1(k)\ \ \underline{\alpha}_2(k)\ \ \cdots\ \ \underline{\alpha}_q(k)\ ]^T$, and replacing the process noise term with an exponential data window, we get

$$\underline{z}(k+1) = \underline{z}(k); \tag{4.1}$$

$$\beta^{-\frac{k}{2}} r(k) = \beta^{-\frac{k}{2}} \underline{z}^H(k)\underline{F}(k)\underline{y}(k) + \epsilon(k). \tag{4.2}$$

In Equation (4.2),

$$\underline{F}(k) \equiv [\ f_1^*(k)\underline{I}\ \ f_2^*(k)\underline{I}\ \ \cdots\ \ f_q^*(k)\underline{I}\ ]^T,$$

can be interpreted as modulating the sensor data vector $\underline{y}(k)$ to generate q sets of sensor outputs with different time variations. Writing $\underline{\tilde{y}}(k) = \underline{F}(k)\underline{y}(k)$ we obtain the following measurement equation,

$$\beta^{-\frac{k}{2}} r(k) = \beta^{-\frac{k}{2}} \underline{z}^H(k)\underline{\tilde{y}}(k) + \epsilon(k). \tag{4.3}$$

This defines a state-space model which can be solved recursively using the RLS algorithm, provided that the basis functions can be computed. In choosing the basis functions prior knowledge of the system dynamics can be brought to bear. A variety of different functions has been suggested in the literature including terms from a power series [3], Legendre polynomials, and terms from a Fourier series [4], [10]. When the system dynamics are periodic and the signal statistics are cyclostationary Equation (2.2) shows that the time-varying Wiener solution will also be periodic, and terms from a Fourier series are then a sensible choice of basis. It has been shown that many communications signals exhibit cyclostationarity; a fact which is exploited in the beamformer described in [11]. While it can be argued that most external processes will lead to periodically varying statistics, when the period is unknown the use of terms from a power series may be more appropriate for modelling low frequency fluctuations.

The contribution of this paper is to show how such basis functions can be computed recursively to ensure that they remain bounded and linearly independent as new data are incorporated into the estimate. Linear independence of the basis functions at each time step is required for the existence of a unique solution and consequently the stability of the algorithm. Ideally the functions should be orthogonal on each successive exponentially fading window.

5 Recursive solution for power series expansion

Consider the following set of basis functions

$$\begin{aligned} f_1(k) &= 1, \\ f_2(k) &= k, \\ &\vdots \\ f_q(k) &= \tfrac{k^{q-1}}{q!}. \end{aligned} \tag{5.1}$$

A simple recursion exist whereby the value of each of the functions at time $k+1$ can be written as a linear combination of the values at time k. In other words we can write

$$f_i(k+1) = \sum_{j=1}^{i} \frac{f_j(k)}{(i-j)!}, i = 1, q. \tag{5.2}$$

We can make use of these recursions to compute the basis functions implicitly through the application of the state-transition matrix $\underline{\Phi}(k+1,k)$, by defining a modified state vector as follows

$$\underline{\tilde{z}}(k) \equiv \begin{bmatrix} \underline{\alpha}_1(k)f_1(k) & \underline{\alpha}_2(k)f_2(k) & \cdots & \underline{\alpha}_q(k)f_q(k) \end{bmatrix}^T. \tag{5.3}$$

The elements of $\underline{\Phi}(k+1,k)$ are determined by the recursions of Equation (5.2). In fact we can apply any full-rank linear transformation to $\underline{\tilde{z}}(k)$ providing that there is a corresponding transformation of $\underline{\Phi}(k+1,k)$ and $\underline{F}(k)$. One option is to find such a transformation which reduces $\underline{F}(k)$ to the form $\underline{F}(k) \equiv \begin{bmatrix} \underline{I} & \underline{0} & \cdots & \underline{0} \end{bmatrix}^T$. This results in the first m elements of $\underline{\tilde{z}}(k)$ representing the instantaneous weight vector. As an example, when $q = 2$, we get the following state-space model

$$\underline{\tilde{z}}(k+1) = \begin{bmatrix} \underline{I} & \underline{I} \\ \underline{0} & \underline{I} \end{bmatrix} \underline{\tilde{z}}(k); \tag{5.4}$$

$$\beta^{-\frac{k}{2}}r(k) = \beta^{-\frac{k}{2}}\underline{\tilde{z}}^H(k)\left[\begin{array}{c} I \\ \underline{0} \end{array}\right]\underline{y}(k) + \epsilon(k). \tag{5.5}$$

The form of the state-space model defined by Equations (5.4) and (5.5) avoids the need to explicitly generate the modulated sensor data, and allows the use of the well known KF equations to recursively compute the LS solution for $\tilde{z}(k)$ and consequently $\underline{w}_a(k)$. We can exploit the vast literature on the KF to obtain numerically efficient solutions. In the latter part of this paper we describe the results of simulation experiments that make use of a square-root form of the KF equations taken directly from [12] to compute a time-varying beamforming weight for the case of $q = 2$.

The algorithm defined here is essentially that suggested by Hudson in [13], where he derives recursions equivalent to the KF equations for the first 3 terms of a Taylor series expansion. As he points out this choice of basis corresponds to estimating the first q derivatives of the weight vector at each time k, the assumption being that the qth derivative is quasi-stationary. When the value of β is close to one the algorithm has a long memory, and the values taken by the basis functions become large enough to affect the conditioning of the state-error covariance matrix. By using a normalisation factor which is appropriate to the value of β the ill-conditioning can be avoided, and the algorithm is found to be numerically stable in practice.

6 Beam pattern stabilisation techniques

By estimating the projection coefficients of the time-varying weight vector onto a set of appropriate basis functions we can reduce the bias in the estimate, but in general increasing the number of independent parameters in an estimation problem leads to an increase in the variance of the estimation error, unless more data is available to help smooth the estimates. In the adaptive beamforming problem this effect manifests itself as weight or pattern "jitter". For the problem of estimating $\underline{\tilde{z}}(k)$ in Equation (5.4) using the measurements of Equation (5.5), the number of data samples needed to achieve a given mean-square error is approximately proportional to q. To get round this problem we can use prior knowledge of the solution to reduce the variance without significantly increasing the bias. Both subspace methods and soft constraint methods have been suggested as a solution. Subspace projection is a way of getting rid of "nuisance" parameters. By assuming that the data lies in a space which can be divided into signal and noise subspaces, and that the noise singular values are known, the estimation problem can be restricted to parameters associated with the signals (see for example [14]). While the use of subspace tracking techniques might permit a recursive subspace approach, a simpler solution is to make use of a soft constraint method, such as that described in [15], for reducing the estimation error and controlling the level of the sidelobes. The use of fictitious data or "pseudo-measurements" for constraining the KF solution has been described previously in the context of beamforming [16] and target tracking [17]. Here we develop a way of applying a soft constraint to KF solutions. Referring to the signal model of Equation (2.1) we can write a LS cost-function, including a penalty term, as

$$J(\underline{w}) = \underline{w}^H \hat{\underline{\mathrm{M}}}\, \underline{w} + \rho \underline{w}^H \underline{\mathrm{X}}\, \underline{w} \tag{6.1}$$

where $\underline{\mathrm{X}}$ is a weighting matrix and ρ is a positive scalar. The time-invariant LS weight solution $\underline{w}_{LS}$, subject to a linear gain constraint, is

$$\underline{w}_{LS} = \frac{(\hat{\underline{\mathrm{M}}} + \rho \underline{\mathrm{X}})^{-1}}{\underline{c}^H (\hat{\underline{\mathrm{M}}} + \rho \underline{\mathrm{X}})^{-1} \underline{c}}. \tag{6.2}$$

Restricting attention to linear arrays for ease of exposition, a suitable choice for the matrix $\underline{\mathrm{X}}$ is

$$\underline{\mathrm{X}} = \frac{1}{2} \int_{-1.0}^{1.0} \underline{\tilde{s}}(u) \underline{\tilde{s}}^H(u), \tag{6.3}$$

where $\underline{\tilde{s}}(u) \equiv \underline{s}(u) - \underline{w}_q \frac{\underline{w}_q^H \underline{s}(u)}{\underline{w}_q^H \underline{w}_q}$ is the component of the array manifold vector orthogonal to a desired quiescent weight vector $\underline{w}_q$. Typically $\underline{w}_q$ is chosen to have uniformly low-sidelobes. When $\underline{\mathrm{X}}$ is defined in this way $\underline{w}_{LS}$ has the following properties

- $\underline{w}_{LS} \to \underline{w}_q$ as $\rho \to \infty$
- $\underline{w}_{LS}$ reverts to the usual LS solution as $\rho \to 0$.

By changing the value of ρ the solution is made to vary between that of a fully adaptive beamformer and a fixed beamformer. To see how this approach is applicable to the KF solution we first observe that if $\underline{\mathrm{X}}$ has rank j we can write $\rho \underline{\mathrm{X}} = \sum_{i=1}^{j} \underline{\zeta}_i \underline{\zeta}_i^H$. It is then easy to see that the penalty term in Equation (6.1) is equivalent to adding the j dummy data vectors to the usual LS cost function. We know that the KF equations can be used to solve the LS problem, and without going into the details (see [5] for a general description of the KF equations) at each iteration a time update and a measurement update are performed. All we have to do to include the penalty function term is to compute j pseudo-measurement updates without the corresponding time updates at each iteration. To avoid excessive computation we exploit the fact that the effectiveness of the penalty term is relatively insensitive to the exact value of ρ, and only perform the pseudo-measurement updates once every b iterations, where b is the effective length of the exponentially fading data window, given by $b = \frac{1+\beta}{1-\beta}$. As a result the effective value of ρ decays by a factor of approximately 6 before being refreshed.

7 Simulation examples

To illustrate the effectiveness of the estimation procedure described above some simple simulation experiments using a 16 element equispaced linear array are described. The sensor outputs each consist of a target signal corrupted by additive white Gaussian thermal noise and jamming. The jamming is modelled as a succession of plane waves arriving at the array with Gaussian distributed random amplitudes having zero mean and variance 40dB greater than that of thermal noise at each sensor. The target signal power was set well below that of thermal noise. The following jamming scenarios were modelled:

(a) no jamming, no array rotation

(b) 8 jammers, evenly distributed across the sidelobes of the quiescent pattern, array rotation of 4.0x10-5 radians per sample

(c) mainbeam jammer with direction of arrival $u = 0.08$ (0.64 beamwidths), array rotation of 4.0x10-5 radians per sample.

Scenario (a) is included as a reference, while the others are known to be difficult for conventional adaptive beamforming algorithms to cope with; scenario (b) because the array motion leads to an increase in the effective number of jammers, and scenario (c) because the proximity of the jammer to the constraint direction can lead to severe pattern distortion. For comparison with the KF algorithm we have implemented the conventional RLS algorithm using a QR decomposition technique with a sliding data window. Both algorithms are soft constrained via pseudo-measurements, but only the KF solution has explicit time variation through the use of basis functions $f_1(k) = 1$ and $f_2(k) = k$. The two criteria that have been used to compare the algorithms are the shape of the adapted beampatterns in terms of the deviation from the desired quiescent pattern, and the output signal-to-noise-plus-interference ratio (SNIR). The adapted beampatterns are compared in Figures 1 and 2, where superficially there is little difference between the two.

The desired quiescent pattern is shown for reference. Both sets of patterns show sidelobes close to the -40db level set by the soft constraint. When viewed in detail the KF solution for scenario (c) is found to produce a narrow null which tracks the jammers, while the RLS solution forms a broad shallow null.

The SNIR has been measured as a function of the effective number of snapshots processed. Measurements of signal and interference power were made once the initial conditions had been forgotten, so they represented only the steady-state tracking performance of the algorithms. In the presence of sidelobe jamming (Figure 3) the RLS beamformer achieves a SNIR which peaks at about 22 samples, and then deteriorates as the algorithm is unable to track the motion of the jammers. In contrast the performance of the KF increases to within 1 dB of the jammer-free scenario after about 80 samples. In the mainbeam jammer experiment (Figure 4) we see a more rapid fall-off in the performance of the

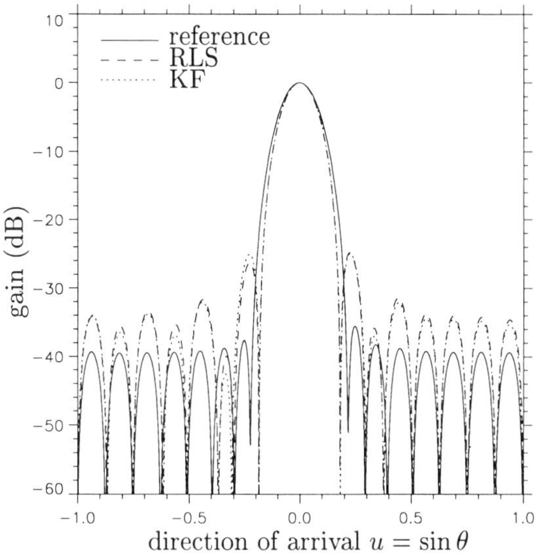

Figure 1. Beampatterns adapted against sidelobe jamming

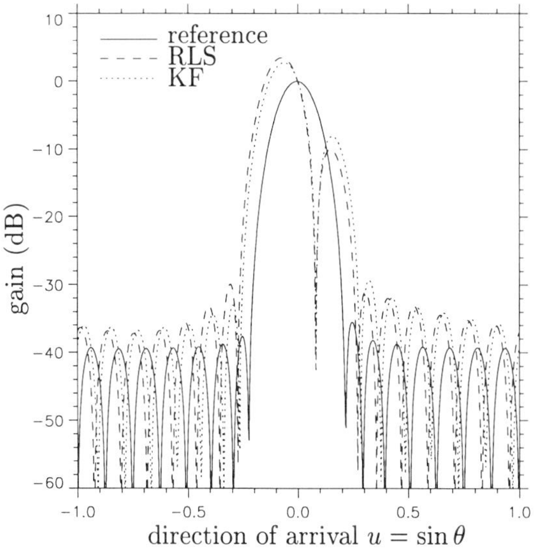

Figure 2. Beampatterns adapted against mainbeam jamming

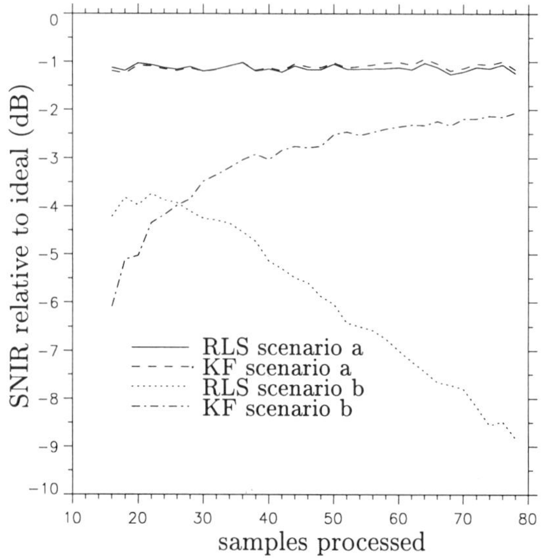

Figure 3. Convergence statistics for sidelobe jammer scenario

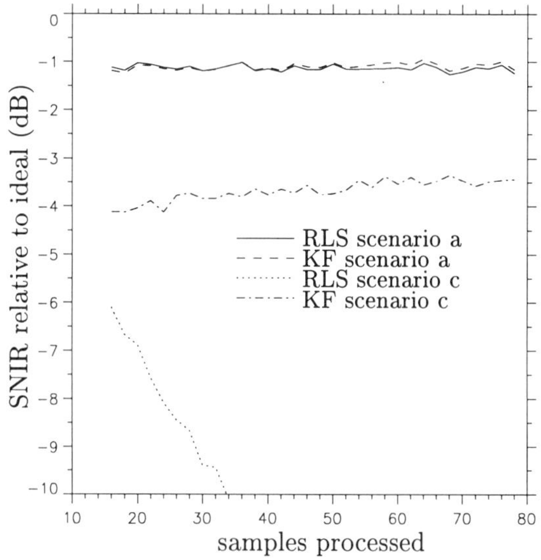

Figure 4. Convergence statistics for mainbeam jammer scenario

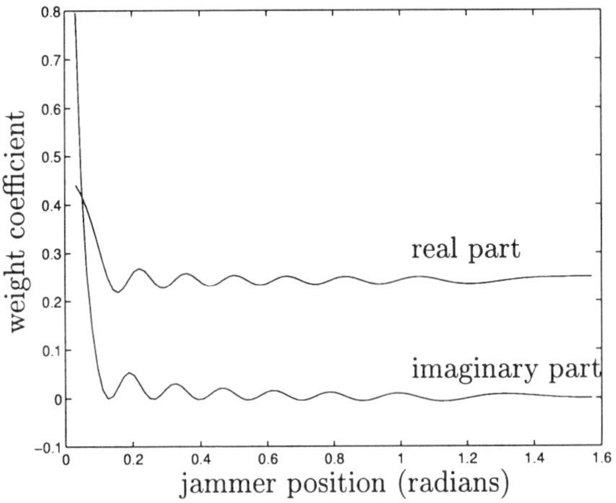

Figure 5. Optimum weight behaviour as a function of jammer position

RLS beamformer, while the KF is able to recover the SNIR achievable when the array is stationary. The comparison is only approximate since the relationship between β and the size of the data window, b, is not exact.

It is interesting to see how closely the actual behaviour of the optimum weight vector matches the model. In Figure 5 the periodic variation of the optimum weight vector is revealed by plotting the real and imaginary parts of one element of the weight vector required to cancel an interferer as a function of the position of that interferer. For the array rotation problem we are interested in very small rotations, for which the linear model is a good approximation to the sinusoidal variation.

8 Observations and conclusions

In summary we have defined a state-space model for an adaptive beamformer and applied a parametric technique from ARMA process estimation to transform a time-varying estimation problem into a time-invariant one. This involves projecting the time-varying weight solution onto a set of known time-varying basis functions, and leads immediately to a RLS solution to the problem. We show that for the special case of a basis comprising the first q terms in a Taylor series there is a Kalman filter solution that avoids the need to explicitly compute the basis functions. The q-fold increase in the dimension of the parameter space means that more data is required to achieve the same degree of convergence, but we show how this problem can be overcome by using soft constraint techniques to incorporate prior information. The resulting constrained Kalman filter is shown to compensate for array motion in simulation experiments for which the standard RLS approach shows rapid degradation. By improving the modelling of the dynamics of the weight solution in this way it should be possible to make the beamformer more efficient by reducing the measurement update rate and relying on time updates to compensate for motion.

Bibliography

1. Hayward, S. D., (1997) Effects of motion in adaptive beamforming, *IEE Proc. Radar, Sonar and Navigation*, to be published.

2. McLaughlin, S., Mulgrew, B., Cowan, C. F. N., Performance bounds for exponentially windowed RLS algorithm in a nonstationary environment, Mathematics in Signal Processing II, Clarendon Press, Oxford, 1990.

3. Subba Rao, T., (1970) The fitting of non-stationary time-series models with time-dependent parameters, *J. Royal Stat. Soc.*, series B, **32**(2), 312-322.

4. Grenier, Y., (1983) Time dependent ARMA modelling of non-stationary signals, *IEEE Trans. ASSP*, **31**(4), 899-911.

5. Haykin, S., (1996) Adaptive Filter Theory, Third Edition, Prentice Hall, New Jersey, 1996.

6. Kovac, S., Piasco, J. M., Guglielmi, M., (1994) Linear frequency modulation signal parameters estimation using time-dependent modelling, *Proceedings of EUSIPCO-94*, Edinburgh, 502-505.

7. Hayward, S. D., (1996) Adaptive beamforming for rapidly moving arrays, Int. Radar Conf., Radar 96, Beijing, 480-483.

8. Griffiths, L. J., Jim, C. W., (1982) An alternative approach to linearly constrained adaptive beamforming, *IEEE Trans. on Ant. and Prop.*, **AP30**, 27-34.

9. Goodwin, G. C., Payne, R. L., (1977) Dynamic System Identification: Experiment and Data Analysis, Academic Press, New York.

10. Wenhua, W., Hongyu, W., (1996) Recursive Least Squares Algorithm for Nonstationary Random Signal, *Proceedings of 3rd international conference on Signal Processing*, Beijing, 197-200.

11. Petrus, P., Reed, J. H., (1995) Time dependent adaptive arrays, *IEEE Signal Processing Letters*, **2**(12), 219-222.

12. Kaminski, P. G., Bryson, A. E., Schmidt, S. F., (1971) Discrete square root filtering: a survey of current techniques, *IEEE Trans. Automatic Control*, **AC-16**(6), 727-735.

13. Hudson, J. E., (1979) A Kalman-type algorithm for adaptive radar arrays and modelling of non-stationary weights in Case studies in advanced signal processing, IEE Conf. Pub., 180.

14. Owsley, N., (1985) Array Signal Processing, Ed. S Haykin, Prentice Hall, 168-177.

15. Hughes, D. T., McWhirter, J. G., (1995), Penalty function method for side-lobe control in least squares adaptive beamforming, *SPIE Proceedings; Advanced Signal Processing Algorithms*, San Diego, 170-181.

16. Chen, Y. C., Chiang, C. T., (1993) Adaptive beamforming using the constrained Kalman Filter, *IEEE Trans. Antennas and Propagation*, **41**(11), 1576-1580.

17. Tahk, M., Speyer, J. L., (1990) Target tracking subject to kinematic constraints, *IEE Trans. on Automatic Control*, **35**(3), 324-326.

Using Algorithmic Engineering to Derive a Fast Parallel Weight Extraction Algorithm Based on QRD Recursive Least Squares

M. Harteneck*, R.W Stewart*, J.G. McWhirter and I.K. Proudler****

**Signal Processing Division, Department of Electrical and Electronic Engineering, University of Strathclyde, Glasgow, Scotland, and* ***DERA Malvern, Worcestershire*

Abstract

In this paper the technique of algorithmic engineering is used to reduce the computational complexity of a QR-RLS adaptive filtering algorithm with parallel weight extraction. Simple, but mathematically rigorous, transformations of a signal flow graph representation are used for the derivation, so that the need for complex matrix algebra is almost completely avoided.

1 Introduction

Adaptive finite impulse response (FIR) filters are used in many real world applications where the environment in which the filter has to operate is time-varying or not known a-priori. Such problems include acoustic echo cancellation, acoustic noise control and on-line modelling of unknown or time-varying plants.

An Nth-order adaptive FIR filter aims to linearly combine N delayed samples of an input signal $x(k)$, which could originate from an input to an unknown plant, in such a way that the corresponding output $\hat{y}(k)$ matches as best as possible a desired signal $y(k)$, which could be the output of the unknown plant. Adaptive system identification tries to determine the parameters of such an unknown and/or time-varying transfer function and, once these parameters are specified, give a complete characterization of the plant. These parameters can then be used for on/off-line controller design or as models of acoustic transfer paths in active noise control systems [1].

Usually, the gradient-descent based least mean squares (LMS) algorithm [2] is used because of its simplicity and low computational complexity of $O(N)$. However, if a rapid convergence is required then a recursive least squares (RLS) algorithm [3] is more advisable. Some drawbacks of the standard RLS algorithm are its high computational complexity of $O(N^2)$ and that it operates on the autocorrelation matrix of the input signal $x(k)$ and therefore needs a high dynamic range which can be problematic in a fixed-point environment. Another method for minimizing the least squares criterion is the QR-RLS algorithm [3] which performs a QR decomposition [4] of the input data matrix, and is therefore better conditioned for use in a fixed point environment, but still needs about the same level of computational complexity.

One approach to reduce the computational complexity is to develop a "fast" RLS algorithm with $O(N)$ complexity. The existence of such a solution relies on the fact that a certain matrix in the problem formulation is Toeplitz. The existence of this structure can be used to avoid performing any calculations upon matrices - which in general require $O(N^2)$ operations. Transforming the problem into such a form has, in the past, required the manipulation of a large number of complicated equations. However recently a more simple approach to deriving fast algorithms has been developed based on signal flow graph representations of the algorithm: algorithmic engineering [5].

Algorithmic engineering is the hybrid discipline of deriving numerical algorithms which are suitable for parallel computation and mapping them onto parallel processing architectures capable of performing the computation efficiently at the required throughput rate. An essential aspect of algorithmic engineering is the compact but accurate representation of established parallel algorithms and architectures which can then be treated as high level building blocks in the development of more complicated processing structures. For the purposes of algorithmic engineering, McWhirter suggested that each building block be treated as a mathematical operator with its parallel structure and interconnections represented in terms of a signal flow graph (SFG) [6]. A signal flow graph does not include the detailed timing features required to specify a practical parallel implementation. However, if required, this detail can easily be inserted at a later stage.

Representing the high level building blocks as well-defined mathematical operators provides a compact specification of the parallel algorithm and hence a much simpler means of proving that it performs the desired computation. However, it is also possible to manipulate the blocks in a rigorous manner using, for example, the associative, commutative and orthogonality properties determined by the matrix (or other) algebra associated with the corresponding operators. This can be used to discover novel processing structures [7, 8].

In this paper algorithmic engineering techniques are used to reduce the computational complexity of the QR-RLS algorithm with parallel weight extraction [3]. The savings are achieved by simple transformations which make the triangular "postprocessor" superfluous while maintaining the same input-output relationship. In Section 2, the QR-RLS algorithm is reviewed and some important properties are explained. Finally, in Section 3 the derivation of the fast algorithm is presented.

2 Review of the adaptive QR-RLS algorithm

An adaptive algorithm for finite impulse response filters tries to predict a desired signal $y(k)$, where k is the discrete time index, with a linear combination of delayed versions of the input signal $x(k)$. The prediction $\hat{y}(k)$ is therefore formed as

$$\hat{y}(k) = \sum_{i=0}^{N-1} w_i(k)x(k-i), \tag{2.1}$$

where $\{w(k)\}$ is the set of adaptive weights and N is the number of taps.

The problem is to find the optimum set of weights such that the least-squares performance criterion $\xi(k)$ is minimized at every time step k. This is defined as

$$\xi(k) = \sum_{i=0}^{k-1} \lambda^i e^2(k-i), \tag{2.2}$$

where the error $e(k)$ is defined as

$$e(k) = y(k) - \hat{y}(k) \tag{2.3}$$

and λ is a *forgetting factor* that allows the algorithm to adapt to changing environments and to ensure convergence in a fixed-point environment. The forgetting factor is usually chosen to be slightly less than 1.

One way to solve this problem is to write it as an Nth-order linear regression and then solve the problem by applying a QR decomposition [4]. To write the above problem in a matrix-vector notation, the following definitions are necessary:

$$\mathbf{w}(k) = [w_0(k)\ w_1(k)\ \ldots\ w_{N-1}(k)]^T \tag{2.4a}$$
$$\mathbf{x}(k) = [x(k)\ x(k-1)\ \ldots\ x(k-N+1)]^T \tag{2.4b}$$
$$\mathbf{y}(k) = [y(1)\ y(2)\ \ldots\ y(k)]^T \tag{2.4c}$$
$$\mathbf{e}(k) = [e(1)\ e(2)\ \ldots\ e(k)]^T \tag{2.4d}$$
$$\mathbf{A}(k) = [\mathbf{x}(1)\ \mathbf{x}(2)\ \ldots\ \mathbf{x}(k)]^T \tag{2.4e}$$
$$\Lambda_k = \text{diagonal}\left(\lambda^{k-1}, \lambda^{k-2}, \ldots, \lambda, 1\right), \tag{2.4f}$$

where the superscript "T" denotes matrix transpose. Using these definitions, (2.1) can be written as

$$\hat{y}(k) = \mathbf{x}^T(k)\mathbf{w}(k) \tag{2.5}$$

and (2.2) can be rewritten as

$$\begin{aligned}\xi(k) &= \sum_{i=0}^{k-1} \lambda^i e^2(k-i) = \mathbf{e}^T(k)\Lambda_k\mathbf{e}(k) = \|\Lambda_k^{\frac{1}{2}}\mathbf{e}(k)\|^2 = \\ &= \|\Lambda_k^{\frac{1}{2}}\mathbf{y}(k) - \Lambda_k^{\frac{1}{2}}\mathbf{A}(k)\mathbf{w}(k)\|^2,\end{aligned} \tag{2.6}$$

To minimize (2.6), the vector inside the Euclidean norm ($||.||$) is premultiplied with an orthogonal rotation matrix $\mathbf{Q}(k)$ which is calculated in such a way that

$$\mathbf{Q}(k)\Lambda_k^{\frac{1}{2}}\mathbf{A}(k) = \begin{bmatrix} \mathbf{R}(k) \\ \mathbf{0}(k) \end{bmatrix}, \tag{2.7}$$

where $\mathbf{R}(k)$ is an $N \times N$ upper-triangular matrix and $\mathbf{0}(k)$ is a $(k-N) \times N$ zero matrix, i.e. $\mathbf{Q}(k)$ performs a QR decomposition [4]. Note that the 2-norm of a vector does not change if it is premultiplied with an orthogonal matrix. By premultiplying the error vector with $\mathbf{Q}(k)$ we get

$$\begin{aligned} \xi(k) &= ||\mathbf{Q}(k)\Lambda_k^{\frac{1}{2}}\mathbf{y}(k) - \mathbf{Q}(k)\Lambda_k^{\frac{1}{2}}\mathbf{A}(k)\mathbf{w}(k)||^2 \\ &= \left\| \begin{bmatrix} \mathbf{p}(k) \\ \mathbf{v}(k) \end{bmatrix} - \begin{bmatrix} \mathbf{R}(k) \\ \mathbf{0}(k) \end{bmatrix} \mathbf{w}(k) \right\|^2, \end{aligned} \tag{2.8}$$

where the vector $\mathbf{Q}(k)\Lambda_k^{\frac{1}{2}}\mathbf{y}(k)$ is partitioned in a suitable way into the vectors $\mathbf{p}(k)$ and $\mathbf{v}(k)$.

If required, the optimal weight vector can now be calculated from (2.8) via backsubstitution as

$$\mathbf{R}(k)\mathbf{w}(k) = \mathbf{p}(k) \Leftrightarrow \mathbf{w}(k) = \mathbf{R}^{-1}(k)\mathbf{p}(k). \tag{2.9}$$

As shown in [3], it is possible to perform the decomposition of (2.7) in a time-recursive fashion and extract the weight vector $\mathbf{w}(k)$, i.e. perform the backsubstitution (2.9), by using the canonical least squares processor, as shown in Figure 1 for the case of 4 adaptive weights ($N = 4$). The processor consists

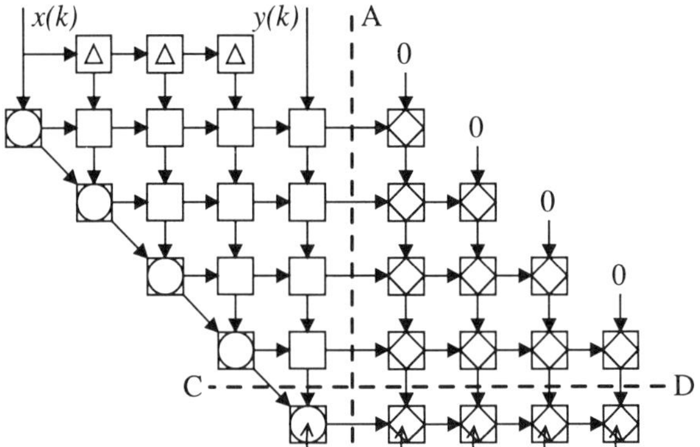

Figure 1. Signal-Flow-Graph representation of the canonical least squares processor with parallel weights extraction for 4 weights

of the interconnection of three components with distinct but coupled functions. The first is the well-known triangular array processor for recursive least squares [9] (left of the line A–B and above the line C–D), consisting of boundary and internal cells of the types (a) and (d) as shown in Figure 2. This array, whose input signals are $x(k)$ and $y(k)$, calculates

$$\begin{bmatrix} \mathbf{R}(k) & \mathbf{p}(k) \\ \mathbf{0} & \alpha(k) \end{bmatrix} = \hat{\mathbf{Q}}(k) \begin{bmatrix} \lambda^{\frac{1}{2}}\mathbf{R}(k-1) & \lambda^{\frac{1}{2}}\mathbf{p}(k-1) \\ \mathbf{x}^T(k) & y(k) \end{bmatrix}. \tag{2.10}$$

The second part is a triangular "postprocessor" [3] (right of the line A–B and above the line C–D) consisting of internal cells type (b) only. The calculation performed by this part is

$$\begin{bmatrix} \mathbf{R}^{-T}(k) \\ \mathbf{g}^T(k) \end{bmatrix} = \hat{\mathbf{Q}}(k) \begin{bmatrix} \lambda^{-\frac{1}{2}}\mathbf{R}^{-T}(k-1) \\ \mathbf{0} \end{bmatrix}, \tag{2.11}$$

where $\hat{\mathbf{Q}}(k)$ is the rotation matrix from the triangular processor (Equation (2.10)) and $\mathbf{g}(k)$ is a scalar multiple of the Kalman gain vector. Note that in the postprocessor the matrix $\mathbf{R}^{-1}(k-1)$ is multiplied by $\lambda^{-\frac{1}{2}}$ which is a value greater than 1 and therefore gives rise to instability as errors in the matrices and vectors are amplified. The third component is the row of cells below the line C–D which stores the least squares weight vector in the form $\tilde{\mathbf{w}}(k-1) = -r^{-1}(k-1)\mathbf{w}(k-1)$ and updates it to $\tilde{\mathbf{w}}(k)$ using the (scaled) Kalman gain vector $\mathbf{g}(k)$ which it receives from above.

One observation which is necessary for the following derivation of the fast algorithm in Section 3 is that if the elements of each input vector $\mathbf{x}(k)$, and hence

(a) *Internal Cell 1:*

$$u_o = c \cdot u_i - s \cdot \lambda^{\frac{1}{2}} \cdot r;$$
$$r = s \cdot u_i + c \cdot \lambda^{\frac{1}{2}} \cdot r;$$

(b) *Internal Cell 2:*

$$u_o = c \cdot u_i - s \cdot \lambda^{-\frac{1}{2}} \cdot r;$$
$$r = s \cdot u_i + c \cdot \lambda^{-\frac{1}{2}} \cdot r;$$

(c) *Downdating Cell:*

$$u_o = \frac{1}{c}(u_o + s\lambda^{\frac{1}{2}}r);$$
$$r = s \cdot u_o + c \cdot \lambda^{\frac{1}{2}}r;$$

(d) *Boundary Cell:*

if $u_i = 0$ then
$\quad c = 1; s = 0; r = \lambda^{\frac{1}{2}} \cdot r;$
$\quad \delta_o = c \cdot \delta_i;$
else
$\quad c = \lambda^{\frac{1}{2}} \cdot r \cdot (\lambda \cdot r^2 + u_i^2)^{-\frac{1}{2}};$
$\quad s = u_i \cdot (\lambda \cdot r^2 + u_i^2)^{-\frac{1}{2}};$
$\quad r = (\lambda \cdot r^2 + u_i^2)^{-\frac{1}{2}};$
$\quad \delta_o = c \cdot \delta_i;$
endif

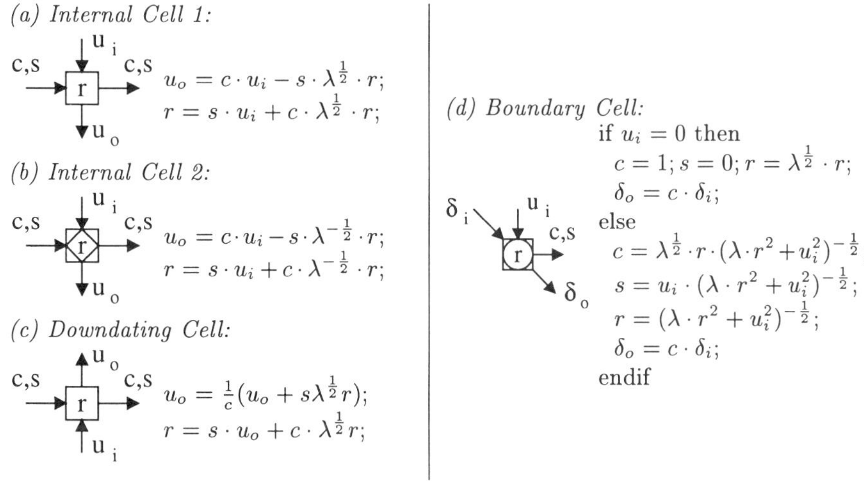

Figure 2. Processing cells

the columns of $\mathbf{A}(k)$ are permuted by a permutation matrix $\mathbf{\Pi}$, then the elements of the resulting weight vector $\mathbf{w}(k)$, and hence the corresponding (scaled) Kalman gain vector $\mathbf{g}(k)$ will be permuted in the same way, i.e.

$$\mathbf{A}(k) \rightarrow \mathbf{w}(k), \mathbf{g}(k) \Leftrightarrow \mathbf{\Pi A}(k) \rightarrow \mathbf{\Pi w}(k), \mathbf{\Pi g}(k) \tag{2.12}$$

where "$\rightarrow$" reads as "leads to" and $\mathbf{\Pi}$ is a $N \times N$ permutation matrix.

3 Fast algorithm

In this section it is shown how the triangular postprocessor (right of the line A–B and above the line C–D in Figure 1) can be made redundant thereby reducing the computational complexity considerably, while maintaining the same input-output relationship of the overall processor.

The derivation consists of three algorithmic engineering transformations, starting with the structure shown in Figure 1. These transformations are shown in Figures 3 to 5. For simplicity, the transformations are demonstrated for a low order example (N=4) but can be easily extended to any higher order problem or indeed a multichannel processor (i.e. vector signals).

The first diagram in Figure 3 is identical to Figure 1 except that cells in the last row of the SFG have been marked to show the index of the associated weight and hence the ordering of the weight vector $\mathbf{w}(k)$ (0-1-2-3). The input vector $\mathbf{x}(k)$ is then reordered and as a result the weights in the bottom row of the SFG are reordered in the same manner (1-2-3-0) leading to the SFG in Figure 3(b). The part of this SFG above the line $A_1 - B_1$ is reproduced as Figure 4(a). The outputs from the 3×3 triangular post processor have been labelled 1, 2, 3 corresponding to the weight vector indexing in Figure 3(b). Note that these outputs constitute the (scaled) Kalman gain vector for an adaptive filtering problem equivalent to that in Figure 1 but of lower order (N = 3). In what follows, the SFG in Figure 4(a) is transformed into a more efficient form. The first step is to duplicate part of this SFG as shown in Figure 4(b). In order to generate the left hand SFG in Figure 4(c), we note that every input to the left hand SFG in Figure 4(b) goes through a common delay element which can therefore be moved from the input of this graph to each of its outputs without affecting the output values. If the inputs to this SFG are then permuted as shown, the left hand SFG in Figure 4(c) results. As noted above, the effect of this permutation is to permute the output (scaled) Kalman gain vector (of order N = 3) as indicated. The right hand SFG in Figure 4(c) is identical to that in Figure 4(b).

Now, comparing the left and right hand SFGs in Figure 4(c), it can be seen that there exists a common sub-graph (i.e. a common computation) within the two SFGs. One of these redundant parts can be removed and the two SFGs recombined as shown in Figure 4(d) where the common subgraph appears above the line $A_2 - B_2$. The next step of the derivation is to show that the 2×2 array of

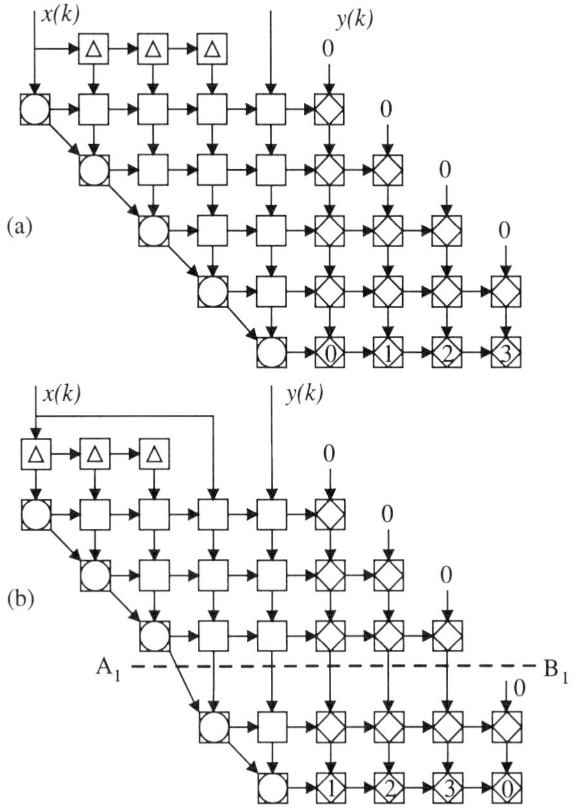

Figure 3. 1st Step of the derivation from a 4 × 4 array to a 3 × 3 array

type (b) cells is also redundant. This is achieved by considering the part of this SFG below the line $A_2 - B_2$. This is reproduced for convenience as Figure 5(a).

Now recall that the output vectors labelled 2-3-1 and 1-2-3 in Figure 5(a) are identical apart from the reordering. Therefore if the operator represented by the type (b) cells on the right side could be "inverted" such that they calculate the input signal u_i given the output signal u_o and the rotation parameters c and s, then these values could be fed back as the required input to the cells on the left hand side. As a consequence, the signals from the postprocessor above the line $A_2 - B_2$ would no longer be necessary. These "inverse" cells are easily derived and have been presented as the type (c) cells in Figure 2. The resulting transformation is shown in Figure 5(b).

Finally, in Figure 6, we assemble from the transformed components a processor which is equivalent to that in Figure 1 except that the least squares weight vector in Figure 6 is reordered with respect to that in Figure 1. In Figure 6 there are only two rows of cells above the line $A_2 - B_2$. However, it is easy to see that

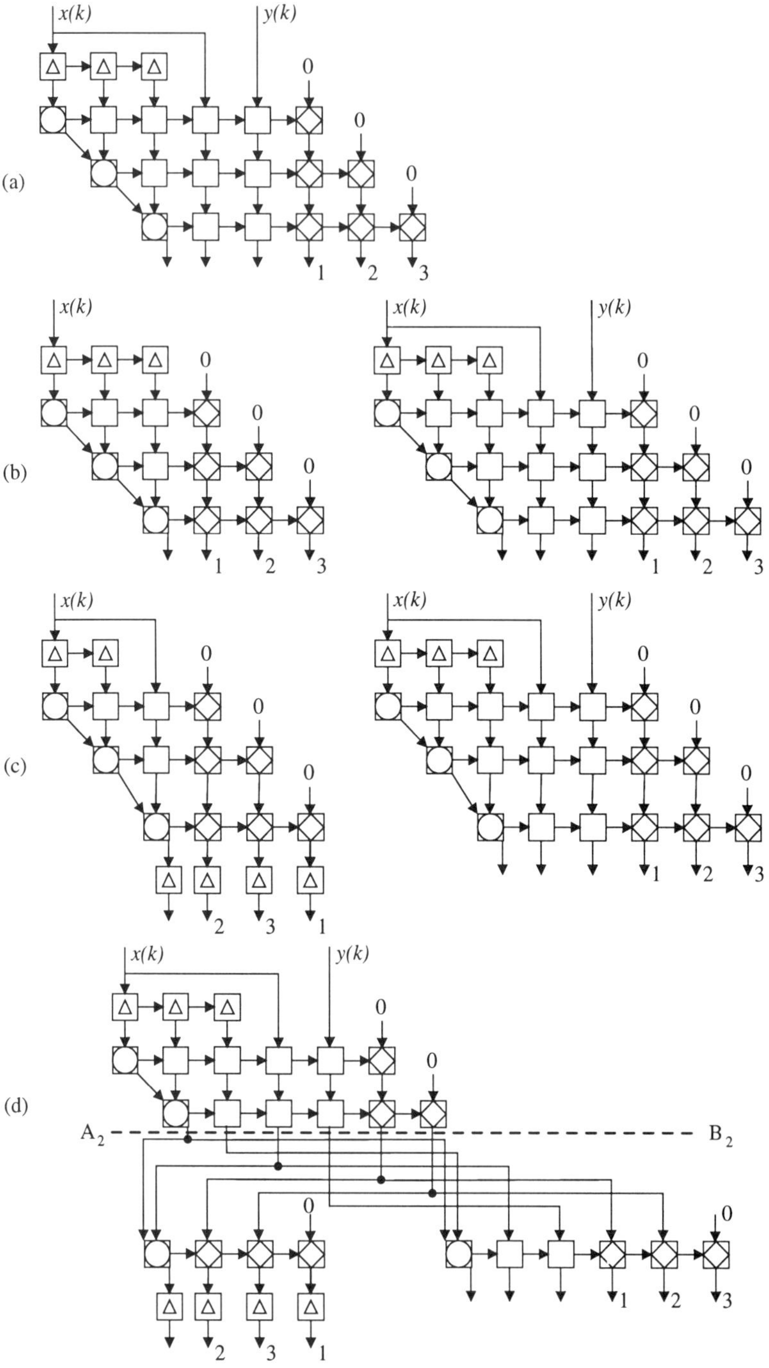

Figure 4. 2nd Step of the derivation from a 3×3 array to a 2×2 array

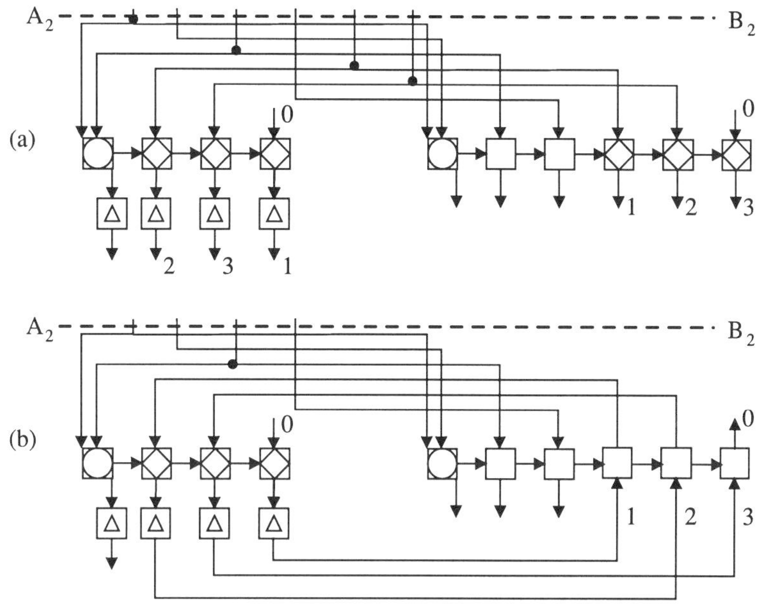

Figure 5. 3rd Step of the derivation: transformation of the additional stage

if we had begun with a Nth order problem instead of a 4th order one (i.e. N weights instead of 4), this part of the SFG would have $N - 2$ rows. Note that the SFG in Figure 6 does not require a triangular postprocessor of $O(N^2)$ type (b) cells above the line $A_2 - B_2$ as in Figure 1. Furthermore, as shown in [7] that part of the original SFG which remains above the line $A_2 - B_2$ can be reduced in complexity from $O(N^2)$ as shown here to $O(N)$ by transforming it into a least squares lattice processor. Thus, the SFG in Figure 6 can be used as the basis of a "fast" recursive least squares algorithm which includes weight extraction.

4 Conclusions

In this paper we have shown via, algorithmic engineering, how to transform a QR-RLS algorithm with parallel weight extraction to an equivalent algorithm with a lower computational complexity. The derivation is easily extendable to a higher order triangular array and even to multichannel systems. The algorithm was derived by transforming a signal flow graph representation of the algorithm. The transformations are based on a high-level, operator-based characterization of the component parts of the original algorithm and are thus mathematically rigorous.

One drawback of the algorithm, from a practical point of view, is its inherent numerical instability. The Cholesky factor of the autocorrelation matrix $\mathbf{R}(k)$ and the inverse of the Cholesky factor $\mathbf{R}^{-1}(k)$ are propagated independently

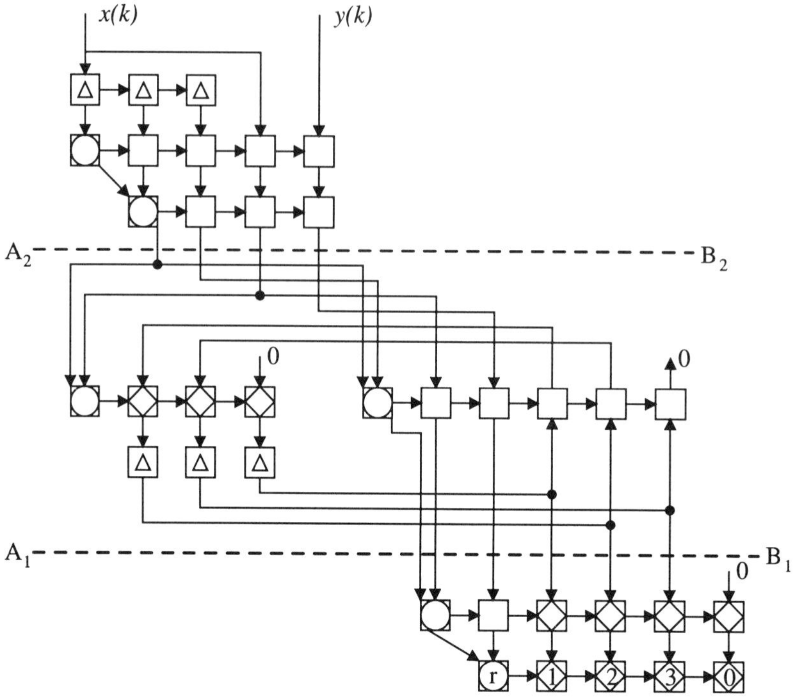

Figure 6. Final structure without postprocessor above line $A_2 - B_2$

i.e. there is no feedback mechanism to ensure that $\mathbf{R}(k)\mathbf{R}^{-1}(k) = \mathbf{I}$. Furthermore, assuming that $\lambda < 1$, any errors in the inverse Cholesky factor will be amplified by the term $\lambda^{-\frac{1}{2}}$ and, lacking a feedback mechanism, the algorithm will diverge. Nevertheless, the algorithm could still be useful for block processing (i.e. processing a predetermined number of data snapshots) with $\lambda = 1$. The main purpose of this paper is, however, not to promote the algorithm but provide a novel illustration of algorithmic engineering.

Bibliography

1. S J Elliot and P A Nelson. Active Noise Control. *IEEE Signal Processing Magazine*, 10(4):12–35, October 1993.

2. B Widrow and S D Stearns. *Adaptive Signal Processing.* Prentice Hall, Englewood Cliffs, 1985.

3. S Haykin. *Adaptive Filter Theory.* Prentice Hall, 2nd edition, 1991. ISBN 0-13-012236-5.

4. G H Golub and C F Van Loan. *Matrix Computations.* John Hopkins University Press, 1989.

5. J G McWhirter. Algorithmic Engineering in Adaptive Signal Processing. *IEE Proceedings-F*, 139(3):226–232, June 1992.

6. S.Y.Kung. *VLSI Array Processors.* Prentice Hall Information and Systems Sciences Series, 1988.

7. I K Proudler and J G McWhirter. Algorithmic Engineering in Adaptive Signal Processing: Worked Examples. *IEE Proceeding - Vis. Image Signal Process.*, 141(1):19–26, February 1994.

8. I K Proudler. Fast Time-Series Adaptive-Filtering Algorithm Based on the QRD Inverse-Updates Method. *IEE Proceeding - Vis. Image Signal Process.*, 141(5):325–333, October 1994.

9. W.M.Gentleman and H.T.Kung. *Matrix Triangularisation by Systolic Arrays.* Proc. SPIE, Vol.298, Real Time Signal Processing IV, 1981, pp.19-26.

The Simultaneous Strict Positive Real Problem: Similarities with the Simultaneous Stability Problem and Applications in Adaptive Recursive Algorithms

C. Mosquera and F. Pérez González

Departamento Tecnologías de las Comunicaciones, ETSI Telecomunicación, Universidad de Vigo, Spain

Abstract

The robust strict positive real (SPR) problem arises in identification and adaptive control, where strict positive realness of a certain transfer function is a sufficient condition for ensuring convergence of a number of adaptive recursive algorithms. The SPR condition involves the denominator of the unknown system, so the problem becomes a robustness question provided some a priori knowledge regarding the unknown system is available. This is a quite extended case, since the physics of the system can justify in many cases assumptions on the parameters under study. The similarities between the robust stabilization problem and the so-called robust SPR problem are analyzed here, and used for designing appropriate compensators making a set SPR.

1 Introduction

This paper addresses the robust Strict Positive Real (SPR) problem: trying to make a set of polynomials simultaneously SPR. Its immediate application is the convergence of adaptive Infinite Impulse Response (IIR) filters. IIR filters are desirable in many situations, for their reduced complexity and improved performance, as a counterpart to Finite Impulse Response (FIR) filters. Adaptive schemes are well understood for the latter class, but they still present many problems in the case of recursive filters, in terms of convergence, stability, undermodelling, etc. A family of adaptive recursive schemes, known as hyperstability-based algorithms, offer guaranteed convergence and assured stability as long as an SPR property concerning the denominator of the system under study is satisfied. This allows us to take advantage in some situations of partially known systems, becoming a robustness problem: designing a compensator making SPR a whole set of polynomials, in the so-called robust SPR problem.

Some important analogies can be traced with the classical problem of robust stability, in which a fixed controller must stabilize a set of systems. These similarities will be exploited to solve the robust SPR problem in some important cases, and thus guarantee convergence of a family of adaptive recursive schemes in some typical signal processing contexts. For the sake of clarity we will reserve uppercase letters for denoting transfer functions when referring to the SPR problem; for the stabilization problem, lowercase letters will be used.

2 Preliminaries

A detailed approach to the simultaneous stabilization problem is presented in [1]. The central question in simultaneous stabilization is about the conditions that make it possible the existence of a controller $c(z)$ stabilizing each of a given family of LTI systems $a_i(z)$ $(i = 1, \cdots, k)$. Those $a_i(z)$ may represent different operating conditions of a given system, and the objective is to stabilize them with a fixed controller. A seemingly more generic problem, the stabilization of a continuum of systems, is easier to solve, at least in terms of testing whether the uncertain set is stabilizable by a fixed controller. Motivated by the robust SPR problem, we will focus on the case of a finite number of systems, the so-called simultaneous stabilization problem, since many different uncertainties in the SPR problem can be handled through a finite number of extreme systems.

An algebraic framework for the simultaneous stabilization problem is sustained on the coprime fractional factorization [2]. First we need some preliminary definitions. By stabilization we mean to place all the poles outside a certain region of instability, denoted as Λ, with Λ a region in the complex plane $\mathbf{C}$. A rational function is Λ-stable if it has no poles inside Λ. We denote by $\mathbf{S}(\Lambda)$ the set of Λ-stable rational functions. The ring of rational functions $\mathbf{R}(z)$ is a field: every nonzero element in $\mathbf{R}(z)$ is invertible.

Theorem 1. *If the subset Λ does not contain the whole extended real line, then the set $\mathbf{S}(\Lambda)$ is a principal ideal domain whose field of fractions is the set of rational functions $\mathbf{R}(z)$.*

The importance of this theorem stated in [1] lies on the fact that a principal ideal domain (p.i.d.) allows a coprime fractional factorization:

Lemma 2. *If $\mathcal{R}$ is a principal ideal domain, any rational function in $\mathbf{R}(z)$ can be factored as a ratio of two coprime rational functions in $\mathcal{R}$.*

This definitions will prove to be very useful later. Next we characterize the family of SPR transfer functions, and describe the robust SPR problem. We will work first with arbitrary regions Λ, although the results will be particularized later for discrete-time systems, i.e., $\Lambda = \{z \in \mathbf{C}, |z| \geq 1\}$.

Definition 3. *Let Λ be any general simply connected domain in* $\mathbf{C}$. *Let $\partial\Lambda$ denote the boundary of Λ. A rational function $T(z)$ is Λ-SPR if:*

$$\begin{aligned} &T(z) \text{ is analytic in } \Lambda \\ &Re\{T(z)\} > 0 \,\forall\, z \in \partial\Lambda. \end{aligned} \tag{2.1}$$

When no mention is done with respect to Λ, the discrete-time case will be assumed, and the term SPR will be used.

In certain output error based identification schemes, convergence is ensured by making $\frac{C(z)}{A(z)}$ SPR, with $A(z)$ the denominator of the unknown system transfer function which is to be identified, and $C(z)$ a transfer function to be designed. The robust SPR problem can be posed as choosing $C(z)$ such that the SPR condition holds for a given minimum-phase family $\mathcal{A}(z)$, which is an uncertain description of the unknown system denominator. A necessary and sufficient condition for the existence of such a $C(z)$, in the discrete-time case, is provided in [3], namely

$$\max_{\omega\in[0,2\pi)} |\max_{A(z)\in\mathcal{A}(z)} \arg(A(e^{j\omega})) - \min_{A(z)\in\mathcal{A}(z)} \arg(A(e^{j\omega}))| < \pi, \tag{2.2}$$

In that case the set $\mathcal{A}(z)$ is said to be SPRizable. In the same paper a design method for $C(z)$ is given. However, this is not a practical method, since it consists of an approximation procedure with non a priori bounds on the degree of the solution. Next we undertake the problem from a different perspective, which will allow us to obtain practical procedures for designing $C(z)$.

3 Synthesis results

First we present the simultaneous stabilization problem from an avoidance approach, which will also be the approach taken in the simultaneous SPR problem. It will be done for a general instability region Λ. The results regarding stabilization are mainly taken from [1] and [4].

Definition 4. *Let $a_1(z), a_2(z)$ be two functions for all $z \in \mathbf{C}$. $z_0 \in \mathbf{C}$ is a point of intersection between $a_1(z)$ and $a_2(z)$ in $\mathbf{C}$ if $a_1(z_0) = a_2(z_0)$. $a_1(z)$ avoids $a_2(z)$ in $\mathbf{C}$ if $a_1(z)$ and $a_2(z)$ have no points of intersection in $\mathbf{C}$.*

Lemma 5. *$a_i(z), i = 1,\ldots,k$, are simultaneously Λ-stabilizable if and only if there exists a rational function $q(z)$ that avoids $a_i(z)$ on Λ. In such a case the controller defined by $c(z) = -\frac{1}{q(z)}$ is the stabilizing controller.*

Now, let $\mathcal{A}(z)$ be a set of polynomials in z, such that for $A(z) \in \mathcal{A}(z)$, $A(z_0) = 0$ implies $z_0 \notin \Lambda$. Without loss of generality we assume that, with $C(z)$ the compensator, $Re\{C(z_r)/A(z_r)\} > 0$ for some z_r and some $A(z) \in \mathcal{A}(z)$. Thus, the robust SPR problem can be presented as an avoidance problem in the following form.

Lemma 6. *Let $C(z)$ be a transfer function with neither zeros nor poles in Λ. $C(z)$ makes $\mathcal{A}(z)$ Λ-SPR, i.e., $Re\{C(z)/A(z)\} > 0, \forall\ z \in \partial\Lambda, A(z) \in \mathcal{A}(z)$, if and only if $tan(arg(C(z)))$ avoids $tan(arg(A(z)) \pm \frac{\pi}{2}), \forall\ A(z) \in \mathcal{A}(z)$, and $\forall\ z \in \partial\Lambda$.*

Proof. The proof is straightforward, since for $b(z)$ a Λ-SPR transfer function, we have that $|arg(b(z))| < \frac{\pi}{2}, \forall\ z \in \partial\Lambda$.

The equivalence between the solvability of both problems in terms of avoidance is evident, and gives a powerful geometrical insight. Next we formulate the avoidance in terms of equations whose solution requires the satisfaction of interpolation and avoidance conditions.

If $a_i(z) = \frac{n_i(z)}{d_i(z)}$, $n_i(z), d_i(z) \in S(\Lambda), i = 1, \ldots, k$ for each system, the following equations must be satisfied so as all the systems are simultaneously stabilized:

$$\left.\begin{array}{rcl} n_1 n_c + d_1 d_c & = & u_1 \\ n_2 n_c + d_2 d_c & = & u_2 \\ & \cdots & \\ n_k n_c + d_k d_c & = & u_k \end{array}\right\} \tag{3.1}$$

with u_i Λ-units, i.e., rational functions with neither zeros nor poles in Λ. We have omitted the argument z for simplicity. From the first two equations, the solution for the controller c is of the form

$$c = \frac{\frac{u_1 d_2 - u_2 d_1}{a_{12}}}{\frac{u_2 n_1 - u_1 n_2}{a_{12}}} = \frac{u_1 d_2 - u_2 d_1}{u_2 n_1 - u_1 n_2} \tag{3.2}$$

with $a_{12} = n_1 d_2 - n_2 d_1$. The following interpolation requirements must then be satisfied:

$$\frac{u_1}{u_2} = \frac{n_1}{n_2} = \frac{d_1}{d_2} \tag{3.3}$$

at the roots of $a_{12} = n_1 d_2 - n_2 d_1$ in Λ. For $k > 2$, there are also avoidance conditions:

$$\frac{u_1}{u_2} \neq \frac{a_{i1}}{a_{i2}}, \forall\ i = 3, \ldots, k \text{ whenever } a_{12} \neq 0 \text{ in } \Lambda \tag{3.4}$$

with $a_{ij} = n_i d_j - n_j d_i$.

Next we specify the set of equations that must be fulfilled in the robust SPR problem for k polynomials. We will work hereafter in the discrete-time case, given its greater interest for signal processing applications. The results can be easily extended to the continuous-time case by means of the bilinear transform. First we define k symmetric transfer functions, $Q_1(z), \ldots, Q_k(z)$, for a set of k polynomials $A_1(z), \ldots, A_k(z)$:

Definition 7.

$$\left.\begin{array}{rcl} Q_1(z) & = & A_1(z)C(z^{-1}) + A_1(z^{-1})C(z) \\ Q_2(z) & = & A_2(z)C(z^{-1}) + A_2(z^{-1})C(z) \\ & \cdots & \\ Q_k(z) & = & A_k(z)C(z^{-1}) + A_k(z^{-1})C(z) \end{array}\right\}. \tag{3.5}$$

It can be readily checked that $Q_i(z) = Q_i(z^{-1}), i = 1, \ldots, k$. The avoidance conditions which arise in the robust SPR problem can be rephrased as follows.

Lemma 8. *Assume that $A_1(z), \ldots, A_k(z)$ are k Schur polynomials (i.e., with all their zeros inside the unit circle). Then there exists a compensator solving the robust SPR problem for $A_1(z), \ldots, A_k(z)$ if and only if there exists a stable $C(z)$ with all its zeros inside the unit circle such that $Q_1(z), \ldots, Q_k(z)$ as in (3.5) verify $Q_i(e^{j\omega}) > 0, \omega \in [0, 2\pi), i = 1, \ldots, k$.*

Proof. The proof follows from the consideration that $Q_i(z), i = 1, \ldots, k$ are symmetric transfer functions with the same sign as $Re\{\frac{C(z)}{A_i(z)}\}$ on the unit circle.

By comparing Equations (3.1) and (3.5) the resemblance can be immediately noticed. The algebraic manipulations follow those for (3.1), and are described in the following steps. From the first two equations, we have

$$C(z) = \frac{Q_2(z)A_1(z) - Q_1(z)A_2(z)}{D_{12}(z)} \tag{3.6}$$

where a new polynomial, $D_{12}(z)$, has been introduced, responding to the following general definition:

$$D_{ij}(z) = A_i(z)A_j(z^{-1}) - A_i(z^{-1})A_j(z). \tag{3.7}$$

Notice the parallelism with a_{ij} above. In this case, the zeros of $D_{ij}(z)$ on the unit circle denote those points at which the phases of $A_i(e^{j\omega})$ and $A_j(e^{j\omega})$ differ by an integer multiple of π, whereas a_{ij} cancels at those points at which $a_i = a_j$. The interpolation conditions follow from (3.6) in the same way as in (3.3):

$$\frac{Q_1(z)}{Q_2(z)} = \frac{A_1(z)}{A_2(z)} \tag{3.8}$$

at the roots of $D_{12}(z)$ on the unit circle. Now the interpolation algorithm should yield symmetric positive functions. One such algorithm is described in [5] as an adaptation of the classic Youla-Saito algorithm [6] for strictly positive real functions. As proved there, the existence of a solution is guaranteed by the positivity of the values to be interpolated, or equivalently, by the verification of condition (2.2).

For k systems, $k > 2$, we have the same pattern as for the stabilization problem above:

$$\frac{Q_1(z)}{Q_2(z)} = \frac{A_1(z)}{A_2(z)} \tag{3.9}$$

whenever $D_{12}(z) = 0$ on the unit circle and

$$\frac{Q_1(z)}{Q_2(z)} \neq \frac{D_{1i}(z)}{D_{2i}(z)} \tag{3.10}$$

whenever $D_{12}(z) \neq 0$ on the unit circle, for $i = 3, \ldots, k$. A final step is necessary in order to guarantee that all the zeros of $C(z)$ lie inside the unit circle. Thus, if $C(z) = C_{in}(z)C_{out}(z)$, where $C_{in}(z)$ and $C_{out}(z)$ include the zeros inside and outside the unit circle respectively, then $C_0(z) = \frac{C(z)}{C_{out}(z)C_{out}(z^{-1})}$ is stable and minimum-phase, and such that $sgn(Re\{\frac{C_0(e^{j\omega})}{A_i(e^{j\omega})}\}) = sgn(Re\{\frac{C(e^{j\omega})}{A_i(e^{j\omega})}\})$, $i = 1, \ldots, k$, where $sgn(\cdot)$ denotes the usual sign function.

The avoidance conditions in the simultaneous stabilization problem must be enforced in the whole instability region, whereas the simultaneous SPR problem only needs to be worked out in one dimension, for example, $|z| = 1$ for discrete-time systems. Regardless the type of uncertainty, a necessary and sufficient condition for the existence of a solution in the robust SPR problem exists, namely, that expressed in (2.2). By the other hand, and for a finite number of systems, there are no tractable necessary and sufficient conditions in the general simultaneous stabilization problem for more than two systems. For two systems, it is a classical result that the necessary and sufficient condition for the existence of a controller $c(s)$ is the verification of a *parity interlacing condition (p.i.p)*:

Theorem 9. *Let $b = n_2x + d_2y$, with x, y Λ-stable solutions of the equation $n_1x + d_1y = 1$. Then a_1 and a_2 are simultaneously Λ-stabilizable if and only if the the number of real zeros of b between each pair of real zeros in Λ of a_{12} is even. This is the so-called parity interlacing condition (p.i.p.).*

The condition under which there exists a solution to the SPR problem for two polynomials follows a similar pattern to the condition exposed above in the stabilization problem. If we define, for the discrete-time case,

$$E_{ij}(z) = A_i(z)A_j(z^{-1}) + A_j(z)A_i(z^{-1}) \tag{3.11}$$

then we have the following result.

Theorem 10. *A stable and minimum-phase compensator $C(z)$ making $A_1(z)$ and $A_2(z)$ SPR exists if and only if the polynomial $E_{12}(z)$ is real and positive at the roots of $D_{12}(z)$ on the unit circle.*

Proof. This is clearly a *p.i.p.* condition on the unit circle. Condition (2.2) will be satisfied if and only if the Nyquist plot of $A_1(z)/A_2(z)$ does not touch the negative real axis, or equivalently, iff $Re\{A_1(e^{j\omega})/A_2(e^{j\omega})\} > 0$ whenever $Im\{A_1(e^{j\omega})/A_2(e^{j\omega})\} = 0$. It is trivial to show that this is only true if $E(e^{j\omega}) > 0$ whenever $D(e^{j\omega}) = 0$.

The reasons for considering the robust SPR problem for two polynomials are manifold. First, it is the first step for solving the problem for a higher number of polynomials. In addition, different types of uncertainty can be tackled by making two polynomials SPR, for example, convex combinations of two polynomials, uncertainty in the root domain, etc. See [5] for a more detailed survey. This problem can be solved by means of interpolation, as shown earlier. For a higher number of systems, a method for avoiding a certain set of functions while interpolating a given set of points must be devised. The stabilization of three systems, for example, is studied in [7], and solved only for $\Lambda = \mathbf{R}_{+\infty}$, the real positive axis. Again, the similarities with our problem make it possible to apply those ideas to finding a compensator making three systems SPR. A so called *strong interlacing property* in [7] arises in the SPR problem as well, and can be proved to be equivalent to the necessary and sufficient condition presented in [3]. See [7] for more details.

Finally, we can stress the relation between the simultaneous SPR problem for k polynomials and the simultaneous SPR problem for $k-1$ transfer functions with an SPR compensator C. This is a well-known result in simultaneous stabilization: the simultaneous stabilization of k systems is equivalent to the simultaneous stabilization of $k-1$ systems with a stable controller.

Lemma 11. *Let $A_1(z), \cdots, A_k(z)$ be k Schur polynomials. Then they are simultaneously SPRizable if and only if the $k-1$ transfer functions $A_i(z)/A_j(z)$, $i = 1, \cdots, k$, $i \neq j$ are SPRizable with an SPR compensator.*

Proof. $C(z)/A_1(z), \cdots, C(z)/A_k(z)$ are SPR if and only if $\frac{C(z)/A_j(z)}{A_1(z)/A_j(z)}$, $\cdots$, $\frac{C(z)/A_j(z)}{1}$, $\cdots$, $\frac{C(z)/A_j(z)}{A_k(z)/A_j(z)}$ are. The SPR compensator would be $C(z)/A_j(z)$.

4 Numerical example

For the purpose of demonstration the results of a simulation are shown. This is a typical application of adaptive IIR filtering in noise canceling [8]. Figure 1 shows the diagram. The adaptive algorithm used was SHARF [9], with $\mu = 0.001$. The desired signal $s(n)$ is corrupted by an additive noise process $g(n)$, considered as a filtered version of a white sequence $v(n)$. The reference noise $u(n)$ is supposed to be the same white sequence filtered by a different transfer function ($H_u(z) = 1$ in this example). The task of the adaptive noise canceler is to identify the transfer function between $u(n)$ and $g(n)$, thus extracting the signal from the additive noise.

The exact transfer function is:

$$H(z) = \frac{1 - 0.2z^{-1} + 0.3z^{-2}}{1 - 1.6z^{-1} + 0.64z^{-2}}.$$

The actual poles are in 0.8, and the only available information is an uncertainty region for these poles, centered at 0.6, and with radius 0.35 (Figure 2).

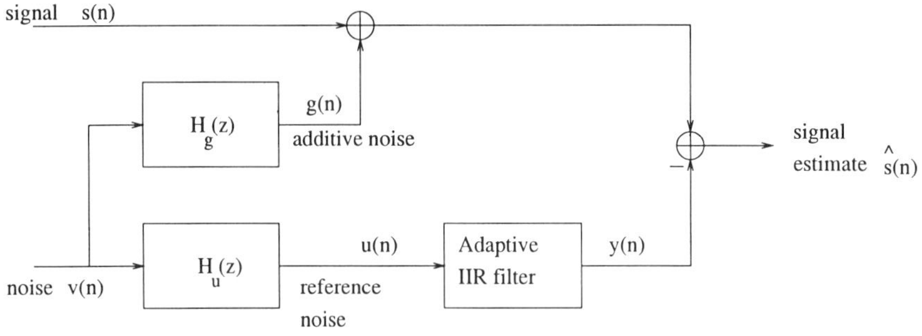

Figure 1. Adaptive noise canceling

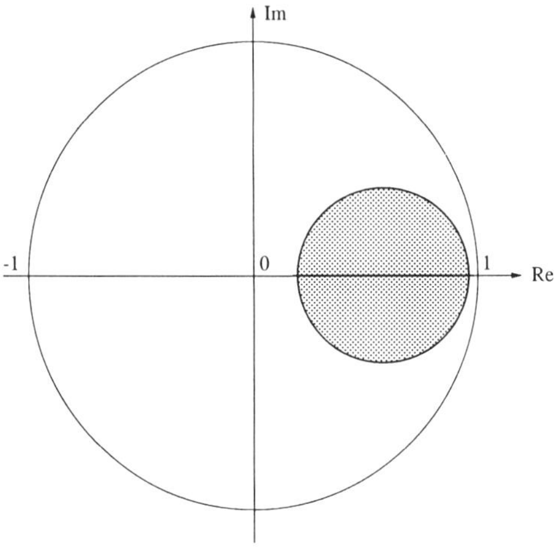

Figure 2. Uncertain root location

Solving the robust SPR problem for this type of uncertainty in the roots amounts to finding a compensator $C(z)$ making SPR the polynomials with both roots at the real extremes of the uncertainty set, 0.25 and 0.95, respectively [5]. In this case, one possible $C(z)$ is seen to be

$$C(z) = 1 - 1.0013z^{-1} + 0.0619z^{-2}.$$

Convergence to the correct parameters is achieved when using that compensator, after 30,000 iterations; otherwise, the scheme fails, as shown in Figure 3.

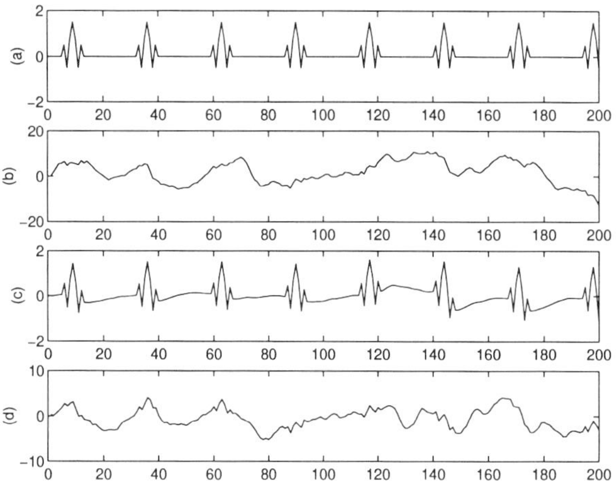

Figure 3. Performance of the adaptive canceler. (a) Desired signal, (b) noise corrupted signal, (c) signal estimate with compensator, and (d) signal estimate without compensator

5 Concluding remarks

The simultaneous SPR problem aims to build robust compensators, with applications in the convergence of adaptive recursive algorithms in signal processing and control. This problem finds its counterpart in the design of robust controllers in the simultaneous stabilization field, by the hand of analogous formulations in terms of interpolation and avoidance. We exploited these analogies so as to obtain algebraic procedures for the design of compensators for some important cases. Research is in progress to obtain new analytical procedures which allow to extend the results presented to more complex types of uncertainties in the description of the unknown system.

Bibliography

1. V. Blondel. (1994) *Simultaneous Stabilization of Linear Systems.* Lecture Notes in Control and Information Sciences. Springer-Verlag, Berlin.

2. M. Vidyasagar. (1985). *Control System Synthesis: A Factorization Approach.* The MIT Press.

3. B.D.O. Anderson, S. Dasgupta, P. Khargonekar, F.J. Kraus, and M.Mansour. (1990). Robust strict positive realness: Characterization and construction. *IEEE Trans. on Circuits and Systems*, *37*:869–876.

4. M. V. Bredemann. (1995). *Feedback Controller Design for Simultaneous Stabilization.* PhD thesis, University of New Mexico-Albuquerque.

5. F. Pérez and C. Mosquera. (1996). Characterization and algebraic solution to the extreme-point robust SPR problem. In *Proc. 13th World Congress IFAC, San Francisco, CA*, pages 391–396.

6. D.C. Youla and M. Saito. (1967). Interpolation with positive-real functions. *Journal of The Franklin Institute*, *284*:77–108.

7. K. Wei. (1992). The solution of a transcendental problem and its applications in simultaneous stabilization problems. *IEEE Transactions on Automatic Control*, *37*:1305–1315.

8. C.R. Johnson Jr. (1984). Adaptive IIR filtering: Current results and open issues. *IEEE Transactions on Information Theory*, *30*:237–250.

9. M.G.Larimore, J.R. Treichler, and Jr. C.R.Johnson. (1980). SHARF: An algorithm for adapting IIR digital filters. *IEEE Trans. on Acoustics, Speech and Signal Processing*, *28*:428–440.

Asymptotic Performance Analysis of Yang's Subspace Tracking Algorithm

Jean-Pierre Delmas* **and Jean-François Cardoso****

**Institut National des Télécommunications, Evry, France, and* ***Ecole Nationale Supérieure des Télécommunications, Paris, France*

Abstract

This paper proposes a complete asymptotic performance analysis of a simple adaptive algorithm for tracking of dominant subspaces, proposed by B. Yang [1]. Our approach is to associate to this algorithm, a second stochastic algorithm that governs the projection matrix on a dominant invariant subspace. This enables us to apply a general result of Gaussian approximation theory to derive the asymptotic covariance of the estimated projection matrix in constant step size situation. Close form expressions are given for this covariance in case of independent observations. It is found that the asymptotic distribution has a structure very similar to those describing batch estimation technique. Finally, the accuracy of the asymptotic analysis is checked by numerical simulations and is found to be valid not only for a "small" step size but in a very large domain.

1 Introduction

Subspace tracking. Over the past decade, adaptive estimation of subspaces of covariance matrices has been applied successfully to high resolution spectral analysis in signal processing and principal component analysis in data compression and feature extraction. The interest for these methods, a tool of growing importance in many fields of signal processing, has recently been renewed by the subspace approach used in blind identification of multichannel FIR filters [2]. Numerous solutions have been proposed to recursively updating subspaces of covariance matrices (see for example the references in [1]), but there are relatively few performance analysis concerning stochastic gradient algorithms derived from constrained or unconstrained optimization problems. Among them, Larimore and Calvert [3] presented a convergence study of the Thompson algorithm, while Yang and Kaveh [4] made an analysis of convergence rate and stability of their constrained gradient search procedure resorting to the classical independence assumption. A study of convergence with the help of the associated ordinary differential equation (ODE) and an evaluation of the performance by computing the covariances of the estimated eigenvectors of the parameterized stochastic gradient algorithm by Regalia [5], was proposed in [6]. A deflation algorithm for tracking dominant or minorant subspaces was presented and studied in [7] by the same approach. Some algorithms tracking dominant subspaces from a least square-like approach are presented in [1] and analyzed in [8], [9]. It is the purpose of this paper to provide a thorough study of the behavior of an attractive algorithm presented by Yang in [1].

Yang's algorithm. For a given $n \times n$ covariance matrix $\mathbf{R}$, denote $\lambda_1 \geq \ldots \geq \lambda_n$ the eigen-values of $\mathbf{R}$ and $\mathbf{v}_1, \ldots, \mathbf{v}_n$ corresponding eigen-vectors. The r-dimensional dominant invariant subspace of $\mathbf{R}$ is the span of $\mathbf{v}_1, \ldots, \mathbf{v}_r$ and it is well defined if, as assumed throughout the paper, $\lambda_r > \lambda_{r+1}$. Denote $\mathbf{\Pi}$ the orthogonal projector onto this subspace. One has:

$$\mathbf{R}\mathbf{v}_a = \lambda_a \mathbf{v}_a \qquad \mathbf{\Pi}\mathbf{v}_a = \pi_a \mathbf{v}_a \qquad 1 \leq a \leq n, \tag{1.1}$$

where we have defined $\pi_1 = \cdots = \pi_r = 1$ and $\pi_{r+1} = \cdots = \pi_n = 0$. Defining the rank-one projection matrices $\mathbf{\Pi}_a$ onto each eigen-vector, one can also write

$$\mathbf{\Pi}_a \stackrel{\text{def}}{=} \mathbf{v}_a \mathbf{v}_a^H \qquad \mathbf{R} = \sum_{a=1,n} \lambda_a \mathbf{\Pi}_a \qquad \mathbf{\Pi} = \sum_{a=1,n} \pi_a \mathbf{\Pi}_a = \sum_{1 \leq a \leq r} \mathbf{\Pi}_a. \tag{1.2}$$

Subspace tracking consists in recursively updating at time t an (approximately) orthonormal basis of this subspace upon reception of sample $\mathbf{x}_t$ of a stationary process with covariance $\mathbf{R} = E\mathbf{x}_t\mathbf{x}_t^H$.

There are several interesting algorithms described in Yang's paper [1], based on the unconstrained minimization of the objective function:

$$J(\mathbf{W}) \stackrel{\text{def}}{=} E\| \mathbf{x}_t - \mathbf{W}\mathbf{W}^H \mathbf{x}_t \|_{\text{Fro}}^2$$

with respect to the $n \times r$ matrix $\mathbf{W}$. In this contribution, we consider the stochastic gradient algorithm for the minimization of $J(\mathbf{W})$. This yields the following algorithm:

$$\mathbf{W}_{t+1} = \mathbf{W}_t + \gamma_t (2\mathbf{x}_t\mathbf{x}_t^H - \mathbf{x}_t\mathbf{x}_t^H\mathbf{W}_t\mathbf{W}_t^H - \mathbf{W}_t\mathbf{W}_t^H\mathbf{x}_t\mathbf{x}_t^H)\mathbf{W}_t. \tag{1.3}$$

The stationary points of the stochastic algorithm (1.3) are $\mathcal{W} = \{\mathbf{W} | \mathbf{W} = \mathbf{V}_r\mathbf{U}\}$ where $\mathbf{V}_r$ is any r-dimensional eigenvector basis of $\mathbf{R}$ and $\mathbf{U}$ is an $r \times r$ arbitrary unitary matrix; the set $\mathcal{W}$ is globally asymptotically stable [1]. Furthermore, it is proved in [1] that the Hessian matrix of $J(\mathbf{W})$ calculated for $\mathbf{W}$ in $\mathcal{W}$ is positive semi-definite if and only if $\mathbf{W} = \mathbf{V}_S\mathbf{U}$ with $\mathbf{V}_S \stackrel{\text{def}}{=} [\mathbf{v}_1, \ldots, \mathbf{v}_r]$. Therefore the matrix set $\mathcal{W}_S = \{\mathbf{W} | \mathbf{W} = \mathbf{V}_S\mathbf{U}\}$ of $\mathcal{W}$ is globally asymptotically stable. Thus the stochastic algorithm (1.3) converges almost surely to $\mathcal{W}_S$.

Projector tracking. A difficulty arises in the study of the behavior of $\mathbf{W}_t$ because the set $\mathcal{W}_S$ forms a *continuum* of attractors: the column vectors of $\mathbf{W}_t$ do not in general tend to the eigenvectors $\mathbf{v}_1, \ldots, \mathbf{v}_r$ and we have no proof of convergence of $\mathbf{W}_t$ to a particular orthonormal basis of their span Therefore, the approach followed in this paper is to study the trajectory of matrix $\mathbf{P}_t$

$$\mathbf{P}_t \stackrel{\text{def}}{=} \mathbf{W}_t \mathbf{W}_t^H \tag{1.4}$$

whose dynamic is governed by the stochastic equation:

$$\begin{aligned} \mathbf{P}_{t+1} &= \mathbf{P}_t + \gamma_t f(\mathbf{P}_t, \mathbf{x}_t\mathbf{x}_t^H) + \gamma_t^2 h((\mathbf{P}_t, \mathbf{x}_t\mathbf{x}_t^H) & (1.5)\\ f(\mathbf{P}, \mathbf{M}) &\stackrel{\text{def}}{=} \mathbf{P}(2\mathbf{M} - \mathbf{MP} - \mathbf{PM}) + (2\mathbf{M} - \mathbf{MP} - \mathbf{PM})\mathbf{P} & (1.6)\\ h(\mathbf{P}, \mathbf{M}) &\stackrel{\text{def}}{=} (2\mathbf{M} - \mathbf{MP} - \mathbf{PM})\mathbf{P}(2\mathbf{M} - \mathbf{MP} - \mathbf{PM}) & (1.7) \end{aligned}$$

obtained by combining (1.3) and (1.4). A remarkable feature of (1.5) is that the field f and the complementary term h actually depend only on $\mathbf{P}_t$ and *not* on $\mathbf{W}_t$. This fortunate circumstance makes it possible to study the evolution of $\mathbf{P}_t$ without determining the evolution of the underlying matrix $\mathbf{W}_t$. The characteristics of $\mathbf{P}_t$ are indeed the most interesting since they completely characterize the estimated subspace.

Outline of the paper. This paper is organized as follows. In Section 2, after presenting a brief review of a general Gaussian approximation result, we obtain in closed form the asymptotic covariance of $\mathbf{P}_t$ for the case where $\mathbf{x}_t$ is a white complex circular Gaussian sequence. In Section 3, we extend this result to real signals and we compare the asymptotic performance of the algorithm with the performance of batch algorithms for subspace estimation. Section 4 presents some simulation results and investigates the validity of the asymptotic approach.

2 Asymptotic performance analysis

2.1 A short review of a general Gaussian approximation result

In this section, we evaluate the asymptotic distributions of eigenvector and subspace projection matrix estimators given by the previous algorithms. For this purpose, we shall use the following result ([10] theorem 2, p. 108). Let a constant step size recursive stochastic algorithm:

$$\Theta_{t+1} = \Theta_t + \gamma f(\Theta_t, \mathbf{x}_t) + \gamma^2 h_t(\Theta_t, \mathbf{x}_t) \tag{2.1}$$

with $\mathbf{x}_t = g(\xi_t)$, where ξ_t is a Markov chain independent of Θ_t and with $h_t(\Theta, \mathbf{x})$ an uniformly bounded function for $(\Theta, \mathbf{x})$ in some fixed compact set. Suppose that the vector parameter Θ_t converges almost surely to the unique asymptotically stable point Θ_* in the corresponding decreasing step size algorithm. Consider the continuous Lyapunov equation:

$$\mathbf{D}\mathbf{C}_\Theta + \mathbf{C}_\Theta\mathbf{D}^T + \mathbf{G} = \mathbf{O} \tag{2.2}$$

and where $\mathbf{D}$ and $\mathbf{G}$ are respectively the derivative of the mean field and the covariance of the field of the algorithm (2.1)

$$\mathbf{D} \stackrel{\text{def}}{=} E[\frac{\partial f}{\partial \Theta}(\Theta, \mathbf{x}_t)]_{\Theta=\Theta_*} \tag{2.3}$$

$$\mathbf{G} \stackrel{\text{def}}{=} \sum_{t=-\infty}^{\infty} \text{Cov}[f(\Theta_*, \mathbf{x}_t), f(\Theta_*, \mathbf{x}_0)]. \tag{2.4}$$

If all the eigenvalues of the derivative of the mean field $\mathbf{D}$ have strictly negative real parts, then when $\gamma \to 0$ and $t \to \infty$ in stationary situation:

$$\frac{1}{\sqrt{\gamma}}(\Theta_t - \Theta_*) \stackrel{\mathcal{L}}{\to} \mathcal{N}(0, \mathbf{C}_\Theta) \tag{2.5}$$

where $\mathbf{C}_\Theta$ is the unique symmetric solution of the Lyapunov Equation (2.2).

2.2 Local characterization of the field

According to previous section, one needs to characterize two local properties of the field $f(\mathbf{P}, \mathbf{x}\mathbf{x}^H)$: the mean value of its derivative and its covariance, both evaluated at point $\mathbf{P} = \mathbf{\Pi}$. To proceed, it will be convenient to define the following set of orthonormal Hermitian matrices:

$$\mathbf{H}_{ab} = \begin{cases} \mathbf{v}_a \mathbf{v}_a^H & a = b \\ \frac{\mathbf{v}_a \mathbf{v}_b^H + \mathbf{v}_b \mathbf{v}_a^H}{\sqrt{2}} & a < b \\ \frac{\mathbf{v}_a \mathbf{v}_b^H - \mathbf{v}_b \mathbf{v}_a^H}{i\sqrt{2}} & a > b. \end{cases} \tag{2.6}$$

With this definition, a first order approximation of the mean field in the neighborhood of $\mathbf{\Pi}$ and the eigenstructure of the covariance matrix of the field are given by the following lemma (the proof is available from the authors):

Lemma 1. *If $\{\mathbf{x}_t\}$ is an i.i.d. sequence of zero-mean complex circular Gaussian random vectors, then, for $1 \leq a, b \leq n$,*

$$Ef(\mathbf{\Pi} + \epsilon\, \mathbf{H}_{ab},\ \mathbf{x}_t\mathbf{x}_t^H) = \epsilon\ \mu_{ab}\ \mathbf{H}_{ab}\ +\ O(\epsilon^2) \tag{2.7}$$

$$\text{Cov}(\text{Vec}(f(\mathbf{\Pi}, \mathbf{x}\mathbf{x}^H)))\ \text{Vec}(\mathbf{H}_{ab}) = \nu_{ab}\ \text{Vec}(\mathbf{H}_{ab}) \tag{2.8}$$

with respectively

$$\mu_{ab} \stackrel{\text{def}}{=} 2\lambda_a(1-\pi_a) + 2\lambda_b(1-\pi_b) - (\lambda_a + \lambda_b)(\pi_a + \pi_b) \tag{2.9}$$

$$\nu_{ab} \stackrel{\text{def}}{=} (\pi_a - \pi_b)^2 \lambda_a \lambda_b. \tag{2.10}$$

2.3 Real parameterization

To adapt the results recalled in the previous section to our needs, the $n \times n$ rank-r complex Hermitian matrix $\mathbf{P}$ should be parameterized by a vector Θ of real parameters. Counting degrees of freedom shows that the set of $n \times n$ rank-r complex Hermitian matrices is a $r(2n-r)$-dimensional manifold. This section introduces a parameterization of this manifold in a neighborhood of $\mathbf{\Pi}$ by a $r(2n-r) \times 1$ vector Θ of real parameters.

For an $n \times n$ matrix $\mathbf{M}$ and any pair $1 \leq a, b \leq n$ of indices, define

$$\theta_{ab}(\mathbf{M}) \stackrel{\text{def}}{=} \operatorname{Tr}\{\mathbf{H}_{ab}(\mathbf{M} - \mathbf{\Pi})\}. \tag{2.11}$$

These are *real* scalars if $\mathbf{M}$ is Hermitian. Since $\{\mathbf{H}_{ab} \mid 1 \leq a, b \leq n\}$ is an orthonormal basis for the linear space of $n \times n$ matrices, the scalars $\theta_{ab}(\mathbf{M})$ are the coordinates of $\mathbf{M} - \mathbf{\Pi}$ on this basis. Thus any $n \times n$ matrix is parameterized by the values of $\theta_{ab}(\mathbf{M})$ according to:

$$\mathbf{M} = \mathbf{\Pi} + \sum_{1 \leq a,b \leq n} \theta_{ab}(\mathbf{M})\, \mathbf{H}_{ab}. \tag{2.12}$$

Matrices close to $\mathbf{\Pi}$ are parameterized by small values of these parameters: by definition $\theta_{ab}(\mathbf{M}) = O(\|\mathbf{M} - \mathbf{\Pi}\|)$ for any pair (a, b). The relevance of these parameters is shown by this lemma (the proof is available from the authors):

Lemma 2. *If $\mathbf{P}$ is an $n \times n$ rank-r Hermitian matrix, then*

$$\mathbf{P} = \mathbf{\Pi} + \sum_{(a,b) \in P_h} \theta_{ab}(\mathbf{P})\, \mathbf{H}_{ab} \; + \; O(\|\mathbf{P} - \mathbf{\Pi}\|^2) \tag{2.13}$$

where P_h is the complement of $\{(a, b) \mid r < a, b \leq n\}$, i.e. $P_h \stackrel{\text{def}}{=} \{(a, b) \mid 1 \leq a \leq r \text{ or } 1 \leq b \leq r\}$.

In other words, a rank-r Hermitian matrices lying less than ϵ away from $\mathbf{\Pi}$ (*i.e.* $\|\mathbf{P} - \mathbf{\Pi}\| < \epsilon$) have negligible (of order ϵ^2) components in the direction of $\mathbf{H}_{ab}$ for $r < a, b \leq n$. Equation (2.13) is more compactly expressed by using an $n^2 \times r(2n-r)$ matrix $\mathcal{H}$:

$$\mathcal{H} \stackrel{\text{def}}{=} [\ldots, \operatorname{Vec}(\mathbf{H}_{ab}), \ldots], \quad (a, b) \in P_h \tag{2.14}$$

so that the $r(2n-r) \times 1$ vector $\Theta(\mathbf{P})$ defined by

$$\Theta(\mathbf{P}) \stackrel{\text{def}}{=} \mathcal{H}^H \operatorname{Vec}(\mathbf{P} - \mathbf{\Pi}) \tag{2.15}$$

contains the values of $\theta_{ab}(\mathbf{P})$ for $(a, b) \in P_h$ and Equation (2.13) reads, after vectorization

$$\operatorname{Vec}(\mathbf{P}) \;=\; \operatorname{Vec}(\mathbf{\Pi}) \;+\; \mathcal{H}\Theta(\mathbf{P}) \;+\; O(\|\mathbf{P} - \mathbf{\Pi}\|^2). \tag{2.16}$$

Note that the particular ordering of the pairs in the set P_h is irrelevant in expressions like $\mathcal{H}\Theta$.

There are $n^2 - (n-r)^2 = r(2n-r)$ pairs in P_h and this is exactly the dimension of the manifold of $n \times n$ rank-r Hermitian matrices. This point, together with Equation (2.13), shows that the matrix set $\{\mathbf{H}_{ab} \mid (a,b) \in P_h\}$ in fact is a *basis* of the tangent plane to this manifold at point $\mathbf{\Pi}$. It follows that, in a neighborhood of $\mathbf{\Pi}$, Hermitian rank-r matrices are uniquely determined by the (real) values of Θ. This is the required real (local) reparameterization of a rank-r Hermitian matrix by a $r(2n-r)$-dimensional vector Θ: we denote $\mathcal{P}(\Theta)$ the unique (for small enough $||\Theta||$) Hermitian matrix with rank r such that $\mathcal{H}^H\text{Vec}(\mathcal{P}(\Theta) - \mathbf{\Pi}) = \Theta$. It is not necessary to express explicitly $\mathcal{P}(\Theta)$: as will turn out, it is sufficient to use the property derived from (2.16):

$$\text{Vec}(\mathcal{P}(\Theta)) = \text{Vec}(\mathbf{\Pi}) + \mathcal{H}\Theta + O(||\Theta||^2). \tag{2.17}$$

2.4 Solution of the Lyapunov equation

We are now in position to solve the Lyapunov equation in the new parameter Θ defined in the previous section. The stochastic equation governing the evolution of this vector parameter is obtained by applying the transformation $\mathbf{P}_t \rightarrow \Theta_t = \mathcal{H}^H\text{Vec}(\mathbf{P}_t - \mathbf{\Pi})$ to the original Equation (1.5).

$$\Theta_{t+1} = \Theta_t + \gamma_t\phi(\Theta_t, \mathbf{x}_t) + \gamma_t^2\psi(\Theta_t, \mathbf{x}_t) \tag{2.18}$$

where functions ϕ and ψ appear to be

$$\phi(\Theta, \mathbf{x}) \stackrel{\text{def}}{=} \mathcal{H}^H\text{Vec}(f(\mathcal{P}(\Theta), \mathbf{x}\mathbf{x}^H)), \tag{2.19}$$

$$\psi(\Theta, \mathbf{x}) \stackrel{\text{def}}{=} \mathcal{H}^H\text{Vec}(h(\mathcal{P}(\Theta), \mathbf{x}\mathbf{x}^H)). \tag{2.20}$$

We need to evaluate the derivative matrix $\mathbf{D}$ of $E\phi(\Theta, \mathbf{x})$ at point $\Theta = \mathbf{0}$ and, since we consider only the case of independent observations, the covariance matrix $\mathbf{\Gamma}$ of $\phi(\mathbf{0}, \mathbf{x})$. With these notations, the results of Section 2.2 are recycled as follows.

$$\begin{aligned}
E\phi(\Theta, \mathbf{x}) &= \mathcal{H}^H\text{Vec}Ef(\mathcal{P}(\Theta), \mathbf{x}\mathbf{x}^H) \\
&= \mathcal{H}^H\text{Vec}Ef\left(\mathbf{\Pi} + \sum \theta_{ab}\mathbf{H}_{ab} + O(||\Theta||^2),\ \mathbf{x}\mathbf{x}^H\right) \\
&= \mathcal{H}^H\text{Vec}\left(\sum \theta_{ab}\mu_{ab}\mathbf{H}_{ab}) + O(||\Theta||^2)\right) \\
&= \mathcal{H}^H(\mathcal{H}\mathbf{\Delta}_\mu\Theta) + O(||\Theta||^2) = \mathbf{\Delta}_\mu\theta + O(||\Theta||^2),
\end{aligned}$$

where the above summations are over $(a,b) \in P_h$. First equality uses definition (2.19) and the linearity of the Vec operation; second equality stems from property (2.16) of the reparameterization; third equality uses Lemma 2 and the differentiability of f; fourth equality is by definitions (2.14) and (2.21); last equality is by orthonormality of the basis $\{\mathbf{H}_{ab}\}$. Hence,

$$\mathbf{D} \stackrel{\text{def}}{=} \left.\frac{\partial E\phi(\Theta, \mathbf{x})}{\partial \Theta}\right|_{\theta=0} = \mathbf{\Delta}_\mu \text{ with } \mathbf{\Delta}_\mu \stackrel{\text{def}}{=} \mathrm{Diag}(\ldots, \mu_{ab}, \ldots) \quad (a,b) \in P_h. \tag{2.21}$$

The covariance of the field at $\Theta = \mathbf{0}$ is:

$$\begin{aligned} \mathrm{Cov}(\phi(\mathbf{0}, \mathbf{x})) &= \mathrm{Cov}(\mathcal{H}^H \mathrm{Vec}(f(\mathbf{\Pi}, \mathbf{x}\mathbf{x}^H))) = \mathcal{H}^H \ \mathrm{Cov}(\mathrm{Vec}(f(\mathbf{\Pi}, \mathbf{x}\mathbf{x}^H)))\ \mathcal{H} \\ &= \mathcal{H}^H \mathcal{H} \mathbf{\Delta}_\nu = \mathbf{\Delta}_\nu. \end{aligned}$$

First equality is by definition of ϕ; second equality is by bilinearity of the Cov operator; third equality is by noting that Lemma 2 also reads $\mathrm{Cov}(\mathrm{Vec}(f(\mathbf{\Pi}, \mathbf{x}\mathbf{x}^H)))\,\mathcal{H} = \mathcal{H}\mathbf{\Delta}_\nu$ with $\mathbf{\Delta}_\nu$ defined by (2.22); last equality is due to the orthonormality of the basis $\{\mathbf{H}_{ab}\}$. We conclude that for independent observations

$$\mathbf{\Gamma} \stackrel{\text{def}}{=} \mathrm{Cov}(\phi(\mathbf{0}, \mathbf{x})) = \mathbf{\Delta}_\nu \quad \text{with} \quad \mathbf{\Delta}_\nu \stackrel{\text{def}}{=} \mathrm{Diag}(\ldots, \nu_{ab}, \ldots) \quad (a,b) \in P_h. \tag{2.22}$$

Thus both $\mathbf{\Gamma}$ and $\mathbf{D}$ are diagonal matrices. In this case, the Lyapunov Equation (2.2) reduces to $r(2n-r)$ *uncoupled* scalar equations so that its solution trivially is

$$\mathbf{C}_\theta = -\frac{1}{2}\mathbf{\Delta}_\nu \mathbf{\Delta}_\mu^{-1}. \tag{2.23}$$

According to (2.5), $\gamma^{-1/2}\Theta_t \stackrel{\mathcal{L}}{\to} \mathcal{N}(\mathbf{0}, -\frac{1}{2}\mathbf{\Delta}_\nu \mathbf{\Delta}_\mu^{-1})$. By Equation (2.16), we have $\mathrm{Vec}(\mathbf{P}_t) = \mathrm{Vec}(\mathbf{\Pi}) + \mathcal{H}\Theta_t + O(\|\Theta_t\|^2)$. We conclude that

$$\frac{1}{\sqrt{\gamma}}(\mathrm{Vec}(\mathbf{P}_t) - \mathrm{Vec}(\mathbf{\Pi})) \stackrel{\mathcal{L}}{\to} \mathcal{N}(\mathbf{0}, \mathbf{C}_P) \quad \text{with} \quad \mathbf{C}_P = \mathcal{H}\mathbf{C}_\theta \mathcal{H}^H = -\frac{1}{2}\mathcal{H}\mathbf{\Delta}_\nu \mathbf{\Delta}_\mu^{-1}\mathcal{H}^H \tag{2.24}$$

for $\gamma \to 0$ and $t \to +\infty$. Expression (2.24) of the covariance matrix $\mathbf{C}_P$ in the asymptotic distribution of $\mathrm{Vec}(\mathbf{P}_t)$ may be written as an explicit sum:

$$\mathbf{C}_P = \sum_{(a,b) \in P_h} \frac{\nu_{ab}}{-2\mu_{ab}} \mathrm{Vec}(\mathbf{H}_{ab}) \mathrm{Vec}(\mathbf{H}_{ab})^H. \tag{2.25}$$

Definitions (2.9) of μ_{ab} and (2.10) of ν_{ab} show that these quantities are symmetric and also that $\nu_{ab} = 0$ for $1 \leq a, b \leq r$. Using this fact and the identity:

$$\mathrm{Vec}(\mathbf{H}_{ab})\mathrm{Vec}(\mathbf{H}_{ab})^H + \mathrm{Vec}(\mathbf{H}_{ba})\mathrm{Vec}(\mathbf{H}_{ba})^H = \mathbf{\Pi}_a \otimes \mathbf{\Pi}_b + \mathbf{\Pi}_b \otimes \mathbf{\Pi}_a, \tag{2.26}$$

expression (2.25) may also be rewritten as

$$\mathbf{C}_P = \sum_{1 \leq a \leq r < b \leq n} \frac{\lambda_a \lambda_b}{2(\lambda_a - \lambda_b)} \left(\mathbf{\Pi}_a \otimes \mathbf{\Pi}_b + \mathbf{\Pi}_b \otimes \mathbf{\Pi}_a\right). \tag{2.27}$$

2.5 Analysis

Several simple properties can be derived from the regular structure of the covariance matrix $\mathbf{C}_P$ as expressed by (2.25) or (2.27). A simple global measure of performance is the MSE between $\mathbf{P}_t$ and $\mathbf{\Pi}$. To give MSE expressions, we assume, as is customary, that the first and second asymptotic moments of $\mathbf{P}_t$ are those of its asymptotic distribution. It follows that:

$$\| E(\mathbf{P}_t) - \mathbf{\Pi} \|^2_{\text{Fro}} = o(\gamma), \qquad \text{Cov}(\text{Vec}(\mathbf{P}_t)) = \gamma \mathbf{C}_P + o(\gamma). \tag{2.28}$$

In particular, the MSE between $\mathbf{P}_t$ and $\mathbf{\Pi}$ is given by the trace of the covariance matrix in the asymptotic distribution of $\mathbf{P}_t$. The trace being invariant under orthonormal change of basis and $\{\text{Vec}(\mathbf{H}_{ab}) | 1 \leq a, b \leq n\}$ being an orthonormal basis, we obtain from Equation (2.25):

$$E\| \mathbf{P}_t - \mathbf{\Pi} \|^2_{\text{Fro}} = \gamma \sum_{1 \leq a \leq r < b \leq n} \frac{\lambda_a \lambda_b}{\lambda_a - \lambda_b} + o(\gamma). \tag{2.29}$$

A finer picture is obtained by decomposing the error into three terms

$$\mathbf{P}_1 \stackrel{\text{def}}{=} \mathbf{\Pi}(\mathbf{P} - \mathbf{\Pi})\mathbf{\Pi}, \quad \mathbf{P}_2 \stackrel{\text{def}}{=} \mathbf{\Pi}\mathbf{P}\mathbf{\Pi}^\perp + \mathbf{\Pi}^\perp\mathbf{P}\mathbf{\Pi}, \quad \mathbf{P}_3 \stackrel{\text{def}}{=} \mathbf{\Pi}^\perp\mathbf{P}\mathbf{\Pi}^\perp. \tag{2.30}$$

Using $\mathbf{I}_n = \mathbf{\Pi} + \mathbf{\Pi}^\perp$, this is easily seen to be an orthogonal decomposition:

$$\mathbf{P} = \mathbf{\Pi} + \mathbf{P}_1 + \mathbf{P}_2 + \mathbf{P}_3, \qquad \|\mathbf{P} - \mathbf{\Pi}\|^2 = \|\mathbf{P}_1\|^2 + \|\mathbf{P}_2\|^2 + \|\mathbf{P}_3\|^2. \tag{2.31}$$

The relevance of this orthogonal decomposition stems from this lemma (the proof is available from the authors):

Lemma 3. *Let* $\mathbf{P}$ *be a rank-*r *Hermitian matrix, and let* $\hat{\mathbf{\Pi}}$ *be the orthogonal projection matrix on the range space of* $\mathbf{P}$. *Then*

$$\mathbf{P}_1 = \mathbf{P}^2 - \mathbf{P} + O(\|\mathbf{P} - \mathbf{\Pi}\|^2), \tag{2.32}$$

$$\mathbf{P}_2 = \hat{\mathbf{\Pi}} - \mathbf{\Pi} + O(\|\mathbf{P} - \mathbf{\Pi}\|^2), \tag{2.33}$$

$$\mathbf{P}_3 = O(\|\mathbf{P} - \mathbf{\Pi}\|^2). \tag{2.34}$$

Thanks to this lemma, each term can be given a simple interpretation. The term $\mathbf{P}_1$ represents the first order of the deviation of $\mathbf{P}$ from orthogonality since if $\mathbf{P}$ was an orthogonal projector, we would have $\mathbf{P}^2 = \mathbf{P}$. Maybe more strikingly, it is not difficult to establish that $\|\mathbf{P}_1\|^2 = \|\mathbf{W}^H\mathbf{W} - \mathbf{I}_r\|^2 + o(\|\mathbf{P} - \mathbf{\Pi}\|^2)$. According to this lemma, the term $\mathbf{P}_2$ represents the deviation between the subspace of interest and the one estimated by $\mathbf{P}$. Finally, the last term $\mathbf{P}_3$ is of order $O(\|\mathbf{P} - \mathbf{\Pi}\|^2)$ because $\mathbf{P}$ have rank r (we already expressed this property in Lemma 2).

The above decomposition is purely geometric. Statistical results are obtained by combining it with expressions (2.25) or (2.27) of the asymptotic covariance matrix $\mathbf{C}_P$ of $\mathbf{P}_t$. In doing so, massive simplifications occur due to orthogonality. This is summarized by

$$\text{Vec}(\mathbf{v}_a\mathbf{v}_b^H)^H(\mathbf{\Pi}_c \otimes \mathbf{\Pi}_d)\text{Vec}(\mathbf{v}_e\mathbf{v}_f^H) = \delta_{c,a}\delta_{c,e}\delta_{d,b}\delta_{d,f} \quad 1 \leq a, b, c, d, e, f \leq n,$$

where $\delta_{i,j}$ denotes the Kronecker notation: $\delta_{i,j} = 1$ if $i = j$ and $\delta_{i,j} = 0$ elsewhere. Many terms cancel by these orthogonality relations. The resulting asymptotic variances are

$$E\|\mathbf{P}_{1,t}\|^2_{\text{Fro}} = E\|\mathbf{W}_t^H\mathbf{W}_t - \mathbf{I}_r\|^2_{\text{Fro}} + o(\gamma) = O(\gamma), \tag{2.35}$$

$$E\|\mathbf{P}_{2,t}\|^2_{\text{Fro}} = E\|\mathbf{P}_t - \mathbf{\Pi}\|^2_{\text{Fro}} + o(\gamma) = \gamma \sum_{1\leq a\leq r<b\leq n} \frac{\lambda_a\lambda_b}{\lambda_a - \lambda_b} + o(\gamma), \tag{2.36}$$

$$E\|\mathbf{P}_{3,t}\|^2_{\text{Fro}} = o(\gamma). \tag{2.37}$$

A very striking result is observed here: the deviation of $\mathbf{W}_t$ from orthonormality, as quantified by $\|\mathbf{P}_1\|$, has a stochastic order lower than $\gamma^{\frac{1}{2}}$! This results from the fact that the summation in (2.27) only is over pairs such that $a \leq r < b$. This is a remarkable feature of this algorithm that there seems to be no price to pay for not constraining matrix $\mathbf{W}$ to have orthonormal columns, at least in the stationary setting considered herein. What is then the order of the deviation from orthonormality? This question cannot be answered by first order performance analysis, but the order can be determined experimentally. We show in Section 4 that the MSE of orthonormality is, in first approximation, proportional to γ^2.

3 Further investigations

In this section, we briefly indicate how the previous results extend to the case of real signals and how they compare to off-line subspace estimation.

3.1 The real case

To address the case of real signals, only slight modifications are needed. They stem from $\text{Cov}(\text{Vec}(\mathbf{x}\mathbf{x}^T)) = \mathbf{R} \otimes \mathbf{R} + (\mathbf{R} \otimes \mathbf{R})\mathbf{K}$ where $\mathbf{K}$ is an $n^2 \times n^2$ block matrix, acting as a permutation operator, *i.e.* $\mathbf{K}\text{Vec}(\mathbf{x}\mathbf{y}^T) = \text{Vec}(\mathbf{y}\mathbf{x}^T)$ for any vectors $\mathbf{x}$ and $\mathbf{y}$. Proceeding as in the complex case, one will easily find that

$$\mathbf{C}_P = \sum_{1\leq a\leq r<b\leq n} \frac{\lambda_a\lambda_b}{2(\lambda_a - \lambda_b)}(\mathbf{\Pi}_a \otimes \mathbf{\Pi}_b + \mathbf{\Pi}_b \otimes \mathbf{\Pi}_a)(\mathbf{I} + \mathbf{K}). \tag{3.1}$$

Because of the similarity between the asymptotic covariance matrices for $\mathbf{P}_t$ in the real case (3.1) and in the complex case (2.27), similar conclusions can be drawn. In particular, the error terms $\|\mathbf{\Pi}(\mathbf{P}_t - \mathbf{\Pi})\mathbf{\Pi}\|^2$ and $\|\mathbf{\Pi}^\perp(\mathbf{P}_t - \mathbf{\Pi})\mathbf{\Pi}^\perp\|^2$ are of order $o(\gamma)$, and the MSE is given by

$$E\|\mathbf{P}_t - \mathbf{\Pi}\|^2_{\text{Fro}} = \gamma\text{Tr}(\mathbf{C}_P) + o(\gamma) = \gamma \sum_{1\leq a\leq r<b\leq n} \frac{\lambda_a\lambda_b}{\lambda_a - \lambda_b} + o(\gamma). \tag{3.2}$$

3.2 Comparison with batch estimation

Denote $\mathbf{P}_m$ the orthogonal projection matrix on the eigenspace of the sample correlation matrix $\mathbf{R}_m$ associated with the r dominant eigenvalues when $\mathbf{R}_m$ is estimated off-line from m independent vectors $\mathbf{x}_1, \ldots, \mathbf{x}_m$. Using a classic result of perturbation theory of linear operators (see for instance [14], theorem 5.4 p.111), we obtain by continuity (for example theorem 6.2a in [15], p. 386) when $m \to \infty$: $\sqrt{m}(\text{Vec}(\mathbf{P}_m) - \text{Vec}(\mathbf{\Pi})) \xrightarrow{\mathcal{L}} \mathcal{N}(0, \mathbf{C}_P)$ with:

$$\mathbf{C}_P = \sum_{1 \leq a \leq r < b \leq n} \frac{\lambda_a \lambda_b}{(\lambda_a - \lambda_b)^2} (\mathbf{\Pi}_a \otimes \mathbf{\Pi}_b + \mathbf{\Pi}_b \otimes \mathbf{\Pi}_a). \tag{3.3}$$

Note the close formal similarity to expression (2.27) for the on-line estimation.

4 Simulations results

In these experiments, we consider the case of the projection matrix $\mathbf{P}_t$ on the eigenspace spanned by the first two eigenvectors of a 3×3 covariance matrix $\mathbf{R}$ which is that of an AR(1) model with parameter $a = 0.3$

Figure 1 shows the learning curve averaged (over 100 runs) of $\| \mathbf{P}_t - \mathbf{\Pi} \|^2_{\text{Fro}}$ for $\gamma = 0.005$. The limiting value is in perfect agreement with the theoretical values predicted by Equation (2.29). This figure also shows the evolution of "small terms" *i.e.* terms with scale $o(\gamma)$ as predicted by Equations (2.28), (2.35), (2.36) and (2.37). We display $\| E(\mathbf{P}_t) - \mathbf{\Pi} \|^2_{\text{Fro}}$, $E\| \mathbf{W}_t^H \mathbf{W}_t - \mathbf{I}_r \|^2_{\text{Fro}}$, $E\| \mathbf{P}_{1,t} \|^2_{\text{Fro}}$, $E\| \mathbf{P}_{2,t} \|^2_{\text{Fro}}$ and $E\| \mathbf{P}_{3,t} \|^2_{\text{Fro}}$ whose significance has been discussed in Section 2.5.

Figure 2 shows the ratio of the estimated mean square error $E\| \mathbf{P}_t - \mathbf{\Pi} \|^2_{\text{Fro}}$ to the theoretical asymptotic mean square error $\gamma \text{Tr}(\mathbf{C}_P)$ as a function of γ. Our present asymptotic analysis is seen to be valid over a large range of γ ($\gamma < 0.01$). Even for values of γ driving the algorithm close to its stability limit (this is about $\gamma < 0.065$ here), this ratio remains close to 1, indicating that the asymptotic analysis is essentially valid over the whole range of possible step size.

Figure 3 reveals something which cannot be determined from our (first-order) asymptotic analysis: the true order of deviation from orthonormality. Indeed, our analysis only yields $E\| \mathbf{W}_t^H \mathbf{W}_t - \mathbf{I}_r \|^2_{\text{Fro}} = O(\gamma)$. In this figure, we plot on a log-log scale $E\| \mathbf{W}_t^H \mathbf{W}_t - \mathbf{I}_r \|^2_{\text{Fro}}$ as a function of γ. We find a slope equal to 2 meaning that, experimentally: $E\| \mathbf{W}_t^H \mathbf{W}_t - \mathbf{I}_r \|^2_{\text{Fro}} \propto \gamma^2$.

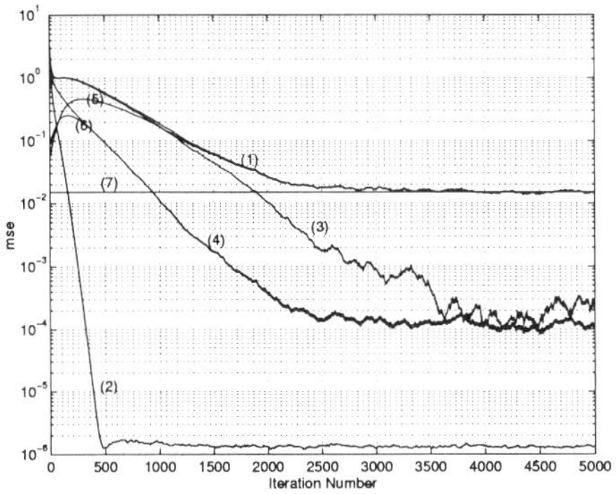

Figure 1. The plots are: **1**: MSE $E\| \mathbf{P}_t - \mathbf{\Pi} \|_{\text{Fro}}^2$, **2**: $E\| \mathbf{W}_t^H \mathbf{W}_t - \mathbf{I}_r \|_{\text{Fro}}^2$, **3**: $\| E(\mathbf{P}_t) - \mathbf{\Pi} \|_{\text{Fro}}^2$, **4**: $E\| \mathbf{P}_{1,t} \|_{\text{Fro}}^2$, **5**: $E\| \mathbf{P}_{2,t} \|_{\text{Fro}}^2$, **6**: $E\| \mathbf{P}_{3,t} \|_{\text{Fro}}^2$, **7**: $\gamma \text{Tr}(\mathbf{C}_P)$

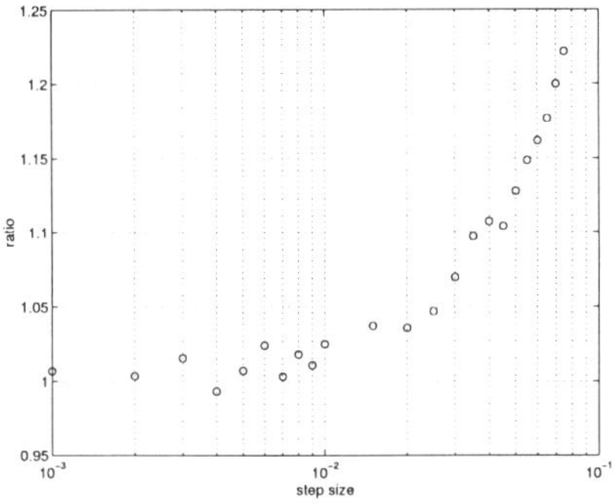

Figure 2. $(\gamma \text{Tr}(\mathbf{C}_P))^{-1} \; E\| \mathbf{P}_t - \mathbf{\Pi} \|_{\text{Fro}}^2$ (average over 400 runs) a function of γ

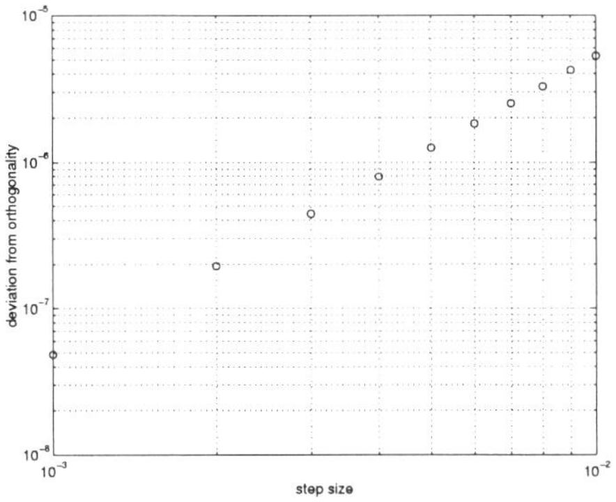

Figure 3. Deviation from orthogonality $E\| \mathbf{W}_t^H \mathbf{W}_t - \mathbf{I}_r \|_{\text{Fro}}^2$ at "convergence" (estimated over 100 independent runs) as a function of γ in log-log scale

Bibliography

1. B. Yang, "Projection approximation subspace tracking," *IEEE Trans. on Signal Processing*, vol. 43, no. 1, pp. 97-105, Jan. 1995.
2. E. Moulines, P. Duhamel, J.F. Cardoso, S. Mayrargue, "Subspace methods for blind identification of multichannel FIR filters," *IEEE Trans. on Signal Processing*, vol. 43, no. 2, pp. 516-525, Feb. 1995.
3. M.G. Larimore, R.J. Calvert, "Convergence studies of Thompson's unbiased adaptative spectral estimator," in *Proc. 14th Asilomar Conf. Circuits Syst. Comput.* Pacific Grove, CA, 1981.
4. J.F. Yang, M. Kaveh, "Adaptive eigenspace algorithms for direction or frequency estimation and tracking," *IEEE Trans. on ASSP*, vol. 36, no. 2, pp. 241-251, Feb. 1988.
5. P.A. Regalia, "An adaptive unit norm filter with applications to signal analysis and Karhunen Love tranformations," *IEEE Trans. on Circuits and Systems*, vol. 37, no. 5, pp. 646-649, May 1990.
6. J.P. Delmas, "Performance analysis of parametrized adaptive eigensubpace algorithms," in *Proc. ICASSP Detroit*, pp. 2056-2059, May 1995.
7. C. Riou, T. Chonavel, PY. Cochet, "Adaptive Subspace Estimation - Application to moving sources localization and blind channel identification," *Proc. ICASSP Atlanta*, pp. 1649-1652, May 1996.
8. B. Yang, F. Gersemsky, "Asymptotic distribution of recursive subspace estimators," *Proc. ICASSP Atlanta*, pp. 1764-1767, May 1996.
9. B. Yang, "Asymptotic convergence analysis of the projection approximation subspace tracking algorithm," *Signal Processing* vol. 50, no. 2, pp. 123-136, April 1996.
10. A. Benveniste, M. Métivier, P. Priouret, *Adaptive algorithms and stochastic approximation.* Springer Verlag, 1990.
11. J.C. Fort, G. Pagès, "Convergence of stochastic algorithms: the Kushner-Clark theorem to the Lyapounov functional method" *Adv. Appl. Prob.* 28, pp. 1072-1094, 1996.
12. D.R. Brillinger, *Times series, data analysis and theory.* Expanded Edition, Holden-Day, Inc., 1980.
13. T.W. Anderson, *An introduction to multivariate statistical analysis.* Second Edition, Wiley and Sons, 1984.
14. T. Kato, *Perturbation Theory for Linear Operators.* Springer Berlin, 1995.
15. C.R. Rao, *Linear statistical inference and its applications.* Wiley and Sons, 1973.

Generalised Prediction Error Approach to Spectral Analysis and Filtering of Irregularly-Sampled Data

R.J. Martin

GEC-Marconi Research Centre, Hertfordshire

Abstract

We discuss time-domain techniques for AR modelling and AR-based signal separation using a generalised prediction error technique in which the prediction coefficients depend on the observation intervals and on the underlying poles. Spectral estimation is effected by minimising a certain energy function with respect to the poles; this coincides with the Covariance method in the regular case. In separation schemes we show that the time-domain removal of features with known spectra is able to allow identification of smaller features made invisible by the spectral "smearing effect" associated with irregular sampling.

1 Introduction

An AR process is a discrete-time linear stochastic process given by

$$x_n + \sum_{j=1}^{p} a_j x_{n-j} = \varepsilon_n$$

in which the driving process on the RHS is uncorrelated and, if the model is useful, small. They are capable of representing data with strong spectral features and exhibit high-resolution spectral estimates even when the data length is short. Methods for fitting such models are well-established and often use the least-squares minimisation of a prediction error energy [1]. When the sampling is irregular the position becomes more complicated. For the missing data problem one can estimate the autocorrelation lags from the available data and then use maximum entropy [2, 3]; alternatively there methods for jointly estimating the coefficients and the missing data [5]. On the other hand these are unsuitable if the sampling is completely general ("random"). In that case one can try to use the sample autocovariance, $R'_m = \mathcal{E} x_n^* x_{n+m}$, but this is made difficult by the dependence of R' on the sampling scheme and on the consequence that reliable estimates of R' require large amounts of data [4]. A more direct, but rather computationally-intensive, procedure is to fit a continuous-time AR process by Kalman recursive estimation [6].

Irregular sampling has a long theoretical history (see [7, 8, 9] for extensive reviews). In practice it occurs in many areas such as biomedicine, geophysics and astronomy where data naturally arises irregularly either in time or space; also, deliberate aperiodic sampling may used to reduce or eliminate aliasing [8, 10], notably in modern synthetic-aperture radars [11] and high-speed digitisers [12].

One of the most serious practical issues in irregular sampling is that the observed spectrum is a smeared version of the underlying one [13, 7, 14]; consequently one only sees the strongest signals in such spectra. Therefore signal identification has to rely heavily on time-domain methods. This has led to the development of sequential component extraction techniques [12] for the estimation of sinusoids in white noise. The filtering methods that we shall present can be viewed as a substantial generalisation, for use when the components are not coherent.

In this paper we do two things. In Section 2 we construct *generalised prediction errors* which depend on the putative model poles, on the sample spacings and of course on the data points; an error energy function is then minimised to produce estimates of the poles. This energy function is shown to generalise the classical Covariance approach in a very natural way. Then in Section 3 we show how to use prediction errors to perform maximum-likelihood separation of two signals of which one is autoregressive and the other is unknown (in particular, of unknown variance). By using generalised prediction errors we are able to perform the filtering whether or not the sampling is regular, and this goes a long way to solving what is generally regarded today as a difficult issue [8]. Although waveform reconstruction followed by resampling is sometimes possible, we shall deal with cases where there are components at frequencies exceeding half the average sampling rate, causing low-pass reconstruction methods to fail.

2 Generalised prediction errors

2.1 Generalities

As we wish to make the prediction coefficients time-varying, it is better to fix our minds on the poles (i.e. the underlying model) and work out how to get the coefficients from the poles. We see immediately that in the case of regular sampling the AR coefficients a_j are determined from the poles α_i by the matrix equation

$$\begin{bmatrix} \alpha_1^p & \alpha_1^{p-1} & \cdots & 1 \\ \vdots & \vdots & \ddots & \vdots \\ \alpha_p^p & \alpha_p^{p-1} & \cdots & 1 \end{bmatrix} \begin{bmatrix} a_0 \\ a_1 \\ \vdots \\ a_p \end{bmatrix} = \mathbf{0}$$

and the "normalisation condition" : $a_0 = 1$. The forward prediction errors are

$$f_n = \sum_{j=0}^{p} a_j y_{n-j}. \tag{2.1}$$

At this point it is convenient to make some notational definitions. For a set of p complex numbers $\boldsymbol{\alpha} = (\alpha_i)_{i=1}^p$ we define $A(z) = \prod_{j=1}^p 1 - \alpha_i/z$ and

$$\begin{aligned} N(\boldsymbol{\alpha}) &:= \frac{1}{2\pi \mathrm{i}} \oint_{|z|=1} A(z)A^*(1/z)\frac{dz}{z} \\ 1/C(\boldsymbol{\alpha}) &:= \frac{1}{2\pi \mathrm{i}} \oint_{|z|=1} \frac{1}{A(z)A^*(1/z)}\frac{dz}{z}. \end{aligned}$$

System-theoretically these are, respectively, the power gains observed by putting white noise through the FIR filter with zeros α_i and the IIR filter with poles α_i. In terms of the coefficients a_j and the reflection coefficients ρ_i :

$$N(\boldsymbol{\alpha}) = 1 + \sum_{j=1}^{p} |a_j|^2 \quad \text{and} \quad C(\boldsymbol{\alpha}) = \prod_{i=1}^{p} 1 - |\rho_i|^2.$$

2.2 Derivation for the coherent case

When the data consist of p pure tones at frequencies ω_i, the poles are at $\alpha_i = \exp(\mathrm{i}\omega_i \delta t)$ and the prediction errors are all zero (independently of the amplitudes and phases of the tones). For irregular sampling let use define $\alpha_i = \exp(\mathrm{i}\omega_i \tau)$ in which τ is reasonably arbitrary; then to make the generalised forward prediction errors,

$$f_n = \sum_{j=0}^{p} r_j^n y_{n-j}, \tag{2.2}$$

zero independently of the amplitudes and phases of the tones, we require

$$\begin{bmatrix} \alpha_1^{(t_n - t_{n-p})/\tau} & \alpha_1^{(t_{n-1} - t_{n-p})/\tau} & \cdots & 1 \\ \vdots & \vdots & \ddots & \vdots \\ \alpha_p^{(t_n - t_{n-p})/\tau} & \alpha_p^{(t_{n-1} - t_{n-p})/\tau} & \cdots & 1 \end{bmatrix} \begin{bmatrix} r_0^n \\ r_1^n \\ \vdots \\ r_p^n \end{bmatrix} = \mathbf{0}. \tag{2.3}$$

Note that a nontrivial solution must exist, as the matrix is $p \times (p+1)$. The normalisation needs to be established and we quickly find that $r_0^n = 1$ will not do. For example, take

$$\begin{gathered} \alpha_1 = \mathrm{i}, \quad \alpha_2 = -\mathrm{i} \\ t_n = 1.5, \quad t_{n-1} = 1, \quad t_{n-2} = 0, \quad \tau = 0.5. \end{gathered}$$

The matrix equation now reads

$$\begin{bmatrix} -\mathrm{i} & -1 & 1 \\ \mathrm{i} & -1 & 1 \end{bmatrix} \begin{bmatrix} r_0^n \\ r_1^n \\ r_2^n \end{bmatrix} = \mathbf{0} \Rightarrow r_0^n = 0.$$

Thus requiring $r_0^n = 1$ would make the r_1^n and r_2^n infinite. Accordingly we constrain the *size* of $\mathbf{r}$ by imposing the following normalisation :

$$\sum_{j=0}^{p} |r_j^n|^2 = N(\boldsymbol{\alpha}). \tag{2.4}$$

Recalling that for regular sampling $N(\boldsymbol{\alpha})$ is just the sum of the mod-squares of the AR coefficients, we see that in that case the definitions (2.2–2.4) produce the same prediction errors as the classical definition (2.1), up to a factor of modulus 1. This does not matter because we will only need the generalised prediction error *energies* $||f_n||^2$.

2.3 The incoherent case

Moving to the incoherent case, we now show that the definition (2.3) is the correct one whether or not the poles are on the unit circle. To do this we consider the Ito diffusion

$$\{\mathcal{L}Y(t)\}dt = dB(t) \tag{2.5}$$

in which $\mathcal{L}$ is a linear differential operator with constant coefficients and $B(t)$ is a Brownian motion on $\mathbb{R}$ or $\mathbb{C}$. The solution of (2.5) is

$$Y(t) = \int_{-\infty}^{\infty} G(t-t')\, dB(t') \tag{2.6}$$

where $G(t)$ is the Green's function satisfying

$$\mathcal{L}G(t) = \delta(t), \qquad t < 0 \Rightarrow G(t) = 0. \tag{2.7}$$

For a stable process, $G(t) \to 0$ exponentially as $t \to \infty$. For an unstable process the integral does not exist. Suppose now that (2.6) is sampled, so $y_n = Y(t_n)$. Let the z-domain poles α_i be related to the Laplace-domain poles β_i of (2.5) by $\alpha_i = \exp(\beta_i \tau)$. For distinct poles the Green's function is constructed thus :

$$G(t) = \begin{cases} 0, & t < 0 \\ \sum_{i=1}^{p} \lambda_i \alpha_i^{t/\tau}, & t > 0 \end{cases}$$

for appropriate coefficients λ_i which we need not find explicitly. Then

$$f_n = \sum_{j=0}^{p} r_j^n \int_{-\infty}^{\infty} G(t_{n-j} - t')\, dB(t') \,.$$

From the construction of $G(t)$ above, and from (2.3), we have that, if either of the following conditions holds,

$$\begin{aligned} 0 \le j \le p \ &\Rightarrow\ t_{n-j} - t' \ge 0 \\ 0 \le j \le p \ &\Rightarrow\ t_{n-j} - t' \le 0 \end{aligned}$$

then

$$\sum_{j=0}^{p} r_j^n G(t_{n-j} - t') = 0.$$

Accordingly, the above integral can be truncated at both its ends :

$$f_n = \sum_{j=0}^{p} r_j^n \int_{t_{n-p}}^{t_n} G(t_{n-j} - t')\, dB(t')$$

allowing us to conclude that f_n is a MA($p-1$) process, i.e. $m \geq p \Rightarrow \mathbb{E} f_n^* f_{n+m} = 0$, and also that f_n is independent of past observations, in the sense that $m \geq p \Rightarrow \mathbb{E} f_n^* y_{n-m} = 0$. This latter property allows f_n to be described as an innovation, or forward prediction, error. It is a standard result [6] that regular sampling of $Y(t)$ produces an ARMA($p, p-1$) process, which implies that after pth order linear prediction the residuals are a MA($p-1$) process; that this latter property is shared by our generalised prediction errors is further evidence that our definitions (2.2) and (2.3) are natural in the incoherent case as well.

2.4 Backward prediction errors

If an AR process is stable then its time-reverse is also stable with complex-conjugate AR coefficients. From (2.3) we have the following expression for the generalised backward prediction errors :

$$b_n = \sum_{j=0}^{p} s_j^{n*} y_{n+j}$$

$$\begin{bmatrix} \alpha_1^{(t_{n+p}-t_n)/\tau} & \alpha_1^{(t_{n+p}-t_{n-1})/\tau} & \cdots & 1 \\ \vdots & \vdots & \ddots & \vdots \\ \alpha_p^{(t_{n+p}-t_n)/\tau} & \alpha_p^{(t_{n+p}-t_{n-1})/\tau} & \cdots & 1 \end{bmatrix} \begin{bmatrix} s_0^n \\ s_1^n \\ \vdots \\ s_p^n \end{bmatrix} = \mathbf{0} \tag{2.8}$$

and the same normalisation as (2.5) is used to fix the size of $\mathbf{s}^n$.

2.5 Model fitting

Now suppose that we have an irregularly-sampled dataset with x_n the observations and t_n the times at which they are observed (we assume $m < n \Rightarrow t_m < t_n$). We choose a time interval τ characteristic of the problem; more specifically as the estimated spectrum

$$P(f) = \left| \prod_{i=1}^{p} 1 - \alpha_i e^{-2\pi i f \tau} \right|^{-2}$$

is periodic with "period" $1/\tau$, we should choose $\tau = 1/2B$ with B the highest frequency that is likely to be present in the data (otherwise aliasing will result).

Then, for a set of poles $\boldsymbol{\alpha}$ we construct the forward and backward prediction coefficients r_j^n and s_j^n and prediction errors f_n and b_n and thence the total forward and backward prediction error energy

$$E(\mathbf{x},\boldsymbol{\alpha}) = \sum_{n=p+1}^{N} ||f_n||^2 + ||b_{n-p}||^2.$$

We then minimise $E(\mathbf{x},\boldsymbol{\alpha})$ with respect to $\boldsymbol{\alpha}$.

Two points are worth noting. From the discussion of the coherent case, we have that E is zero precisely when the α_i are in the correct places ($\alpha_i = e^{\mathrm{i}\omega_i\tau}$); so of course minimising E gives the right answer. Also, when the sampling is regular, there is no difference between this and the Covariance method.

3 Filtering

3.1 The method

In this section we shall consider the following problem

A signal is given consisting of the sum of an AR process with known poles and a "message signal" about which nothing is known. Do the separation.

followed by its generalisation

Now do it when the sampling is irregular.

We remark at this point that we are going to characterise an AR process by means of its forward and backward prediction error energy, which (as seen in Section 2) is generalisable to the irregular case. To start with it is worthwhile to consider the maximum-likelihood separation of two signals $\mathbf{x}$ and $\mathbf{s}$ that have known covariance matrices $\mathbf{C_x}$ and $\mathbf{C_s}$. Their sum is given as $\mathbf{z}$. Assuming the two are independent and Gaussian we maximise

$$\exp -\tfrac{1}{2}\left\{\mathbf{x}^\dagger {\mathbf{C_x}}^{-1}\mathbf{x} + (\mathbf{z}-\mathbf{x})^\dagger {\mathbf{C_s}}^{-1}(\mathbf{z}-\mathbf{x})\right\}$$

with respect to $\mathbf{x}$; this of course can readily be solved to give the Wiener filter,

$$\hat{\mathbf{x}} = \left(\mathbf{I} + \mathbf{C_s}{\mathbf{C_x}}^{-1}\right)^{-1}\mathbf{z}.$$

Now suppose that $\mathbf{x}$ comes from an AR model whose parameters are known but whose driving noise variance is possibly unknown. Suppose also that nothing is known about $\mathbf{s}$. Then we must assume that $\mathbf{C_s}$ is a multiple of the identity matrix, so $\mathbf{C_x} = \sigma_\varepsilon^2 \left(\mathbf{F}^{f\dagger}\mathbf{F}^f\right)^{-1}$ and $\mathbf{C_s} = \sigma_s^2\mathbf{I}$, where $\mathbf{F}^f$ is the forward prediction error matrix,

$$\mathbf{F}^f = \begin{bmatrix} a_p & a_{p-1} & \dots & 1 & & & \\ & a_p & a_{p-1} & \dots & 1 & & \\ & & \dots & \dots & \dots & \dots \end{bmatrix}$$

(the vacant spaces are supposed to be filled with zeros). Then the solution is

$$\hat{\mathbf{x}} = \left(\mathbf{I} + \frac{\sigma_s^2}{\sigma_\varepsilon^2}\mathbf{F}^{f\dagger}\mathbf{F}^f\right)^{-1}\mathbf{z}.$$

We may wish to use the backward prediction error filter matrix $\mathbf{F}^b$ (defined in the obvious way) as well, and if we write

$$\mathbf{F} = \begin{pmatrix} \mathbf{F}^f \\ \mathbf{F}^b \end{pmatrix}$$

then the above solution can be modified to

$$\hat{\mathbf{x}} = \left(\mathbf{I} + \frac{\sigma_s^2}{2\sigma_\varepsilon^2}\mathbf{F}^{\dagger}\mathbf{F}\right)^{-1}\mathbf{z}. \tag{3.1}$$

Notice that the matrix to be inverted in (3.1) is banded, so the equation can be solved quickly using Cholesky factorisation in $O(Np^2)$ operations. The sticking-point is that we do not know the variance ratio $\sigma_s^2/\sigma_\varepsilon^2$, and an attempt to incorporate this into the maximisation seems destined to produce $\hat{\mathbf{x}} = \mathbf{z}$ or $\hat{\mathbf{x}} = \mathbf{0}$, neither of which is very helpful. We therefore try a different approach. Consider, for realisations $\mathbf{v}$ of the given AR process, the quantities

$$||\mathbf{F}\mathbf{v}||^2 \text{ and } ||\mathbf{F}\mathbf{v}||^2/||\mathbf{v}||^2.$$

The first has expected value $2(N-p)\sigma_\varepsilon^2$, which is known if σ_ε^2 is known. The expectation of the second is just $2C(\boldsymbol{\alpha})$ (q.v.). We then consider the following subtractive approach :

$$\text{Minimise } ||\mathbf{z} - \hat{\mathbf{x}}||^2 \text{ w.r.t. } \hat{\mathbf{x}} \text{ s.t. } \hat{\mathbf{x}} \in \mathcal{X}$$

in which $\mathcal{X}$ contains all realisations of the known AR process.

The set $\mathcal{X}$ is then *specified* by the equations

$$\mathcal{X} = \left\{\mathbf{v} \in \mathbb{C}^N \text{ s.t. } ||\mathbf{F}\mathbf{v}||^2 = 2\dot{E}\right\}$$

or

$$\mathcal{X} = \left\{\mathbf{v} \in \mathbb{C}^N \text{ s.t. } ||\mathbf{F}\mathbf{v}||^2/||\mathbf{v}||^2 = 2\dot{C}\right\}$$

in which the values $\dot{E}$ and $\dot{C}$ can be set by the user. When the driving noise variance of the AR process is known, the first definition is used, and $\dot{E} = (N-p)\sigma_\varepsilon^2$. When it is not, the second definition is used and $\dot{C} = C(\boldsymbol{\alpha})$.

Thus in each case there is a theoretical value of the associated constant. We now solve the minimisation by introducing a Lagrange multiplier in the usual way, to get

$$\hat{\mathbf{x}}_E = \left(\mathbf{I} + \mu \mathbf{F}^\dagger \mathbf{F}\right)^{-1} \mathbf{z}, \qquad \|\mathbf{F}\hat{\mathbf{x}}_E\|^2 = 2\dot{E} \tag{3.2}$$

in the first case and

$$\hat{\mathbf{x}}_C = \left((1 - 2\mu\dot{C})\mathbf{I} + \mu \mathbf{F}^\dagger \mathbf{F}\right)^{-1} \mathbf{z}, \qquad \|\mathbf{F}\hat{\mathbf{x}}_C\|^2 / \|\hat{\mathbf{x}}_C\|^2 = 2\dot{C} \tag{3.3}$$

in the second. As the correct value of μ is found by adjustment until the constraint $\mathbf{x} \in \mathcal{X}$ be obeyed, there are no "unknowns" in these equations. So the problem of the unknown variance ratio has disappeared. Also the first expression is the maximum-likelihood estimate,

$$\hat{\mathbf{x}}_E = \hat{\mathbf{x}}_{ML}, \qquad \widehat{\sigma_s^2/\sigma_\varepsilon^2} = \mu,$$

and the second method produces a scalar multiple of the maximum-likelihood estimate :

$$\hat{\mathbf{x}}_C = (1 - 2\mu\dot{C})^{-1} \hat{\mathbf{x}}_{ML}, \qquad \widehat{\sigma_s^2/\sigma_\varepsilon^2} = \frac{\mu}{1 - 2\mu\dot{C}}.$$

So the way in which we propose to solve the first boxed problem is to write

$$\hat{\mathbf{x}} = \hat{\mathbf{x}}_E \text{ or } (1 - 2\mu\dot{C})\hat{\mathbf{x}}_C \tag{3.4}$$

(see (3.2) and (3.3)) depending on whether we know σ_ε^2.

3.2 Computational considerations

To solve these optimisations in practice is straightforward. One simply has to find the value of μ so that (3.2) (or (3.3)) is obeyed. Let us take (3.2). For a chosen value of μ one performs the (quick) matrix inversion to obtain $\hat{\mathbf{x}}_E(\mu)$; then $g_E(\mu) := \|\mathbf{F}\mathbf{x}_E(\mu)\|^2$ is calculated. This process is repeated until $g_E(\mu) \approx 2\dot{E}$. It turns out that $g_E(\mu)$ is monotone-decreasing for $\mu \in [0, \infty)$ (the range of interest) and so the correct μ is unique and can be found by repeated bisection. Similar remarks apply to (3.3), where the range for μ is $[0, 1/2\dot{C}]$. We remark that *the optimisations have unique solutions.*

3.3 Irregular sampling

To answer the second boxed question is now very easy. We simply replace the AR coefficients in $\mathbf{F}$ with the generalised forward and backward prediction coefficients :

$$\mathbf{F} = \begin{bmatrix} r_p^{p+1} & r_{p-1}^{p+1} & \cdots & r_0^{p+1} & & & \\ & r_p^{p+2} & r_{p-1}^{p+2} & \cdots & r_0^{p+2} & & \\ & & \cdots & \cdots & \cdots & & \\ & & & r_p^N & r_{p-1}^N & \cdots & r_0^N \\ s_0^{1*} & s_1^{1*} & \cdots & s_p^{1*} & & & \\ & s_0^{2*} & s_1^{2*} & \cdots & s_p^{2*} & & \\ & & \cdots & \cdots & \cdots & & \\ & & & s_0^{N-p*} & s_1^{N-p*} & \cdots & s_p^{N-p*} \end{bmatrix}.$$

The normalisation described in Section 2 means that the rows of $\mathbf{F}$ are defined only up to multiplication by scalars of modulus 1. This is not a problem because only $\mathbf{F}^\dagger\mathbf{F}$ is used, and that *is* well-defined. Again, the solution is unique.

4 Results and Discussion

To illustrate the spectral estimation and filtering procedure we have considered the following example. To obtain a broad-band process suitable as a test example for the modelling work of Section 2 we have multiplied a sinusoid of frequency 0.21(Hz) — an arbitrary choice — by a continuous-time waveform from the Lorenz attractor. Such a signal is continuous-time, continuous-spectrum and "stationary". A section of the waveform is shown below. We shall later call this the "interferer".

The first thing we did was to sample this signal in various ways and fit an AR(2) model using the techniques described in Section 2. Several sampling schemes were used, two of which were (a) regular sampling, rate 1Hz, and (b) additive-random sampling with intersample times drawn from the rectangular distribution on the interval [0.5,1.5]s. These two schemes are used later for testing the filtering method. The results were almost identical even for short data sets (for example 64 points) showing that the estimated *underlying* spectrum was independent of the choice of sampling scheme.

Figure 2 was produced as follows. A sinusoid of amplitude 1 and frequency 0.61Hz was added and the signal regularly sampled at rate 1Hz. Then the sampled set was DFTd. One sees the two spectral features very clearly ... but the sinusoid is at the "wrong" frequency. This, of course, is a straightforward case of aliasing.

Using the poles obtained from the modelling the coherence $C(\boldsymbol{\alpha})$ was found (actually 0.043) and the filtering scheme of Section 3 was used to remove the interferer. The residue $\hat{\mathbf{x}}$ was then DFTd (Figure 3).

Good separation has been achieved; however it was not really necessary because both spectral features were perfectly obvious in Figure 2. But the position for irregular sampling is very different. When the sampling was irregular the unfiltered and filtered signals were found to have the following DFTs :

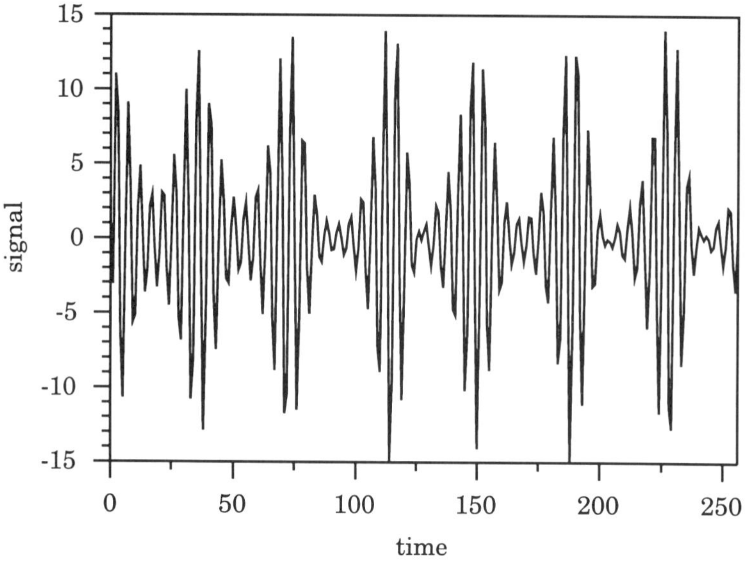

Figure 1. Time series of "broadened sinusoid"

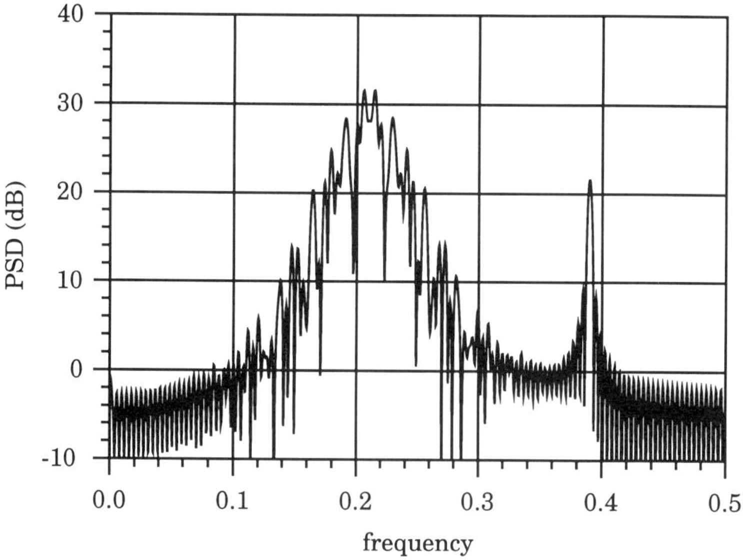

Figure 2. Spectrum of unfiltered signal (regular sampling)

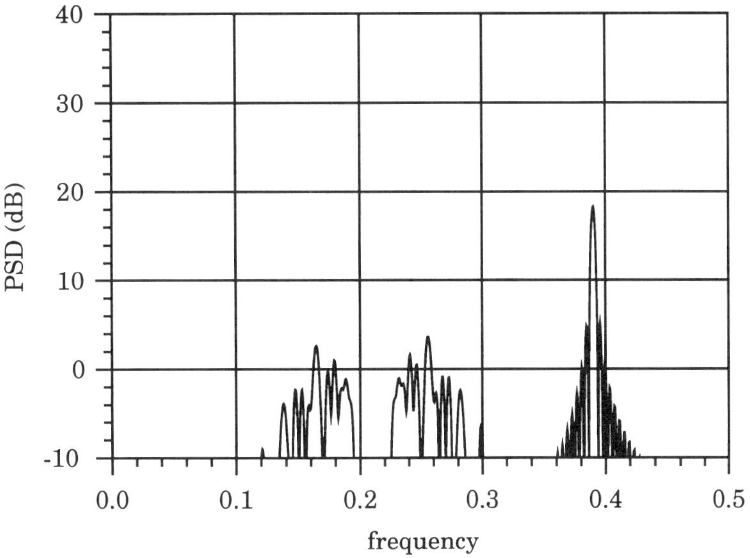

Figure 3. Spectrum of filtered signal (regular sampling)

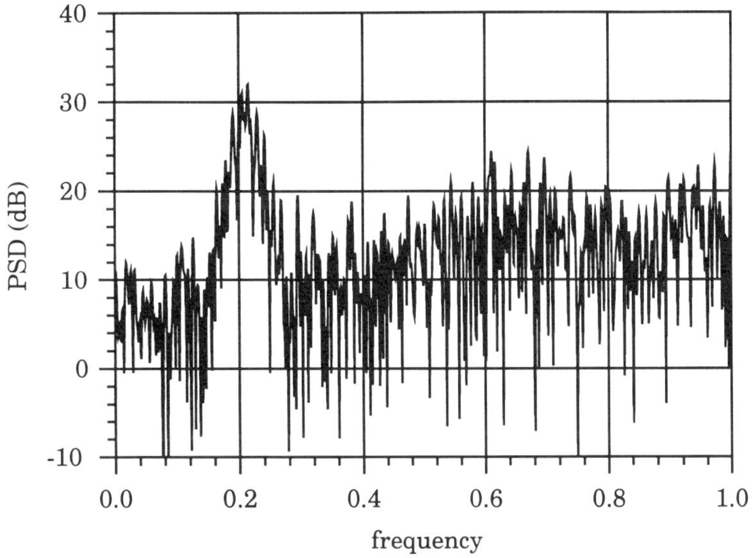

Figure 4. Spectrum of unfiltered signal (irregular sampling)

Figures 4 and 5 show very nicely the difficulties and the benefits of irregular sampling. In Figure 4 the dynamic range is so reduced that one can only see the strongest component; in Figure 5 the "interference" has been removed, and now the sinusoid now appears strongly and *not as an alias.*

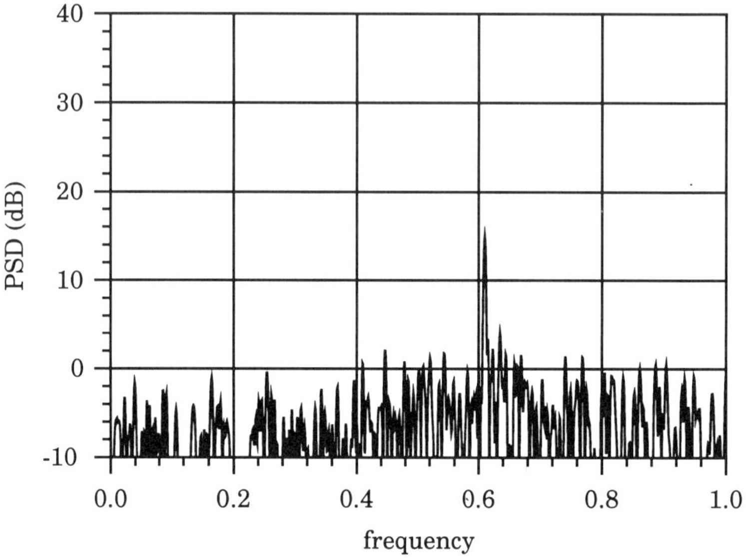

Figure 5. Spectrum of filtered signal (irregular sampling)

Acknowledgements

R.J. Martin would like to thank the Royal Commission for the Exhibition of 1851 for their generous financial support, and Dr. Jaroslav Stark of University College London for helpful discussions.

Bibliography

1. Marple, S. L. Jr. (1987). Digital spectral analysis with applications. Prentice Hall, NJ.

2. Newman, W. I. (1977,1979). Extension to the maximum entropy method. *IEEE Trans. Info. Theory* **23** 89–93 and **25** 705–8.

3. Dowla F. U., McClellan J. H. (1981). MEM spectral analysis for nonuniformly sampled signals. *Proc. IEEE ICASSP'81.* 79–85.

4. Lii, K.-S., Masry, E. (1992). Model fitting for continuous-time stationary processes from discrete-time data. *J. Multivariate Analysis* **41** 56–79.

5. Shumway, R. H. (1983). Some applications of the EM algorithm. *Time series analysis of irregularly-observed data*, Springer, NY. 290–324.

6. Jones R. H. (1983). Fitting multivariate models to unequally-spaced data. *Time series analysis of irregularly-observed data*, Springer, NY. 158–88.

7. Marvasti F. A. (1987). A unified approach to zero-crossings and nonuniform sampling. Nonuniform publications, Oak Park, Illinois.

8. Bilinskis, I., Mikelsons, A. (1992). Randomized signal processing. Prentice Hall, NY.

9. Marks R. J. II. (1992). Advances in Shannon sampling and interpolation theory. Springer, NY.

10. Shapiro, H. S., Silverman, R. A. (1960). Alias-free sampling of random noise. *SIAM J. Appl. Math.* **8** 245–58.

11. Legg J. A. (1997). Synthetic Aperture Radar using Non-Uniform Sampling. PhD thesis, DSTO / Univ. Adelaide.

12. Bilinskis, I. (1995). Deliberately Pseudorandomized DSP — advantages and limitations. *Proc. SAMPTA'95, Jurmala (Riga), Latvia.* 201–7.

13. Brillinger D. R. (1972). The spectral analysis of stationary interval functions. *Proc. Sixth Berkeley Symp. Probab. Statist.* 483–513.

14. Roberts, J. B., Gaster, M. (1980). On the estimation of spectra from randomly sampled signals. *Proc. Roy. Soc. London A* **371** 238–58.

Decaying Exponentials and Signal Decomposition

Franklin T. Luk* **and Sanzheng Qiao****

**Department of Computer Science, Rensselaer Polytechnic Institute, Troy, New York, USA, and **Department of Computer Science and Systems, McMaster University, Ontario, Canada*

Abstract

We consider the problem of decomposing a sequence of signals into a small set of decaying exponentials. It is known that the problem can be solved via calculating the generalized eigenvalues of a pair of closely related Hankel matrices. One of our new results in this paper is the application of a rank-revealing decomposition in place of the more expensive singular value decomposition. Furthermore, we show how to efficiently update the solution when new signals are added. Finally, we discuss the case where the two Hankel matrices are square and we show how to exploit complex symmetry to reduce the work by approximately one-half.

1 Introduction

We consider a signal decomposition problem: Given a sequence $\{s_k\}$ of signals, we decompose them into a small set of decaying complex exponentials z_i $(i = 1, ..., r)$:

$$s_k = \sum_{i=1}^{r} c_i z_i^{k-1}, \tag{1.1}$$

assuming $c_i \neq 0$ for all i and z_i distinct. For example, a signal s_k composed of sinusoids can be expressed by

$$s_k = \sum_{i=1}^{r} a_i e^{j((k-1)\omega_i + \phi_i)}$$

where a_i are amplitudes, ω_i frequencies, and ϕ_i phases. Setting

$$c_i = a_i e^{j\phi_i} \quad \text{and} \quad z_i = e^{j\omega_i},$$

we get the expression (1.1) for s_k. This problem arises from spectral analysis [2]. The decomposition (1.1) is determined by the integer r, the complex coefficients c_i, and the complex exponentials z_i $(i = 1, ..., r)$. We will concentrate on the determination of r and z_i, since c_i can be readily found by the linear least squares data fitting technique. Specifically, let $s = (s_1, ..., s_m)^T$ be the signal vector and

$$V = \begin{pmatrix} e^T \\ e^T D_z \\ \vdots \\ e^T D_z^{m-1} \end{pmatrix}$$

where $e = (1, 1, ..., 1)^T$ is an r-by-1 vector and D_z denotes an r-by-r diagonal matrix $\mathrm{diag}(z_1, ..., z_r)$, then we find the coefficient vector $c = (c_1, ..., c_r)^T$ by minimizing

$$\|s - Vc\|_2.$$

In this paper, we first show an association between the problem of finding r exponentials z_i and the problem of solving the generalized eigenproblem in the following section. Then we present an algorithm for computing the exponentials in Section 3 and an updating scheme in Section 4. In Section 5, we present another algorithm by exploiting the symmetry of the data matrix. This method can lead to an algorithm for the generalized eigenproblem of complex and symmetric matrices.

2 Background

Suppose that we are given m signals: $s_1, ..., s_m$. It is reasonable to assume that $m > 2r$, since we have $2r + 1$ unknowns, namely r, $c_1, ..., c_r$, $z_1, ..., z_r$, to be determined. We arrange these m signals into a Hankel matrix

$$H = \begin{pmatrix} s_1 & s_2 & \cdots & s_n \\ s_2 & s_3 & \cdots & s_{n+1} \\ \vdots & \vdots & \cdots & \vdots \\ s_k & s_{k+1} & \cdots & s_m \end{pmatrix} \tag{2.1}$$

where $k = m - n + 1 > n \geq r$. The first inequality $k > n$ means that H has more columns than rows. The second inequality $n \geq r$ implies that r can be the rank of H. In the following, we will show that r is indeed the rank. Using (1.1), we can verify the following decomposition

$$H = \begin{pmatrix} e^T \\ e^T D_z \\ \vdots \\ e^T D_z^{k-1} \end{pmatrix} D_c(e \; D_z e \; \cdots \; D_z^{n-1} e)$$

where e and D_z are defined previously and $D_c = \mathrm{diag}(c_1, ..., c_r)$. This decomposition shows that r is the rank of H. Partitioning

$$H = \begin{pmatrix} g^T \\ H_1 \end{pmatrix} = \begin{pmatrix} H_2 \\ h^T \end{pmatrix} \tag{2.2}$$

where g and h are n-by-1 vectors and denoting

$$V_1 = \begin{pmatrix} e^T \\ e^T D_z \\ \vdots \\ e^T D_z^{k-2} \end{pmatrix} \quad \text{and} \quad V_2 = \begin{pmatrix} e^T \\ e^T D_z \\ \vdots \\ e^T D_z^{n-1} \end{pmatrix},$$

we get a pair of decompositions:

$$H_1 = V_1 D_c D_z V_2^T \quad \text{and} \quad H_2 = V_1 D_c V_2^T.$$

Note that V_1 and V_2 are Vandermonde matrices. Thus, both V_1 and V_2 are of full column rank, since z_i are distinct. This implies that H, H_1 and H_2 all have rank r. Therefore, r can be determined by applying any rank-revealing techniques to H, H_1, or H_2.

Now we consider the problem of computing the exponentials z_i. We will show an association between the exponentials and the generalized eigenvalues of the matrix pair (H_1, H_2). Let

$$V_1 = Q_1 R_1 \quad \text{and} \quad V_2 = Q_2 R_2$$

be the QR decompositions, where $Q_1 \in C^{(k-1)\times r}$ and $Q_2 \in C^{n\times r}$ have orthonormal columns, i.e., $Q_1^T Q_1 = Q_2^T Q_2 = I_r$, and R_1 and R_2 (r-by-r) are upper triangular and nonsingular. Then we have the following simultaneous decompositions:

$$H_1 = Q_1 R_1 D_c D_z R_2^T Q_2^T \quad \text{and} \quad H_2 = Q_1 R_1 D_c R_2^T Q_2^T.$$

Consequently,

$$H_1 Q_2 R_2^{-T} = H_2 Q_2 R_2^{-T} D_z. \tag{2.3}$$

Recall that if a scalar λ satisfies

$$Ax = \lambda Bx,$$

for some $x \neq 0$, then it is called the generalized eigenvalue of the pair of matrices A and B and x is the associated eigenvector [1]. It then follows from (2.3) that z_i are the r nonzero generalized eigenvalues of the pair (H_1, H_2) and the columns of $Q_2 R_2^{-T}$ are the corresponding eigenvectors. Thus, from the above derivation, we have transformed the problem of finding the exponentials z_i into the problem of computing the nonzero generalized eigenvalues of the matrix pair (H_1, H_2).

So, given H_1 and H_2, how can we compute the generalized eigenvalues of (H_1, H_2)? A basic tenet in matrix computation is that an original problem is first transformed into a simpler problem by decomposing the original matrices into simpler matrices. In our case, we first decompose H_1 and H_2, say,

$$H_1 = FAG \quad \text{and} \quad H_2 = FBG$$

such that F is of full column rank, G is nonsingular, and A and B have simpler structure than H_1 and H_2. Then the generalized eigenproblem of H_1 and H_2 is equivalent to that of A and B. In general, to compute the generalized eigenvalues of (A, B), we solve for a nonsingular Y and an upper triangular T in the matrix equation:

$$AY = BYT. \tag{2.4}$$

Then the diagonal elements of T are the desired generalized eigenvalues. Note that the equation (2.4) may not have a solution. For example, if

$$A = \begin{pmatrix} 2 \\ 3 \end{pmatrix} \quad \text{and} \quad B = \begin{pmatrix} 1 \\ 2 \end{pmatrix},$$

then there do not exist Y and T (scalars) satisfying (2.4).

In conclusion, in the absence of noise, this signal decomposition problem can be solved by techniques for revealing the rank of H, H_1, or H_2 and finding the nonzero generalized eigenvalues of (H_1, H_2). The singular value decomposition (SVD) provides a high-resolution method for solving this problem [2, 4]. However, the SVD method is unsuitable for adaptive processing and makes no use of the symmetric structure of the Hankel matrix H (2.1). In the following section, we present a particular pair of decompositions $H_1 = FAG$ and $H_2 = FBG$ which leads to an adaptive algorithm for computing r and the exponentials z_i. In Section 5, we show another particular pair of decompositions $H_1 = FAG$, and $H_2 = FBG$ which exploits the symmetry of the Hankel matrix H.

3 Incorporating the URV decomposition

The first method for finding r and the exponentials z_i incorporates the rank-revealing URV decomposition [5]. As shown in the previous section, in the absence of noise, the Hankel matrix H in (2.1) has rank r, so any rank-revealing method can be used to find r. Suppose that, the following URV decomposition reveals the rank r

$$H = U \begin{pmatrix} R & 0 \\ 0 & 0 \end{pmatrix} V^H \tag{3.1}$$

where U (k-by-n) has orthonormal columns, i.e., $U^H U = I_n$, R (r-by-r) is upper triangular and nonsingular, and V (n-by-n) is unitary. As described in the previous section, to find the complex exponentials z_i, we first decompose H_1 and H_2. Partitioning

$$U = \begin{pmatrix} v^T \\ U_1 \end{pmatrix} = \begin{pmatrix} U_2 \\ u^T \end{pmatrix}$$

where u and v are n-by-1 vectors, from the partition (2.2) and the decomposition (3.1), we get

$$H_1 = U_1 \begin{pmatrix} R & 0 \\ 0 & 0 \end{pmatrix} V^H \quad \text{and} \quad H_2 = U_2 \begin{pmatrix} R & 0 \\ 0 & 0 \end{pmatrix} V^H.$$

Furthermore, we perform the QR decomposition

$$U_2 = Q \begin{pmatrix} S \\ 0 \end{pmatrix}$$

where Q (($k-1$)-by-($k-1$)) is unitary and S (n-by-n) is upper triangular. If we partition

$$S = \begin{pmatrix} S_{11} & S_{12} \\ 0 & S_{22} \end{pmatrix}$$

where S_{11} is r-by-r, then we have the simultaneous decompositions:

$$H_1 = QQ^H U_1 \begin{pmatrix} I_r & 0 \\ 0 & 0 \end{pmatrix} \begin{pmatrix} R & 0 \\ 0 & I_{n-r} \end{pmatrix} V^H$$

and

$$H_2 = Q \begin{pmatrix} S_{11} & S_{12} \\ 0 & S_{22} \\ 0 & 0 \end{pmatrix} \begin{pmatrix} R & 0 \\ 0 & 0 \end{pmatrix} V^H = Q \begin{pmatrix} S_{11} & 0 \\ 0 & 0 \end{pmatrix} \begin{pmatrix} R & 0 \\ 0 & I_{n-r} \end{pmatrix} V^H.$$

Thus the problem of finding the complex exponentials z_i, i.e., the generalized eigenvalues of H_1 and H_2, is equivalent to the problem of finding the generalized eigenvalues of

$$A = Q^H U_1 \begin{pmatrix} I_r & 0 \\ 0 & 0 \end{pmatrix} \quad \text{and} \quad B = \begin{pmatrix} S_{11} & 0 \\ 0 & 0 \end{pmatrix}. \tag{3.2}$$

We can show that any nonzero generalized eigenvalue of (A, B) is an eigenvalue of the r-by-r matrix

$$X = S_{11}^{-1}(I_r \ 0)Q^H U_1 \begin{pmatrix} I_r \\ 0 \end{pmatrix}. \tag{3.3}$$

Suppose that λ is a nonzero generalized eigenvalue of (A, B), i.e.,

$$Az = \lambda Bz$$

for some vector z. Partitioning

$$z = \begin{pmatrix} z_1 \\ z_2 \end{pmatrix},$$

where z_1 is r-by-1, from (3.2), we get

$$Q^H U_1 \begin{pmatrix} z_1 \\ 0 \end{pmatrix} = \lambda \begin{pmatrix} S_{11} z_1 \\ 0 \end{pmatrix}.$$

It follows that

$$(I_r \ 0)Q^H U_1 \begin{pmatrix} I_r \\ 0 \end{pmatrix} z_1 = \lambda S_{11} z_1.$$

This shows that λ is an eigenvalue of X in (3.3). Denoting

$$\hat{U}_1 = U_1 \begin{pmatrix} I_r \\ 0 \end{pmatrix} \quad \text{and} \quad \hat{Q} = Q \begin{pmatrix} I_r \\ 0 \end{pmatrix},$$

which are composed of the first r columns of U_1 and Q, respectively, we have

$$X = S_{11}^{-1} \hat{Q}^H \hat{U}_1. \tag{3.4}$$

Thus, we have reduced the generalized eigenproblem of (H_1, H_2) to the eigenproblem of the r-by-r matrix X.

Although the above discussion assumes the absence of noise, it is readily amendable to the case when noise is present. In summary, we present the following algorithm for computing r and the exponentials.

Algorithm 1. *Given the Hankel matrix H composed of the signals, this algorithm computes the r exponentials.*

1. *Perform the URVD:*
$$H = U \begin{pmatrix} R_{11} & E_1 \\ 0 & E_2 \end{pmatrix} V^H$$
and determine the rank r;

2. *Partition*
$$U = \begin{pmatrix} v^T \\ U_1 \end{pmatrix} = \begin{pmatrix} U_2 \\ u^T \end{pmatrix};$$

3. *Compute the QRD*
$$U_2 = Q \begin{pmatrix} S \\ 0 \end{pmatrix};$$

4. *Multiply $\hat{Q}^H \hat{U}_1$, where $\hat{Q}$ and $\hat{U}_1$ consist of the first r columns of Q and U_1, respectively;*

5. *The exponentials are the eigenvalues of the r-by-r matrix $X = S_{11}^{-1}(\hat{Q}^H \hat{U}_1)$, where S_{11} is the rth order leading principal submatrix of S.*

4 Adding a row

When a new row is added to the Hankel matrix H, its URV decomposition in step 1 of Algorithm 1 can be efficiently updated, see [5] for details. However, whenever a row is added to H, the U matrix in the URV decomposition (3.1) also grows by one row. As rows are added to H, the matrix U grows indefinitely. As shown by steps 3 and 4 of Algorithm 1, U_1 and U_2, the submatrices of U, are needed for computing the exponentials. Thus, it is undesirable to store the matrices U_1 and U_2. Fortunately, in this application, we are interested in the eigenvalues of X in (3.4), a matrix given by S_{11} and $\hat{Q}^H \hat{U}_1$. The matrix S_{11} is the

rth order leading principal submatrix of S, which is the upper triangular factor in the QR decomposition of U_2 (see Algorithm 1, step 3). Moreover, the product $\hat{Q}^H\hat{U}_1$ is the rth order leading principal submatrix of $Q^H U_1$ (Algorithm 1, step 4), where Q is the unitary factor in the QR decomposition of U_2 (Algorithm 1, step 3). Thus, to update X, it suffices to update S and $Q^H U_1$, both $n \times n$ matrices. We note that it is necessary to update the $n \times n$ matrices S and $Q^H U_1$ instead of the $r \times r$ matrices S_{11} and $\hat{Q}^H\hat{U}_1$, since the rank r can potentially increase up to n as rows are added to H. So, we need all information in S and $Q^H U_1$. In the following we propose schemes for updating S and $Q^H U_1$ without storing U_1 and U_2.

The building blocks in updating the URV decomposition are the application of right- and left-rotations to the triangular factor R in the URV decomposition. When a rotation is applied to the right of R, it is simultaneously applied to the right of the V factor. This is irrelevant to our underlying matrices S and $Q^H U_1$. So, we concentrate our discussion on the application of left-rotations to R. We show how to update R, S, and $Q^H U_1$ without explicitly computing U or U_1 or U_2.

Suppose that a row vector h^T is added to H, we write

$$\hat{H} \stackrel{\text{def}}{=} \begin{pmatrix} H \\ h^T \end{pmatrix} = \begin{pmatrix} U & 0 \\ 0 & 1 \end{pmatrix} \begin{pmatrix} R \\ h^T V \end{pmatrix} V^H \stackrel{\text{def}}{=} \hat{U}\hat{R}\hat{V}^H.$$

Partitioning

$$\hat{U} = \begin{pmatrix} \hat{v}^T \\ \hat{U}_1 \end{pmatrix} = \begin{pmatrix} \hat{U}_2 \\ \hat{u}^T \end{pmatrix} \tag{4.1}$$

and recalling that

$$U = \begin{pmatrix} v^T \\ U_1 \end{pmatrix} = \begin{pmatrix} U_2 \\ u^T \end{pmatrix},$$

we get

$$\hat{U}_1 = \begin{pmatrix} U_1 & 0 \\ 0 & 1 \end{pmatrix}, \quad \hat{U}_2 = \begin{pmatrix} U_2 & 0 \\ u^T & 0 \end{pmatrix}, \quad \text{and} \quad \hat{u} = \begin{pmatrix} 0 \\ \vdots \\ 0 \\ 1 \end{pmatrix}.$$

Consequently, from the step 3 of Algorithm 1, we initialize the QR decomposition

$$\hat{U}_2 = \begin{pmatrix} Q & 0 \\ 0 & 1 \end{pmatrix} \begin{pmatrix} S & 0 \\ 0 & 0 \\ u^T & 0 \end{pmatrix} \stackrel{\text{def}}{=} \hat{Q}\hat{S}. \tag{4.2}$$

Thus, initializing

$$\hat{V} = V, \quad \hat{R} = \begin{pmatrix} R \\ h^T V \end{pmatrix}, \quad \hat{S} = \begin{pmatrix} S & 0 \\ 0 & 0 \\ u^T & 0 \end{pmatrix}, \quad \hat{u} = \begin{pmatrix} 0 \\ \vdots \\ 0 \\ 1 \end{pmatrix}, \tag{4.3}$$

from (4.1) we get the decompositions

$$\hat{H} = \begin{pmatrix} \hat{v}^T \\ \hat{U}_1 \end{pmatrix} \hat{R}\hat{V}^H = \begin{pmatrix} \hat{Q}\hat{S} \\ \hat{u}^T \end{pmatrix} \hat{R}\hat{V}^H. \tag{4.4}$$

In our updating scheme, instead of storing U_1 and U_2, we keep $Q^H U_1$, which has a fixed order n. For updating $Q^H U_1$, we initialize

$$\hat{Q}^H \hat{U}_1 = \begin{pmatrix} Q^H U_1 & 0 \\ 0 & 1 \end{pmatrix}. \tag{4.5}$$

As stated before, in updating the URV decomposition, when we apply a right-rotation to $\hat{R}$, we simply apply this rotation to the right to $\hat{V}$. When we apply a left-rotation P to $\hat{R}$, from (4.4) and (4.5), we apply P^H to the right of $\hat{S}$, $\hat{u}^T$, and $\hat{U}_1$. Since we keep $\hat{Q}^H\hat{U}_1$ instead of $\hat{U}_1$, we update $\hat{Q}^H\hat{U}_1$ by postmultiplying it with P^H. Similarly, in restoring the triangular structure of $\hat{S}$, when we apply a rotation P to the left of $\hat{S}$, from (4.2), we should apply P^H to the right of $\hat{Q}$. Since we store $\hat{Q}^H\hat{U}_1$ instead of $\hat{Q}$, we apply P^H to the left of $\hat{Q}^H\hat{U}_1$.

Algorithm 2. *Suppose that we have the following variables available*

R: the triangular factor in the URV decomposition,
S: the triangular factor in the QR decomposition of U_2,
u^T: the bottom row of the unitary factor U in the URVD,
$Q^H U_1$: see steps 2 and 3 of Algorithm 1 for the definitions of U_1 and Q,

and a new row h^T is added to H, the following steps show how to update these variables and compute the exponentials.

1. *Initialization:*

$$\hat{V} = V; \quad \hat{R} = \begin{pmatrix} R \\ h^T V \end{pmatrix}; \quad \hat{S} = \begin{pmatrix} S & 0 \\ 0 & 0 \\ u^T & 0 \end{pmatrix}; \quad \hat{u} = \begin{pmatrix} 0 \\ \vdots \\ 0 \\ 1 \end{pmatrix};$$

and

$$\hat{Q}^H \hat{U}_1 = \begin{pmatrix} Q^H U_1 & 0 \\ 0 & 1 \end{pmatrix};$$

2. *In updating the URV decomposition,*

   (a) *When $\hat{R}$ is post-multiplied by a rotation P, i.e., $\hat{R} \leftarrow \hat{R}P$, update $\hat{V} \leftarrow \hat{V}P$;*

   (b) *When $\hat{R}$ is pre-multiplied by a rotation P, i.e., $\hat{R} \leftarrow P\hat{R}$, update*

$$\hat{S} \leftarrow \hat{S}P^H, \quad \hat{u}^T \leftarrow \hat{u}^T P^H, \quad \hat{Q}^H\hat{U}_1 \leftarrow \hat{Q}^H\hat{U}_1 P^H;$$

(c) In triangularizing $\hat{S}$, when a rotation P is applied to the left of $\hat{S}$, i.e., $\hat{S} \leftarrow P\hat{S}$, update $\hat{Q}^H\hat{U}_1 \leftarrow P(\hat{Q}^H\hat{U}_1)$;

3. After the URV decomposition is updated, the new rank $\hat{r}$ is determined, and $\hat{S}$ is triangularized, the exponentials are the eigenvalues of $\hat{X} = \hat{S}_{11}^{-1}(\tilde{Q}^H\tilde{U}_1)$ where $\hat{S}_{11}$ and $\tilde{Q}^H\tilde{U}_1$ are the $\hat{r}$th order leading principal submatrices of $\hat{S}$ and $\hat{Q}^H\hat{U}_1$, respectively.

5 Exploiting symmetry

When $m = 2n$, the Hankel matrices

$$H_1 = \begin{pmatrix} s_2 & s_3 & \cdots & s_{n+1} \\ s_3 & s_4 & \cdots & s_{n+2} \\ \vdots & \vdots & \cdots & \vdots \\ s_{n+1} & s_{k+2} & \cdots & s_m \end{pmatrix} \quad \text{and} \quad H_2 = \begin{pmatrix} s_1 & s_2 & \cdots & s_n \\ s_2 & s_3 & \cdots & s_{n+1} \\ \vdots & \vdots & \cdots & \vdots \\ s_n & s_{n+1} & \cdots & s_{m-1} \end{pmatrix}$$

are square and symmetric. While the algorithm presented in the previous section is adaptive, it makes no use of the symmetry. In this section, we propose another algorithm which exploits the symmetry.

In [3], we propose a rank-revealing factorization of symmetric matrices. Here, we apply this technique to H_1 and H_2. To exploit the symmetry, we first decompose a symmetric matrix into a product of a lower triangular matrix and its transpose. Assuming the signals $s_i \neq 0$ for all i and denoting

$$\hat{R}_1 = D_1^{-1/2}\text{triu}(H_1), \quad \text{and} \quad \tilde{R}_1 = \hat{R}_1 - D_1^{1/2}$$

where D_1 is diagonal matrix whose diagonal equals the diagonal of H_1 and $\text{triu}(H_1)$ is the upper triangular part of H_1. Thus $\tilde{R}_1$ is the strict upper triangular part of $\hat{R}_1$. It can be verified that

$$H_1 = \hat{R}_1^T\hat{R}_1 - \tilde{R}_1^T\tilde{R}_1 = \begin{pmatrix} \hat{R}_1 \\ j\tilde{R}_1 \end{pmatrix}^T \begin{pmatrix} \hat{R}_1 \\ j\tilde{R}_1 \end{pmatrix}.$$

Then we find a (complex) orthogonal transformation P such that $P^TP = I$ and

$$P\begin{pmatrix} \hat{R}_1 \\ j\tilde{R}_1 \end{pmatrix} = \begin{pmatrix} R_1 \\ 0 \end{pmatrix}$$

is upper triangularized. Thus we have the symmetric decomposition

$$H_1 = R_1^T R_1$$

where R_1 is upper triangular. Similarly, we compute the decomposition

$$H_2 = R_2^T R_2.$$

The construction of the (complex) orthogonal transformations which introduce zeros into vectors is presented in Appendix.

Next we compute the rank-revealing factorization [3]

$$R_2 = U_2 S_2 V^T \tag{5.1}$$

where U_2, $V \in C^{n\times n}$ are orthogonal, i.e., $U_2^T U_2 = V^T V = I_n$. Thus $S_2^T S_2$ and H_2 have the same eigenvalues. In the absence of noise, the upper triangular matrix S_2 satisfies the rank-revealing condition

$$S_2^T S_2 = \begin{pmatrix} \hat{S}_2^T \hat{S}_2 & 0 \\ 0 & 0 \end{pmatrix}$$

where $\hat{S}_2$ is the rth order leading principal submatrix of S_2. Consequently, we have the symmetric rank-revealing factorization

$$H_2 = V \begin{pmatrix} \hat{S}_2^T \hat{S}_2 & 0 \\ 0 & 0 \end{pmatrix} V^T.$$

This shows that H_2 and $\hat{S}_2^T \hat{S}_2$ have the same set of non-zero eigenvalues, since $V^T = V^{-1}$. Simultaneously, for R_1, we compute the decomposition

$$R_1 = U_1 S_1 V^T \tag{5.2}$$

where $U_1 \in C^{n\times n}$ is orthogonal and S_1 upper triangular. Note that the decompositions (5.1) and (5.2) share the same matrix V. Now the problem of finding the generalized eigenvalues of H_1 and H_2 is equivalent to that of finding the generalized eigenvalues of

$$A = S_1^T S_1 \quad \text{and} \quad B = S_2^T S_2.$$

Similar to (3.3), these generalized eigenvalues are the eigenvalues of the r-by-r matrix

$$(\hat{S}_2^T \hat{S}_2)^{-1} (\hat{S}_1^T \hat{S}_1) \qquad \text{or} \qquad (\hat{S}_1 \hat{S}_2^{-1})^T (\hat{S}_1 \hat{S}_2^{-1})$$

where $\hat{S}_1$ and $\hat{S}_2$ are the rth order leading principal submatrices of S_1 and S_2, respectively. In summary, we present the following algorithm for computing the r exponentials.

Algorithm 3. *Given Hankel matrices H_1 and H_2 where $m = 2n$, this algorithm computes the r exponentials.*

1. *Decompose $H_1 = \hat{R}_1^T \hat{R}_1 - \tilde{R}_1^T \tilde{R}_1$, where $\hat{R}_1 = D_1^{-1/2}\mathrm{triu}(H_1)$, $\tilde{R}_1 = \hat{R}_1 - D_1^{1/2}$, and D_1 is the diagonal part of H_1;*

2. *Obtain the symmetric factorization $H_1 = R_1^T R_1$ through an orthogonal transformation P such that*

$$P^T \begin{pmatrix} \hat{R}_1 \\ i\tilde{R}_1 \end{pmatrix} = \begin{pmatrix} R_1 \\ 0 \end{pmatrix}$$

is upper triangularized;

3. *Similarly, factorize $H_2 = R_2^T R_2$, where R_2 is $n \times n$ and upper triangular;*

4. *Compute the rank-revealing factorization*

$$R_2 = U_2 S_2 V^T$$

where U_2, and V are orthogonal and S_2 is upper triangular and satisfies the rank-revealing condition; simultaneously, decompose

$$R_1 = U_1 S_1 V^T$$

where U_1 is orthogonal and S_1 upper triangular;

5. *Let r be the rank revealed in the previous step and $\hat{S}_1$ and $\hat{S}_2$ are the rth order leading principal submatrices of S_1 and S_2, respectively, then the exponents are the eigenvalues of the r-by-r symmetric matrix $(\hat{S}_1\hat{S}_2^{-1})^T(\hat{S}_1\hat{S}_2^{-1})$.*

Finally, we remark that in step 5, we can further exploit the symmetric structure in $(\hat{S}_1\hat{S}_2^{-1})^T(\hat{S}_1\hat{S}_2^{-1})$ by computing the following decomposition

$$\hat{S}_1\hat{S}_2^{-1} = \hat{U}\hat{\Sigma}\hat{V}^T$$

where $\hat{U}$, $\hat{V} \in C^{n\times n}$ are orthogonal, i.e., $\hat{U}^T\hat{U} = \hat{V}^T\hat{V} = I_r$ and $\hat{\Sigma}$ is diagonal. Then the diagonal elements of $\hat{\Sigma}^2$ are the r eigenvalues of $(\hat{S}_1\hat{S}_2^{-1})^T(\hat{S}_1\hat{S}_2^{-1})$. The whole procedure is analogous to the computation of the singular value decomposition $S = U\Sigma V^H$ to obtain the eigenvalues of the Hermitian and positive definite matrix $S^H S$. The major difference is that here we use the complex orthogonal transformations whereas in the SVD we use unitary transformations. The detailed discussion of the procedure is out of the scope of this paper.

6 Appendix

Algorithm 4. *Complex symmetric reflector.*
Given x such that $x^T x \neq 0$, compute a symmetric reflector $H_S = I - \delta_S^{-1} u u^T$ so that $H_S x = -\alpha e_1$ and $H_S^T H_S = I$.

$\alpha = \sqrt{x^T x}$;
$u = x + \alpha e_1$;
$\delta_S = \alpha u_1$.

Algorithm 5. *Complex symmetric rotation.*
Given a, $b \in C$ such that $a^2 + b^2 \neq 0$, compute the c and s in a symmetric rotation

$$G_S = \begin{pmatrix} c & s \\ s & -c \end{pmatrix}$$

so that

$$G_S \begin{pmatrix} a \\ b \end{pmatrix} = \begin{pmatrix} \alpha \\ 0 \end{pmatrix} \quad \text{and} \quad G_S^T G_S = I_2;$$

if $(|a| > |b|)$
  $\tau = b/a;\ c = 1/\sqrt{1+\tau^2};\ s = c\tau;$
else
  $\tau = a/b;\ s = 1/\sqrt{1+\tau^2};\ c = s\tau.$

Bibliography

1. Golub, G.H. and C.F. Van Loan (1989), *Matrix Computations*, (2nd edn). The Johns Hopkins University Press.

2. Kung, S.Y., K.S. Arun, and D.V. Bhaskar Rao (1983), State-space and singular-value decomposition-based approximation methods for the harmonic retrieval problem, *J. Opt. Soc. Am.*, *73/12*, 1799–811.

3. F.T. Luk and S. Qiao (1996), A symmetric rank-revealing Topelitz matrix decomposition, *J. VLSI Signal Processing*, *14/1*, 19–28.

4. F.T. Luk and D. Vandevoorde (1997), Signal decomposition and Hankel approximation, *Iterative Methods in Scientific Computing*, R.H. Chan, T.F. Chan, and G.H. Golub, Eds., Springer, Singapore, 329–357.

5. Stewart, G.W. (1992), An Updating Algorithm for Subspace Tracking, *IEEE Trans. Signal Proc.*, *40*, 1535–41.

Robust Envelope-Constrained Filter Design

Z. Zang, A. Cantoni, B. Vo and K.L. Teo

Australian Telecommunications Research Institute and Cooperative Research Centre for Broadband Telecommunications and Networking, Curtin University of Technology, Bentley, Australia

Abstract

The envelope-constrained filtering problem is concerned with the design of a time-invariant filter to process a given input signal such that the noiseless output of the filter is guaranteed to lie within a prescribed output mask and the output noise gain is minimized. Mathematically, this problem has been formulated as a constrained optimization problem. Various methods for solving this problem have been reported in the literature. Because of the way in which the problem is posed, the output response corresponding to the optimal filter always lies on the boundary of the feasible region. Consequently, any disturbance at the input of the channel or any implementation error could cause the output to violate the envelope constraints. In this paper, we summarise various approaches to the EC filtering problem and then, we investigate how an optimal filter can be found which maximizes the minimum distance between the feasible output response and the output mask. A new scheme for achieving this is proposed. To illustrate the effectiveness of the scheme, numerical examples are presented.

1 Introduction

In signal processing the design of many filters can often be cast as a constrained optimization problem where the constraints are defined by the specifications of the filter. These specifications can arise either from practical considerations or from the standards set by certain regulatory bodies. The envelope-constrained (EC) filtering problem considered in this paper is a specific constrained optimization problem. We are concerned with the design of a time invariant filter with impulse response $u(t)$ to process a given input pulse $s(t)$ which is corrupted by zero mean white noise $n(t)$, see Figure 1(a). The noiseless output $\psi(t)$ is required to fit into a prescribed pulse shape envelope defined by the lower and upper boundaries $\varepsilon^-(t)$ and $\varepsilon^+(t)$ as shown in Figure 1(b). The optimal EC filter is defined as the filter which minimizes the output noise power while satisfying the pulse shape constraints. If the input noise n is white with constant power spectrum, it can be easily verified that the output noise power is proportional to the squared L_2 norm of the filter. In fact the problem can be generalized to

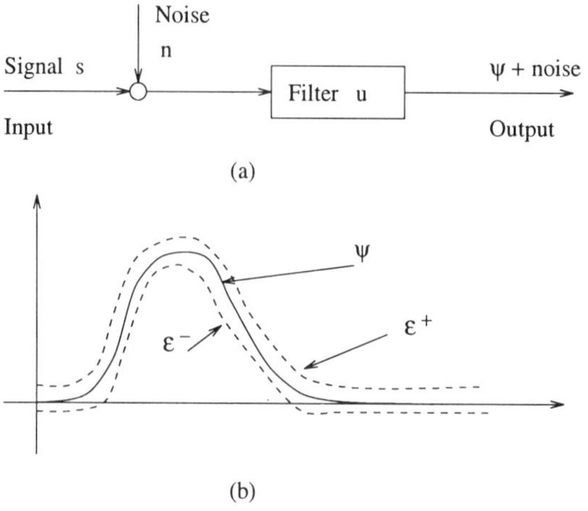

Figure 1. EC filtering problem: (a) Block diagram, and (b) Pulse shape envelope

cope with non-white noise by introducing a generalized filter norm [1]. The EC filtering problem, denoted by (**P**0), can be posed as

$$\min \|u\|^2, \text{ subject to } \varepsilon^-(t) \leq \psi(t,u) \leq \varepsilon^+(t), \ \forall\, t \in [0,T] \tag{1.1}$$

where

$$\|u\| \triangleq \left(\int_0^T |u(t)|^2 dt\right)^{1/2}, \qquad \psi(t,u) = \int_0^t u(\tau)s(t-\tau)d\tau.$$

Traditionally, problems of this type were often treated by minimizing the weighted mean-square difference between ψ and some desired pulse shape. However, in many applications this "soft" least-squares (LS) approach is unsatisfactory because large narrow excursions from the desired shape occur and the norm of the filter can be large and the choice of an appropriate weighting function is not obvious. Moreover, the solution can be sensitive to the detailed structure of the desired pulse, and it is usually not obvious how the shape of the desired pulse should be altered in order to improve on the solution. It was argued in [2] that the EC filtering problem as defined above is more relevant than the "soft" LS approach in a variety of signal processing fields such as robust antenna and filter design [3], communication channel equalization [4,5,6], and pulse compression in radar and sonar [7].

We consider an example to illustrate some of the above issues. In television transmission, a test pulse for example a $\sin^2$ pulse, is sent through a dispersive channel and becomes distorted. If this distorted pulse fits into a prescribed envelope called a K-mask (see Figure 2) and satisfies a specified K-rating, the channel will not subjectively distort transmitted video data [8,9]. Otherwise, it is necessary to insert an equalizing filter to compensate for the distortion,

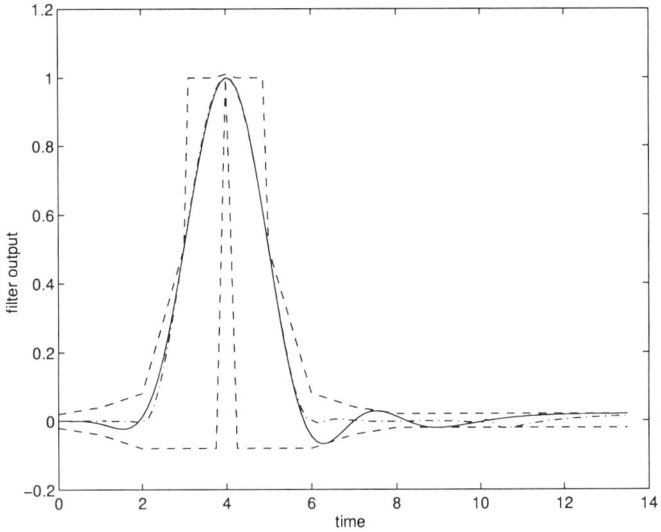

Figure 2. K-rating mask for equalization of TV channel

i.e. fit the distorted pulse to the K-mask. Notice that detailed knowledge of the dispersive channel is not required to design the equalizer, only the distorted pulse is required. Most researchers investigating pulse shaping and equalization problems use LS error and minimax error approximation techniques. These techniques although very powerful do not yield a good solution for the K-mask equalization problem. Figure 2 shows the equalized pulse for the EC (solid line) and LS (dash-dotted line) approach, where the channel under consideration is a coaxial cable. Both methods satisfied the constraints but the filter norm for the LS approach is 9475 times that of the EC. It is important to note that for this particular application there is, apparently, no meaningful formulation of the problem using LS or other "soft" criteria. The equalized pulse must fit into the prescribed envelope.

The EC filtering problem was first posed in early 1970s [10,11]. Since then, various methods for solving it have been reported in the literature (see, for example, [2,12,13,14]). Let u^o denote the solution to the EC filtering problem (**P**0) and $\psi(t, u^o)$ the corresponding output response. As a solution to the constrained minimization problem (**P**0), the output response corresponding to the optimal EC filter always lies on the boundary of the feasible region. This means that

$$\min\{\min_t(\psi(t,u^o) - \varepsilon^-(t)), \min_t(\varepsilon^+(t) - \psi(t,u^o))\} = 0.$$

Consequently, any disturbance at the input of the channel or any implementation error could cause the output to violate the envelope constraints. For a given filter

u which satisfies the envelope constraints as specified in Equation (1.1), let us define its constraint robustness margin as follows

$$\sigma(u) \triangleq \min\{\min_t(\psi(t,u) - \varepsilon^-(t)), \min_t(\varepsilon^+(t) - \psi(t,u))\}.$$

It is clear that if $\sigma > 0$, the minimum distance of the output response $\psi(t,u)$ to the output mask is at least equal to $\sigma(u)$. Therefore, we may say that filter u is robust with constraint robustness margin σ. In this contribution we investigate how an optimal filter can be designed which maximizes the constraint robustness margin.

The rest of the paper is organized as follows: In Section 2, various approaches to the EC filtering problem are summarised. In Section 3, we investigate how an optimal filter can be found which maximizes the minimum distance between the output response and the output mask. A new optimization scheme for achieving this is proposed. In Section 4 we study two numerical design examples to demonstrate the effectiveness of the proposed scheme. Section 5 draws some conclusions regarding the design methods presented.

2 Envelope-constrained filter design

Originally, the EC filtering problem (**P**0) was posed in [10,11] with the additional assumption that both the input signal and the filter to be designed are finite support L_2 space functions. This problem was tackled by the primal dual method, which involves transforming the constrained optimization problem into a non-smooth unconstrained dual problem.

$$\max_\lambda \left\{ -\|\frac{1}{2}\int_0^T s(t-\tau)d\lambda(t)\|^2 - \int_0^T d(t)d\lambda(t) - \sup \Sigma_{i=1}^n |\lambda(E_i)| \int_{E_i} \varepsilon(t)dt \right\}$$

where $d(t) = 0.5(\varepsilon^+(t) + \varepsilon^-(t))$ and $\varepsilon(t) = 0.5(\varepsilon^+(t) - \varepsilon^-(t))$. d can be interpreted as the desired pulse shape and ε as an error tolerance about d. The unconstrained problem is then discretized into a finite dimensional problem to be solved by a steepest ascent type algorithm. The approximate solutions obtained by this technique are not guaranteed to satisfy the envelope constraints. Only in the limiting case would the constraints be satisfied. From analysis [10], the optimal filter u^* can be expressed in terms of the solution λ^* to the dual problem as follows

$$u^*(\tau) = -\frac{1}{2}\int_0^T s(t-\tau)d\lambda^*(t). \tag{2.1}$$

From Equation (2.1) we see that the optimal filter can be interpreted as consisting of a matched filter followed by a second filter (or equalizer) which is determined by the Lagrange multiplier λ^*.

The discretized version of EC filtering problem (**P**0) was considered in [2] and can be stated as

$$\min \|u\|^2 = u'u, \quad u \in R^n, \quad \text{subject to} \quad \mathbf{d} - \varepsilon \leq Su \leq \mathbf{d} + \varepsilon, \tag{2.2}$$

where S is the convolution matrix of the discretized input signal $s(t)$. The primal dual algorithm was also used to solve this problem. This method can be easily implemented on-line due to its iterative nature. However, convergence can be quite slow. The penalty approach [12] allows much faster convergence while preserving the iterative nature. The solution to the finite-dimensional EC filtering problem 2.2 can be expressed as

$$u^* = -\frac{1}{2}S'\lambda^*.$$

where λ^* is the solution of the following dual problem

$$\max_{\lambda}\left\{-\frac{1}{4}\lambda' SS'\lambda - \mathbf{d}'\lambda - \varepsilon'|\lambda|\right\}.$$

As in the continuous time case, the optimal filter consists of a matched filter followed by a second filter (or equalizer) whose weights are the elements of the Lagrange multiplier vector λ^*.

In [15,16], the EC filtering problem 2.2 was generalized to cover the case where the input is not specified exactly, but known to stay within an input mask. This is known as the envelope-constrained with uncertain input (ECUI) filtering problem, denoted as problem (P1), and can be stated as

$$\min \|u\|^2 = u'u, \quad u \in R^n, \quad \text{subject to} \quad \mathbf{d} - \varepsilon \leq Su \leq \mathbf{d} + \varepsilon, \quad \forall s: \ |s - c| \leq \gamma.$$

where c is the nominal input signal. Clearly, the EC filtering problem 2.2 is a special case of the problem (P1) where the input uncertainty (γ) is zero. This non-smooth problem can be solved by the primal-dual approach, however a more efficient method is to transform it into the following standard QP problem, denoted as problem (P2), with affine inequality constraints [17].

$$\min(\|x\|^2 + \|y\|^2)$$

subject to

$$C(x-y) + \Gamma(x+y) \leq \varepsilon^+, \quad -C(x-y) + \Gamma(x+y) \leq -\varepsilon^-,$$

$$-x \leq 0, \quad -y \leq 0,$$

where C is the convolution matrix of the nominal input signal c. It can be proved [1] that if (x^*, y^*) is the solution of problem (P2), $u^* = x^* - y^*$ is the solution of problem (P1).

Recently [13], continuous-time finite-structured orthonormal filters were used to solve the EC filtering problem (**P**0). Unlike the previous approaches, this method does not involve discretization of either the filter or the input signal. Hence, the constraints are satisfied for all t in the constrained interval. A theory for the application of digital techniques to continuous-time EC filtering via hybrid filters has been developed [14] and this problem assumes the same form as the analog formulation.

All the approaches summarised so far assume that the second order statistics of the noise is known. For noise with unknown statistics, a formulation of the EC filtering problem via H_∞ method was developed in [18] and can be stated as

$$\min \|U\|_\infty^2, \text{ subject to } \varepsilon^-(t) \leq \psi(t,u) \leq \varepsilon^+(t), \quad \forall\, t \in [0,T]$$

where $U(s)$ is the Laplace transform of $u(t)$

$$\|U\|_\infty \triangleq \max_\omega |U(j\omega)| = \max_{\Phi_N} \left(\frac{1}{2\pi} \int_{-\infty}^{\infty} \Phi_N(\omega)|U(j\omega)|^2 d\omega \right)^{1/2}$$

$\Phi_N \in \mathcal{B}_N = \{\Phi_N : \frac{1}{2\pi}\int_{-\infty}^{\infty} \Phi_N(\omega)d\omega \leq 1\}$ is the unknown but bounded noise spectrum. This H_∞ optimization approach is based on the theory [19] that minimizing the H_∞ norm of the filter $U(s)$ is equivalent to minimizing the output noise power with respect to the worst case input noise. In cases in which the input signal is subject to random disturbance with unknown but bounded power spectrum, the $\mathcal{H}_\infty$ optimization approach offers a more robust design.

To conclude this section let us reiterate that, due to the way in which the EC filtering problem was posed, the solution always lies on the boundary of the feasible region. Consequently, any disturbance in the input or implementation errors could violate the output constraints. The problem of designing EC filters that are robust to such disturbances or techniques for providing a guard band on the constraint boundary are essential to the implementation of EC filters. But the price to pay is the increased noise gain. In the next section, we investigate how a trade-off between the noise gain and the constraint robustness can be achieved by defining a new optimization problem.

3 Robust envelope-constrained filter design

3.1 Problem formulation

Consider the EC filtering problem (**P**0). Clearly, for any given $u(t)$, if $\sigma(u) \geq 0$, the corresponding output response $\psi(t,u)$ satisfies the output constraints as specified in Equation (1.1). Let us define

$$\mathcal{U} = \{u \in L_2 \text{ such that } \sigma(u) \geq 0\}.$$

Any $u \in \mathcal{U}$ will be called a feasible point of problem (**P**0). Let u^o denote the solution of the EC filtering problem (**P**0). Consider the following constrained maximization problem, denoted as problem (**P**):

$$\max_u \sigma$$

subject to

$$\varepsilon^-(t) + \beta(t)\sigma \leq \psi(t, u) \leq \varepsilon^+(t) - \beta(t)\sigma \tag{3.1}$$

$$\sigma \geq 0, \ \ ||u||^2 \leq (1+\delta)||u^o||^2,$$

where $\beta(t)$ is a positive, piece-wise continuous function and $\delta > 0$ a constant which specifies the allowable amount of increase of the output noise power in the design of the optimal filter with maximum constraint robustness margin.

Remark A. (i). Assume that $\sigma^* > 0$ is the solution of problem (**P**) and u^* is the related filter. If $\beta(t) = 1$, from Equation (3.1) we see that the distance between the output response, denoted as $\psi(t, \sigma^*, u^*)$, and the output mask is at least equal to σ^*. Therefore, we may say that the filter u^* possesses certain robustness with constraint robustness margin being equal to σ^*. (ii). In practical design, it may be the case that output constraint robustness is particularly important in certain time intervals. If this is the case, $\beta(t)$ can be used to specify the weightings in different time intervals. For example, the function $\beta(t)$ can be assigned to larger value in those time intervals where constraint robustness is more important.

3.2 Problem solution

Define $f(u, \sigma) \triangleq -\sigma$. The constrained maximization problem (**P**) can be converted into the following standard minimization problem, again denoted as problem (**P**),

$$\min_{u,\sigma} f(u, \sigma) \tag{3.2}$$

subject to

$$g_1(t, u, \sigma) \triangleq \beta(t)\sigma - \psi(t, u) + \varepsilon^-(t) \leq 0 \tag{3.3}$$

$$g_2(t, u, \sigma) \triangleq \beta(t)\sigma + \psi(t, u) - \varepsilon^+(t) \leq 0 \tag{3.4}$$

$$g_3(t, u, \sigma) \triangleq -\sigma \leq 0 \tag{3.5}$$

$$g_4(t, u, \sigma) \triangleq ||u||^2 - (1+\delta)||u^o||^2 \leq 0. \tag{3.6}$$

Define

$$\mathcal{F} \triangleq \{(u, \sigma) : \ g_j(t, u, \sigma) \leq 0, \ \forall t \in [0, T], \ j = 1, 2, 3, 4\}.$$

Any pair $(u, \sigma) \in \mathcal{F}$ will be called a feasible point of the problem (**P**). In regard to the objective function $f(u, \sigma)$ and the constraints $g_i(t, u, \sigma)$ specified in problem (**P**), the following result can be easily established.

Lemma 1. *Problem* (**P**) *is a convex minimization problem with convex constraints, that is, both the objective $f(u, \sigma)$ and the constraints $g_i(t, u, \sigma) \leq 0$ ($j = 1, 2, 3, 4$) are convex in u and σ.*

From convex optimization theory we know that the solution to problem (**P**) exists and any relative minimum is a global minimum. In fact, with an additional assumption that the amplitude constraint (3.6) is active and nondegenerate (see, for example, [20]), we can establish the following

Theorem 2. *Consider the constrained minimization problem (3.2-3.6). If the constraint (3.6) is active and non-degenerate, the solution is unique and can be expressed as follows*

$$u^*(\tau) = \frac{1}{2\lambda_4^*}\int_0^T s(t-\tau)d\lambda^*(t) \tag{3.7}$$

where λ_4^* and λ^* are the solutions of the following dual problem

$$\max_{(\lambda,\lambda_3\lambda_4)} \min_{\sigma}\{-\sigma + \sigma\int_0^T \beta(t)d|\lambda|(t) + \int_0^T d(t)d\lambda(t) - \int_0^T \varepsilon(t)d|\lambda|(t) \tag{3.8}$$

$$-\lambda_3\sigma - \lambda_4(1+\delta)||u^o||^2 - \frac{1}{4\lambda_4^2}\int_0^T (\int_0^T s(t-\tau)d\lambda(t))^2 d\tau\}. \tag{3.9}$$

Remark B. (i). Theorem 2 gives a structural characterization of the filter which solves the constrained optimization problem (**P**). Similar to the solution (see Equation (2.1)) of the EC filtering problem (**P0**), u^* given by Equation (3.7) can be interpreted as consisting of a matched filter followed by a second filter (or equalizer) which is determined by the Lagrange multiplier λ_4^* and $\lambda^*(t)$. (ii). In practice, the constrained optimization problem (**P**) can be solved using existing standard softwares such MATLAB and its Optimization Toolbox.

4 Numerical examples with Laguerre filters

Let us consider the design of an equalization filter for a digital transmission channel consisting of a coaxial cable on which data is transmitted according to the DX3 standard (see [4]). For simplicity, we shall restrict our attention to those filters which can be expressed as linear combinations of a finite set of orthonormal Laguerre filters.

4.1 Orthonormal Laguerre filters

The time-domain Laguerre polynomials are defined as

$$l_n(t) = \frac{e^t}{n!}\frac{d^n}{dt^n}\left(e^{-t}t^n\right), \quad n = 0,\ 1,\ 2,\ ...$$

For a given $p > 0$, the Laguerre filters with scale factor p are defined as

$$\phi_n^p(t) = \sqrt{2p}\,e^{-pt}l_n(2pt), \quad n = 0,\ 1,\ 2,\ ...$$

It is known that the sequence $\{\phi_n^p(t)\}$ forms a uniformly bounded orthonormal basis for the Hilbert space $L_2(R_+)$ (cf. [21,22]). Therefore, any function in $L_2(R_+)$ can be expanded as a Laguerre-Fourier series. Consider those filters which can be expressed as

$$u(t) = \sum_{n=0}^{N-1} x_n \phi_n^p(t). \tag{4.1}$$

It is easy to verify that the problem (**P**0) can be written as

$$\min \|x\|^2, x \in R^N, \text{ subject to } \varepsilon^-(t) \le \varphi^T(t)x \le \varepsilon^+(t), \ \ \forall\, t \in [0,T] \tag{4.2}$$

where

$$x = [x_0,\ x_1,\ x_2,\ ...,\ x_{N-1}]^T,\ \ \varphi(t) = [y_0(t),\ y_1(t),\ y_2(t),\ ...,\ y_{N-1}(t)]^T$$

and $y_n(t) = \int_0^t \phi_n^p(\tau)s(t-\tau)d\tau$.

Similarly, the problem (**P**) can be written as

$$\min_{x,\sigma} f(x,\sigma) \tag{4.3}$$

subject to

$$\varepsilon^-(t) + \beta(t)\sigma \le \varphi^T(t)x \le \varepsilon^+(t) - \beta(t)\sigma, \tag{4.4}$$

$$\sigma \ge 0,\ \ x^T x \le (1+\delta)\|u^*\|^2. \tag{4.5}$$

4.2 Computer studies

For this filtering problem, the design objective is to find an equalizing filter which takes the impulse response of a coaxial cable with a loss of 30dB at a normalized frequency of $1/\tau$ (where τ is the baud interval) as input, see Figure 3 (dashed line) and produces an output which lies within the envelope given by the DSX-3 pulse template, see Figure 3 (dash-dotted line) (cf. [4]). For computational purpose, we shall discretize both the input signal and the output mask. In our numerical studies, 1024 samples are used over the normalized time interval of $[0, 32\tau]$, i.e. the sampling period is $\tau/32$. Using the filter structure 4.1 with $N = 14$ and $p = 12$, we first solve the optimal filtering problem 4.2. The squared L_2 norm of the optimal filter (i.e., the optimal output noise power gain) is

$$\|u^*\|^2 = 54.2008.$$

The output mask, input signal and the output response are depicted in Figure 3. Obviously, the output response meets the mask at many points. Figure 4 is the plot of the frequency domain input-output signals.

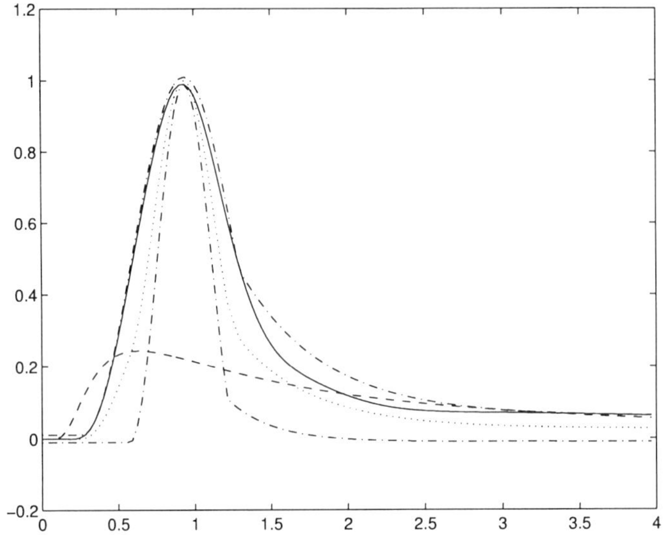

Figure 3. Plot of the sampled time domain output signal (solid line), input signal (dashed line), the output mask (dash-dotted lines), and the center of the output mask (dotted line)

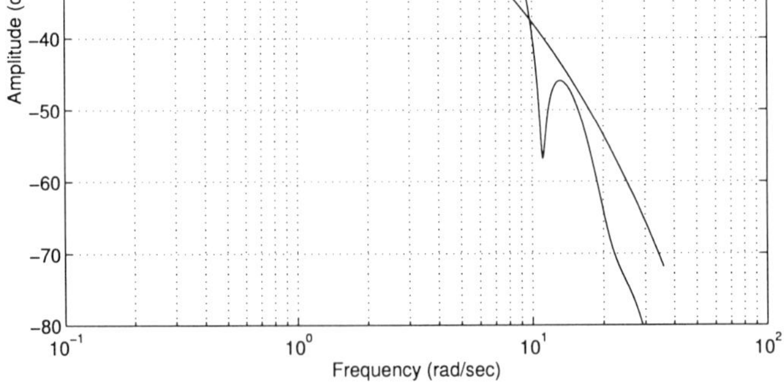

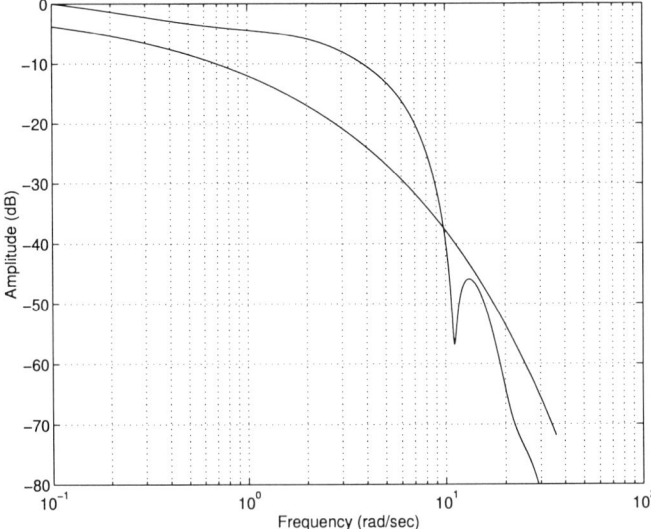

Figure 4. Magnitude spectrum plot of the output signal (solid line) and the corresponding input signal (dashed line)

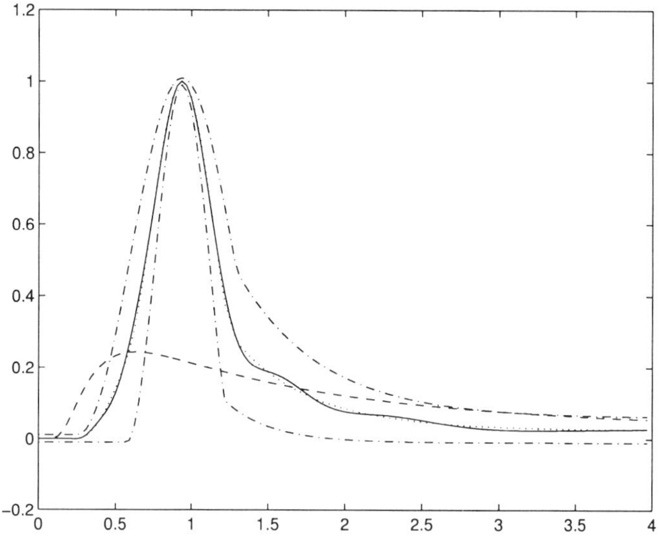

Figure 5. Plot of the sampled time domain output signal (solid line), input signal (dashed line), the output mask (dash-dotted lines), and the center of the output mask (dotted line)

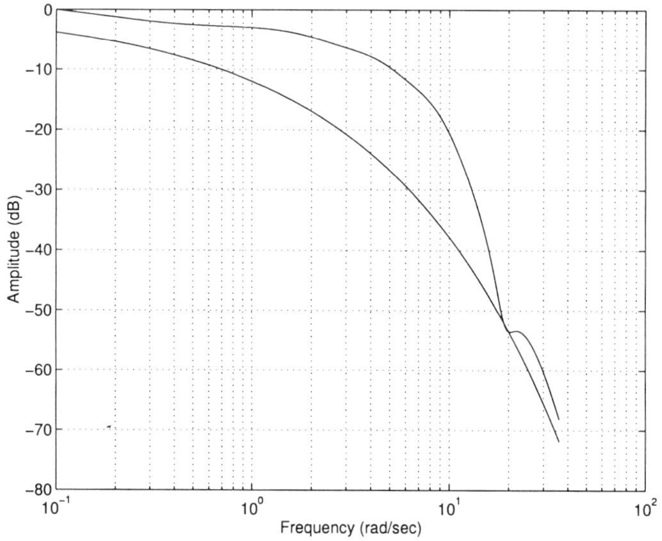

Figure 6. Magnitude spectrum plot of the output signal (solid line) and the corresponding input signal (dashed line)

To achieve maximum output constraint robustness margin, we solve the robust EC filtering problem 4.3-4.5 with $\delta = 1.5$ and

$$\beta(t) = \begin{cases} 3\varepsilon(t), & \text{if } |\varepsilon(t)| \geq 0.05 \\ 0.035, & \text{if } |\varepsilon(t)| \leq 0.03 \\ \varepsilon(t) & \text{elsewhere} \end{cases}$$

where $\varepsilon(t) = 0.5(\varepsilon^+(t) - \varepsilon^-(t))$, $\varepsilon^+(t)$ and $\varepsilon^-(t)$ are given by the DSX-3 pulse template (cf. [4]). The corresponding output mask, input signal and the output response are depicted in Figure 5. The related frequency domain plot of the input-output signals is depicted in Figure 6. From Figure 5 we see that the output signal (solid line) is very close to the center (dotted line) of the output mask. The robustness of this filter with respect to the output constraint is self-evident compared with the optimal filter derived by solving problem (**P**0).

5 Discussions and concluding remarks

In this paper, various approaches to the EC filtering problem have been summarised. Because of the way in which the EC filtering problem (**P0**) was posed, the output response corresponding to the optimal filter always lies on the boundary of the feasible region. This lack of constraint robustness against input uncertainty or implementation error motivates us to investigate how an optimal filter can be found which provides a guard band on the constraint boundary. A new optimization problem has been studied. Given a trade-off between the output noise gain and the constraint robustness, the solution to this problem guarantees that the ouput response lies strictly inside the feasible region. To illustrate the effectiveness of the technique, numerical examples are presented.

Acknowledgements

This work was supported by the Australian Research Council and the Cooperative Research Centre for Broadband Telecommunications and Networking.

Bibliography

1. Vo, B. (1996). Optimum envelope constrained filters. PhD Thesis, ATRI, Curtin University of Technology, Western Australia.

2. Evans, R.J., Cantoni, A. and Fortman, T.E. (1977). Envelope-constrained filter, part I, theory and application; Part II, adaptive structures, *IEEE Trans. Information Theory*, **23,** 421-444.

3. Ahmed, K.M. and Evans, R.J. (1984). An adaptive array processor with robustness and broadband capabilities. *IEEE Trans. Antennas and Propagat.*, **32**, 944-950.

4. Bell Communications. (1988). DSX-3 isolated pulse template and equations. *Technical Reference TR-TSY-000499,* Issue 2, 9-17.

5. Lechleider, J.W. (1991). A new interpolation theorem with application to pulse transmission. *IEEE Trans. Communications*, **39**, 1438-1444.

6. Lucky, R., Salz, J. and Weldon, E.J. (1968). *Principles of data communications,* New York: McGraw-Hill.

7. McAulay, R.J. and Johnson, J.R. (1971). Optimal mismatched filter design for radar ranging detection and resolution. *IEEE Trans. Information Theory,* **17,** 696-701.

8. Seyler, J. and Potter, J. (1960). Waveform testing of television transmission facilities. *Proc. IRE,* Australia, 470-478.

9. Budrikis, Z.L. (1972). Visual fidelity criterion and modeling. *Proc. IEEE,* **60**, 771-779.

10. Fortmann, .E. and Athans, M. (1974). Optimal filter design subject to output sidelobe constraints: Theoretical considerations. *J. Opt. Theory and Appl.* **14,** 179-197.

11. Fortmann, T.E. and Evans, R.J. (1974). Optimal filter design subject to output sidelobe constraints: Computational algorithm and numerical results. *J. Opt. Theory and Appl.* **14,** 271-290.

12. Vo, B., Cantoni, A. and Teo, K.L. (1995). Iterative algorithms for envelope constrained filter design. *Proc. IEEE Int. Conf. Acoustics, Speech and Signal Processing,* Detroit, Michigan, **2**, 1288-1291.

13. Vo, B., Zang, Z., Cantoni, A. and Teo, K.L. (1995). Continuous-time envelope constrained filter design via orthonormal filters. *IEE Proc.-Vis. Image Signal Process.,* **142,** 389-394.

14. Vo, B., Cantoni, A. and Teo, K.L. (1995). Envelope constrained filter with linear interpolator. To appear in *IEEE Trans. Signal Processing.*

15. Evans, J.R., Cantoni, A. and Ahmed, K.M. (1983). Envelope-constrained filters with uncertain input. *Circuits System Signal Processing,* **2,** 131-154.

16. Zheng, W.X., Cantoni, A. and Teo, K.L. (1995). The sensitivity of envelope-constrained filters with uncertain input. *IEEE Trans. Circuits and Systems–I. Fundamental, Theory, and Application*, **42**, 1-9.

17. Teo, K.L., Cantoni, A. and Lin, X.G. (1994). A new approach to optimization of envelope constrained filters with uncertain input. *IEEE Trans. Signal Processing,* **42**, 426-429.

18. Zang, Z., Cantoni, A. and Teo, K.L. (1997). Continuous-time envelope constrained filter design via Laguerre filters and H_∞ optimization methods. To appear in *Proc. IEEE Int. Conf. Acoustics, Speech and Signal Processing,* Munich, Germany.

19. Francis, B.A. (1987). *A Course in H_∞ Control Theory,* Springer-Verlag, Berlin, Heidelberg.

20. Luenberger, D.G. (1983). *Linear and Nonlinear Programming, Wesley Publishing Company,* Reading, Massachusetts.

21. Mäkilä, P.M. (1990). Laguerre series approximation of infinite dimensional systems. *Automatica*, **26,** 985-995.

22. Szegö, G. (1939). *Orthogonal Polynomials*, Amer. Math. Soc. Coll. Publ., **23**.

On the use of Positivity in Reconstructing Images from Noisy Data

G.D. de Villiers

DRA Malvern, Worcestershire

Abstract

We give a method for incorporating positivity of the solution into linear inverse problems. The method uses mathematical programming techniques and involves first solving the dual problem and then obtaining the solution to the primal problem from that of the dual one. It delivers a solution which agrees with the truncated singular function expansion in its first few terms and hence can be viewed as the unique minimum 2-norm positive extension of the truncated singular function expansion. Occasionally the iteration involved in solving the problem fails to converge and we suggest that this is due primarily to inconsistency of the primal problem or, when this is not the case, failure of the constraint qualification. Problems can also occur when a matrix involved in the iteration becomes ill-conditioned. We give a simple approximate solution when this is the case.

1 Introduction

A standard problem in signal processing is analysis of data which is linearly related to some unknown function in order to make some deductions about that function. An example in image processing is the following: Given a blurred image (the data) which has arisen from the convolution of a known point spread function and a clear image one is required to deconvolve the point spread function to retrieve the clear image. This type of problem is known as a linear inverse problem. Often the true image is known to be positive (an intensity distribution, for example) but this property is frequently lost in the reconstruction. Another common problem with reconstruction techniques is extreme sensitivity to noise on the data.

In this paper we describe a novel method, based on mathematical programming, which is insensitive to noise and produces non-negative solutions. Some preliminary details may be found in [1]. The underlying mathematics is discussed in detail in [2]. However their presentation is rather technical and so in this paper, which represents an application of their ideas to the solution of linear inverse problems, we will give a simplified discussion of some of the relevant points. We will restrict ourselves in this paper to discussing one-dimensional problems for simplicity though we will give some results for a two-dimensional one. The extension to higher dimensions is straightforward.

Linear inverse problems (in one dimension) are normally specified by a linear integral equation of the form

$$g(y) = \int_{\alpha}^{\beta} K(y,x)f(x)dx + noise \equiv (Kf)(y) + noise \tag{1.1}$$

where g represents the data (the blurred image), the kernel $K(y,x)$ is known (the point spread function in image processing) and the function f (the object or clear image) is required to be found. f is normally taken to lie in $L^2(\alpha,\beta)$ and g to lie in $L^2(\gamma,\delta)$, for some real numbers $\alpha,\beta,\gamma,\delta$. Such problems usually have a wide range of possible solutions.

Two examples of linear inverse problems which we will consider later are band-limited coherent imaging in one and two dimensions. In one dimension this corresponds to convolution with a sinc function:

$$g(y) = \int_{-1}^{1} \frac{sin[c(y-x)]}{\pi(y-x)} f(x)dx \tag{1.2}$$

and in two dimensions, for a square pupil and square object domain:

$$g(y_1,y_2) = \int_{-1}^{1}\int_{-1}^{1} \frac{sin[c(y_1-x_1)]}{\pi(y_1-x_1)} \frac{sin[c(y_2-x_2)]}{\pi(y_2-x_2)} f(x_1,x_2)dx_1dx_2. \tag{1.3}$$

For further details see [3]. For coherent imaging positivity is perhaps not as useful as for incoherent imaging, where the image to be reconstructed is an intensity distribution (and hence is non-negative) but we have chosen the former since the singular functions are analytic and hence pseudo-Haar - an essential property for the approach in this paper (see Section 5). These singular functions, which are, in fact, prolate spheroidal wave functions, are well known and so we can check whether the singular functions we have calculated are accurate or not.

2 The singular function approach

If, as is often the case, the integral operator $K : L^2(\alpha,\beta) \to L^2(\gamma,\delta)$ in Equation (1.1) is compact then it possesses a singular system

$$Ku_i = \sigma_i v_i, \qquad K^* v_i = \sigma_i u_i, \qquad \sigma_i \geq 0, \qquad i = 0, ..., \infty \tag{2.1}$$

where K^* denotes the adjoint of K, the σ_i are the singular values and the u_i and v_i are respectively the right- and left-hand singular functions. We recall that the u_i and v_i form complete orthonormal bases for $L^2(\alpha,\beta)$ and $L^2(\gamma,\delta)$ respectively and that the $\sigma_i \to 0$ as $i \to \infty$ (except for when K is of finite rank, in which case all the singular values are identically zero except for a finite number of them). The true solution to the linear inverse problem can be found by expanding the data in terms of the left-hand singular functions:

$$g(y) = \sum_{i=0}^{\infty} b_i v_i(y) \tag{2.2}$$

where

$$b_i = \int g(y)v_i(y)dy \qquad (2.3)$$

to yield

$$f(x) = \sum_{i=0}^{\infty} \frac{b_i}{\sigma_i} u_i(x). \qquad (2.4)$$

In the absence of noise on the data b_i for large i is extremely small since it corresponds to the component of f in the direction u_i multiplied by σ_i. However if additive noise is present on the data b_i will become much larger, i.e. to the level of the component of the noise in the direction v_i. Any information contained in b_i will consequently be swamped by the noise and there is no point in including the corresponding term in the expansion in Equation (2.4). If one removes all such terms from the expansion one obtains a standard solution to the linear inverse problem, namely the truncated singular function expansion:

$$\tilde{f}(x) = \sum_{i=0}^{N} \frac{b_i}{\sigma_i} u_i(x). \qquad (2.5)$$

One can view this solution, known also as the numerical filtering solution, as arising from a regularising algorithm with regularisation parameter $\frac{1}{N}$.

In practice the data always has to be sampled and the integral in Equation (2.3) must be replaced by a sum. However we have adopted the philosophy that it is better to solve the ideal problem with continuous data first and then worry about the choice of sampling scheme later on.

3 The problem and its solution

One problem with the truncated singular function expansion is that it often has negative parts even if the true solution is known to be positive. In order to correct this we use a non-linear method based on the scheme in [4] for solving equality-constrained least squares problems where the solution lies in L^2 (see also [2] and [5]). We impose the equality constraints that the first $N + 1$ singular function coefficients of the solution agree with their "known" values (as determined from the data via Equation (2.3)). Hence our solution agrees in the first $N + 1$ terms of its singular function expansion with the truncated singular function expansion. The solution is chosen, from the wide range of positive solutions, to be the one of minimum 2-norm which satisfies the constraints. This is a standard mathematical programming problem. The primal problem, which we have just described, is given by

$$\min_{f \in L^2(\alpha,\beta), f \geq 0} ||f||_2 \qquad (3.1)$$

subject to

$$\int f(x)u_i(x)dx = d_i \qquad i = 0, ..., N \qquad (3.2)$$

where

$$d_i = \frac{b_i}{\sigma_i}. \tag{3.3}$$

In what follows we will denote the mapping from f to d by A.

The solution is obtained by solving the dual problem, which is finite dimensional, and then obtaining the primal solution from the dual one. The relevant form of duality is Fenchel (or conjugate) duality. For a detailed discussion see [6]. The dual problem is given by ([2], page 62)

$$\min_{\lambda \in R^{N+1}} -d^T\lambda + \frac{1}{2}||(\sum_{i=0}^{N} \lambda_i u_i)_+||_2^2 \tag{3.4}$$

where the vector λ contains the dual variables and the + subscript indicates that the positive part of the term in brackets must be taken. We recall that the positive part of a function f is defined by

$$(f)_+ = f(x), \qquad f(x) \geq 0 \tag{3.5}$$

$$(f)_+ = 0, \qquad f(x) < 0. \tag{3.6}$$

The transpose $A^T : R^{N+1} \to L^2(\alpha, \beta)$ of the mapping A defined above is given by

$$(A^T\mu)(x) = \sum_{i=0}^{N} u_i(x)\mu_i, \qquad \mu \in R^{N+1}. \tag{3.7}$$

Hence the dual problem may be written as

$$\min_{\lambda \in R^{N+1}} -d^T\lambda + \frac{1}{2}||(A^T\lambda)_+||_2^2. \tag{3.8}$$

At the minimum the gradient of this function must be zero. The gradient is given by

$$-d + \nabla_\lambda(\frac{1}{2}||(A^T\lambda)_+||_2^2) \tag{3.9}$$

$$= -d + ||(A^T\lambda)_+||_2 \nabla_\lambda(||(A^T\lambda)_+||_2). \tag{3.10}$$

We need the following result for the Gateaux derivative of the norm of a positive part (for $y \in L^p$)

$$\nabla_y(||y_+||_p) = ||y_+||_p^{1-p}(y_+)^{p-1} \tag{3.11}$$

(see [2], page 40, Example 5.7(i)). In this expression we put $y = A^T\lambda$ and $p = 2$ and use

$$(\nabla_\lambda f)(A^T\lambda) = (A(\nabla_{A^T\lambda} f))(A^T\lambda) \tag{3.12}$$

where the left-hand side is the ordinary gradient in the direction λ, the derivative on the right is the Gateaux derivative in the direction $A^T\lambda$ and f is some functional on L^2. Finally we have

$$\nabla_\lambda(\|(A^T\lambda)_+\|_2) = \|(A^T\lambda)_+\|_2^{-1}A(A^T\lambda)_+. \tag{3.13}$$

Substituting this in (3.10) we have that, at the minimum,

$$-d + A(A^T\lambda)_+ = 0. \tag{3.14}$$

Define a matrix $H(\lambda)$ by

$$H(\lambda)_{i,j} = \int_{t:(A^T\lambda)(t)>0} u_i(t)u_j(t)dt. \tag{3.15}$$

Then Equation (3.14) becomes

$$H(\lambda)\lambda = d. \tag{3.16}$$

Suppose we try and solve this using a method of successive approximations

$$H(\lambda^{old})\lambda^{new} = d \tag{3.17}$$

stopping when $\|\lambda^{old} - \lambda^{new}\|_2$ is sufficiently small. Let λ^{opt} be the value of λ delivered by this iteration. Then the solution to the primal problem is given by

$$f^{opt} = (\sum_{i=0}^{N} \lambda_i^{opt} u_i)_+ \tag{3.18}$$

(see [2], page 61, Equation (7.2) with $q = 2$). Note that we must take a positive part to get this solution.

In order for this duality theory to be valid a technical condition known as a constraint qualification must be satisfied. To be more precise, if the constraint qualification is not satisfied one cannot guarantee that the primal and dual optimal values will be the same or that one will be able to obtain the primal solution from the dual one. Let $L^2_+(\alpha,\beta)$ be those functions in $L^2(\alpha,\beta)$ which are non-negative almost everywhere. Then for the case we are considering the constraint qualification reads that there must be a feasible point in the quasi-relative interior of $L^2_+(\alpha,\beta)$. This quasi-relative interior corresponds to those functions which are positive almost everywhere (for details of quasi-relative interiors see [2]). This is equivalent to the following condition

$$b \in interior(A(L^2_+(\alpha,\beta))) \tag{3.19}$$

$A(L^2_+(\alpha,\beta))$ is known as the moment cone. Except in a few special cases this is a very difficult condition to check. We will return to this question later.

In practice when one uses this iterative method there will always be a question as to how close one is to the true solution of minimum norm, since there is no way of knowing if one has found that solution. However as long as the solution is positive, agrees with the equality constraints and agrees with the data to within the noise this is a perfectly acceptable solution to the problem. Indeed though the minimum 2-norm solution has an elegant mathematical structure associated with it, in the absence of additional prior information such as a smoothness constraint there is no reason to suppose it is any more valid than any other positive solution.

Note also that if the truncated singular function expansion solution should turn out to be non-negative then our solution will coincide with it.

4 Simulation of the noiseless problem

We have applied the technique to various linear inverse problems. We show in Figure 1 an example of one-dimensional band-limited coherent imaging. We have included the first six singular functions in the truncated singular function expansion. In Figures 2 and 3 we show the object and reconstructed

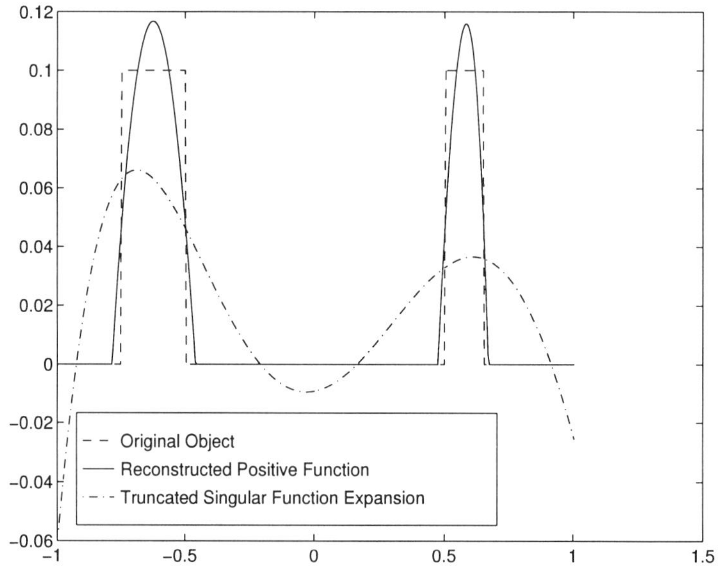

Figure 1. Positive solution versus truncated singular function expansion

object for a two-dimensional coherent imaging problem. We used 36 singular functions in the reconstruction. For both of these problems we chose a value of c (see Equations (1.2) and (1.3)) of less than 1. This is not physically realistic but it was chosen in order to make calculation of the singular functions easier since c corresponds to the number of lowest order singular values which are approximately equal and the closer singular values are to each other the harder it is to find the corresponding singular functions accurately.

An advantage of the method proposed here is that it is not computationally very expensive. For the one-dimensional case the iteration involved in solving the dual problem with N+1 equality constraints in the primal problem typically converges within 3(N+1) steps and the main computational burden in each step involves solving an (N+1)-dimensional linear system.

We chose to use the vector d (given by Equation (3.3)) as an initial value of λ for the iteration. The justification for this was that if the truncated singular function expansion turned out to be non-negative then λ would start with the correct value and the iteration would not be necessary.

As regards resolution our method produces solutions which appear to have better resolution than the truncated singular function expansion, except, of course, when the solutions coincide. This represents an advantage over linear methods for finding positive solutions which appear to have worse resolution [7].

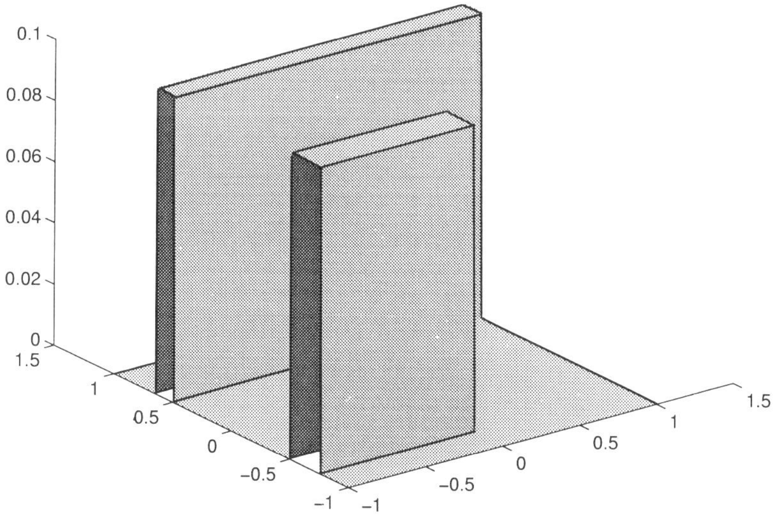

Figure 2. Original object

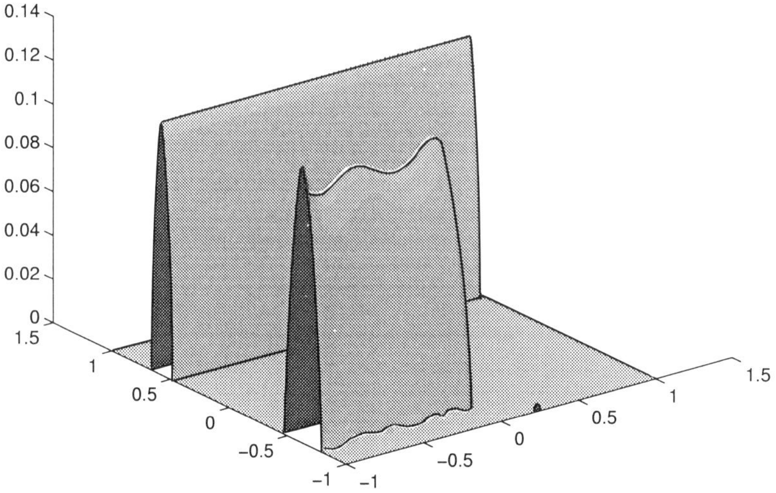

Figure 3. Reconstructed positive function

5 A technical consideration

In order to guarantee a unique solution for the dual problem the singular functions must be pseudo-Haar. This property is defined as follows. A set of functions $\psi_i, i = 1, ..., N$ on some measure space T are *pseudo-Haar* if they are linearly independent on every non-null subset of T. This is true if they are analytic functions (for a proof see [8], page 328).

For the coherent imaging problem and the finite Laplace transform inversion (see [9]) the right-hand singular functions u_i are analytic and hence pseudo-Haar. This is by virtue of the fact that the u_i are eigenfunctions of K^*K and for these problems K^*K has a commuting second order differential operator so that the u_i are also eigenfunctions of this operator (see [10] and [11]). Their analytic nature follows from this.

6 Problems with noise

On adding noise to the data in a linear inverse problem without the positivity constraint the standard difficulty is that the noisy data may lie outside the range of K. This is remedied by projecting the data onto the closure of the range of K

since the closure of the range of K is a subspace and projection onto subspaces is a straightforward matter. For the problem we are considering here things are not quite so simple since $K(L^2_+(\alpha,\beta))$ is not a subspace. To start with we need a few preliminary definitions.

A *cone* is a set C (in some vector space) which is closed under positive scalar multiplication.

A *convex cone* is a cone which is convex.

Let C be a convex cone in some vector space X. The *polar cone* of C, C^o is the set in the dual space X^* defined by

$$C^o = \{y \in X^* : < y, x > \leq 0 \qquad \forall x \in C\}. \tag{6.1}$$

We are dealing with Hilbert spaces exclusively in this paper so that the dual space, X^*, of X can always be identified (as an algebraic set) with X itself. Hence C^o can be thought of as a set in X.

Given a closed convex set S in some normed linear space X and a given element $x \in X$ the *projection of x onto S*, $P_S(x)$ is defined to be the element y in S which minimises

$$\| y - x \| . \tag{6.2}$$

If, in addition the set S is a cone and X is a Hilbert space we have the following result:

Let S be a closed convex cone in some Hilbert space X. Then given an element x in X, an element y in S is the projection of x onto S if and only if

$$y \in S, \qquad x - y \in S^o, \qquad < x - y, y >= 0. \tag{6.3}$$

This form of projection is thus intermediate between projections onto general convex sets and projections onto subspaces.

To see how projection onto closed convex cones applies to our problem it is sufficient to note that we wish the data to lie in $K(L^2_+(\alpha,\beta))$ and that this is a convex cone. Hence we need to project the data onto the closure of $K(L^2_+(\alpha,\beta))$. There is thus an analogue of projecting onto the closure of the range of K when one incorporates positivity into the problem.

Now the foregoing describes how to deal with noise on the inverse problem before one restricts the accurately known coefficients to be the first $N+1$ ones. If one does this the effect of noise in the problem without the positivity constraint is to produce a b which always satisfies $b \in A(L^2(\alpha,\beta))$. With positivity, however, one must project b onto the closure of the convex cone $A(L^2_+(\alpha,\beta))$. It is not clear (to the author at least) how one would determine such a projection in practice. Furthermore for the dual method discussed in this paper to work we require the constraint qualification to be satisfied, i.e. $b \in \textit{interior} A(L^2_+(\alpha,\beta))$.

We have simulated the effects of noise by adding Gaussian noise to the first $N+1$ coefficients of the singular function expansion. The magnitude of this noise was about 5 percent of the largest singular function coefficient in the absence of noise. The main effect of adding this noise was that, for the one-dimensional problem, although two peaks were generally still present in the reconstruction the height and position of these peaks were changed. Another undesirable effect of the noise was that the iteration sometimes failed to converge and cycled round the same sequence of coefficients. One might wonder if this undesirable behaviour could be due to the constraint qualification not being satisfied. Though it is hard to test this some light was shed on the matter by looking at the problem in the next section.

7 A related moment problem

The following moment problem is an example where a condition on the coefficients can be written down which is equivalent to the constraint qualification being satisfied. The problem is given by

$$min\|f\|_2^2 \qquad f \in L^2(0,1) \tag{7.1}$$

subject to

$$\int_0^1 f(x)x^i dx = d_i, \qquad i = 0, 1, ..., N \tag{7.2}$$

and

$$f(x) \geq 0. \tag{7.3}$$

Assume, for the sake of argument, that N is odd. Then the primal problem is consistent if and only if the quadratic forms

$$\sum_{i,j=0}^{(N-1)/2} d_{i+j+1} y_i y_j \tag{7.4}$$

and

$$\sum_{i,j=0}^{(N-1)/2} (d_{i+j} - d_{i+j+1}) y_i y_j \tag{7.5}$$

are positive definite (see [12], page 106). The constraint qualification is then also satisfied. A similar structure exists when N is even. Hence an inspection of the eigenvalues of the matrices in these forms is sufficient for determining whether the constraint qualification is satisfied or not.

However this moment problem is not immediately relevant to our work since functions consisting of simple powers of their argument have little in common with singular functions due to the former not being orthonormal. A set of functions does exist which forms a bridge between typical singular functions and the moment problem. These functions are the Legendre polynomials. If one constructs a problem of the form in (3.1) and (3.2) using Legendre polynomials instead of singular functions then one can write down explicitly the constraint qualification in terms of the constraint qualification for the moment problem. This was done and noise was added in order to reproduce the cycling behaviour in the dual problem iteration. It was found that there was a definite correspondence between the cycling and the failure of the constraint qualification, but also that the problem sometimes occurred when the matrix H defined in Equation (3.15) was ill-conditioned which suggested that some form of regularisation was likely to be needed. The simplest solution was to modify the iteration slightly, for those steps where H was ill-conditioned, to

$$(\epsilon I + H(\lambda^{old}))\lambda^{new} = d \tag{7.6}$$

where ϵ is a small parameter designed to make the ill-conditioned Hessian better conditioned. This iteration scheme seems always to deliver a result (when the constraint qualification is not violated) though, of course, if non-zero values of ϵ are used the solution will be approximate in the sense that the equality constraints are no longer satisfied exactly.

One should note that the results for the Legendre problem only suggest that violation of the constraint qualification is the cause of non-convergence for the other problems. Further work needs to be done on this question. Also note that a basic property of the moment problem is that the condition for consistency of the primal problem is equivalent to the constraint qualification being satisfied. For more general problems there could be situations where the constraint qualification is violated but the primal problem is still consistent so that the primal problem has a meaningful solution whereas the dual one does not. The extent to which this is likely to occur in practice needs to be clarified.

The scheme in this section has been applied to the coherent imaging problem and the problem of finite Laplace transform inversion with encouraging results. However, if the constraint qualification is violated probably the most sensible thing to do is to repeat the experiment (when practical) or add a small amount of simulated noise to the vector b until one arrives at a suitable b for which the iteration converges.

8 Summary

We have looked at a method for finding the minimum 2-norm positive solution to linear inverse problems, the singular function expansion of which agrees with the truncated singular function expansion in its first $N+1$ terms. The method produces solutions with better resolution than the truncated singular function expansion one, except of course, when the two types of solution coincide. When noise is added to the data it is possible to modify the method in order to cope with ill-conditioning of the matrix H, though in this case the equality constraints may have to be slightly relaxed. The problems of the constraint qualification being violated and the primal problem being inconsistent are more serious but the latter should not be taken as a drawback solely of the method proposed here. It is a problem that every method for solving inverse problems based on constrained optimisation will inevitably suffer from when noise is present on the data.

Acknowledgements

The author would like to thank E.R. Pike and B. McNally at King's College, London for useful discussions and also I.K. Proudler at DRA, Malvern for reading and commenting on the manuscript.

Bibliography

1. de Villiers, G.D. (1996). An algorithm for reconstructing positive images from noisy data. *Proceedings of 8th European Signal Processing Conference (EUSIPCO-96)*, LINT Trieste, 895-898.

2. Borwein, J.M. and Lewis, A.S. (1992). Partially finite convex programming, parts I and II. *Mathematical Programming*, **57** , 15-48 and 49-83.

3. Bertero, M. and Pike, E.R. (1982). Resolution in diffraction-limited imaging, a singular value analysis I. The case of coherent illumination. *Optica Acta*, **29**, no.6, 727-746.

4. Borwein, J.M. and Wolkowicz, H. (1986). A simple constraint qualification in infinite dimensional programming. *Mathematical Programming*, **35**, 83-96.

5. Micchelli, C.A., Smith, P.W., Swetits, J. and Ward, J.D. (1985). Constrained L_p approximation. *Constructive Approximation*, **1**, 93-102.

6. Rockafellar, R.T. (1970). Convex analysis, Princeton University Press.

7. Bertero, M., Brianzi, P., Pike, E.R. and Rebolia, L. (1988). Linear regularising algorithms for positive solutions of linear inverse problems. *Proc.Roy.Soc.Lond.*, **A 415**, 257-275.

8. Borwein, J.M. and Lewis, A.S. (1991). Duality relationships for entropy-like minimization problems. *SIAM J. Control and Optimization*, **29**, no.2, 325-338.

9. Bertero, M., Boccacci, P. and Pike, E.R. (1982). On the recovery and resolution of exponential relaxation rates from experimental data: a singular-value analysis of the Laplace transform inversion in the presence of noise. *Proc.R.Soc.Lond.A.*, **383**, 15-29.

10. Slepian, D. and Pollak, H.O. (1961). Prolate spheroidal wave functions, Fourier analysis and uncertainty I. *Bell System Tech. J.*, **40**, 43-64.

11. Bertero, M. and Grunbaum, F.A. (1985). Commuting differential operators for the finite Laplace transform. *Inverse Problems*, **1**, 181-192.

12. Karlin, S. and Studden, W.J. (1966). Tchebycheff systems: with applications in analysis and statistics, Wiley, New York.

Treatment of Noise in Positive Reconstructions in Linear Inverse Problems

E.R. Pike and B. McNally

Department of Physics, King's College, London

Abstract

Any available a priori knowledge should be used when dealing with inverse problems. There are a large class of problems where it is known that the original object could not posses any negative regions. This information can be used to constrain the set of possible reconstructions to those which fit the data *and* are also non-negative. Unfortunately all data measurements are subject to some level of noise, and this perturbation from the true result degrades the reconstruction. Certain noise realisations may even make the construction of a positive, data consistent reconstruction impossible. This paper shows examples from these cases, and also proposes a method to allow for noise by relaxing the data consistency in a way which is commensurate with the noise levels present.

1 Introduction

The background to this work has been presented in a previous paper [1], and it is suggested that this is consulted for a fuller introduction to the problem. A brief introduction is given here. A (discretised) linear inverse problem can be defined by

$$g = Kf + n \tag{1.1}$$

where $\mathbf{f}$ is the unknown (and desired) object, $\mathbf{n}$ is some noise present in the system, $\mathbf{g}$ is the measured data and $\mathbf{K}$ is a transform kernel that operates on $\mathbf{f}$. One method of solving for $\mathbf{f}$ is to use the singular value decomposition (SVD) technique. This defines a set of functions, $\{\mathbf{u}_k\}$ and $\{\mathbf{v}_k\}$ and a spectrum $\{\sigma_k\}$ that satisfy the following relationships

$$\begin{aligned} \mathbf{K}\mathbf{u}_k &= \sigma_k \mathbf{v}_k \\ \mathbf{K}^*\mathbf{v}_k &= \sigma_k \mathbf{u}_k \end{aligned} \tag{1.2}$$

where $\mathbf{K}^*$ is the adjoint of $\mathbf{K}$.

It is a feature of linear inverse problems with compact operators that the singular values, $\{\sigma_k\}$, quickly tend to zero after some index. This corresponds to the matrix operator $\mathbf{K}$ being (almost) rank deficient. From Equations (1.1) and (1.2) it is possible to make the truncated singular value decomposition (TSVD) reconstruction

$$\tilde{\mathbf{f}} = \sum_{k=1}^{R} \frac{\langle \mathbf{g}, \mathbf{v}_k \rangle}{\sigma_k} \mathbf{u}_k \tag{1.3}$$

R is chosen (in an analagous way to a cut-off frequency) as the index which corresponds to the last $\mathbf{v}_k$ function that can be reliably measured in $\mathbf{g}$. This is usually chosen from the form of the singular value spectrum and an estimation of the noise level in the problem. However, it should also be remembered that the expected form of the object can also influence the choice of R. For example, if it is thought that the object (for whatever reason) might posses a lot of $\mathbf{u}_5$, there is little point in truncating the TSVD reconstruction at $R = 4$.

Often, upon forming the TSVD reconstruction, it is noticed that there are negative regions. These are caused by two separate effects

- All the singular function with indicies greater than R are missing in the reconstruction. These contain the "high frequency" information about the object.
- The singular functions with indicies less than or equal to R are corrupted by noise, and hence the function weights ($\langle \mathbf{u}_k, \mathbf{f} \rangle$ is the kth weight of $\mathbf{f}$) are perturbed from what they would be in the absence of noise.

If the cut-off index R is well chosen, by far the larger contribution to the negative regions comes from the missing high order information. In many situations (for example diffraction limited imaging [2], macromolecular particle sizing [3], high temperature superconductivity [4]) it is known a priori that the original object was non-negative. Hence the reconstruction must also have this property. Any reconstruction that does contain negative regions may be regarded as "wrong" in the sense that even though $\mathbf{K}\tilde{\mathbf{f}} \approx \mathbf{g}$ (where $\tilde{\mathbf{f}}$ is the TSVD reconstruction) the information about positivity is not satisfied.

One way to restore positivity, and hopefully get closer to the original object in the process, is to add in the missing high order basis functions $\{\mathbf{u}_k\}, k > R$ by quadratic programming techniques. This approach solves the *primal* problem; an example of solving the equivalent *dual* problem can be found in [5] and [6]. The quadratic programming method has the following features

1. It constrains the relaible ($k \leq R$) weights to be fixed;
2. It adds in weights of $\mathbf{u}_k, k > R$ such that
    - The reconstruction is non-negative;
    - Some aspect of the reconstruction, such as $L^{(1)}$ or $L^{(2)}$ norm, is minimised.

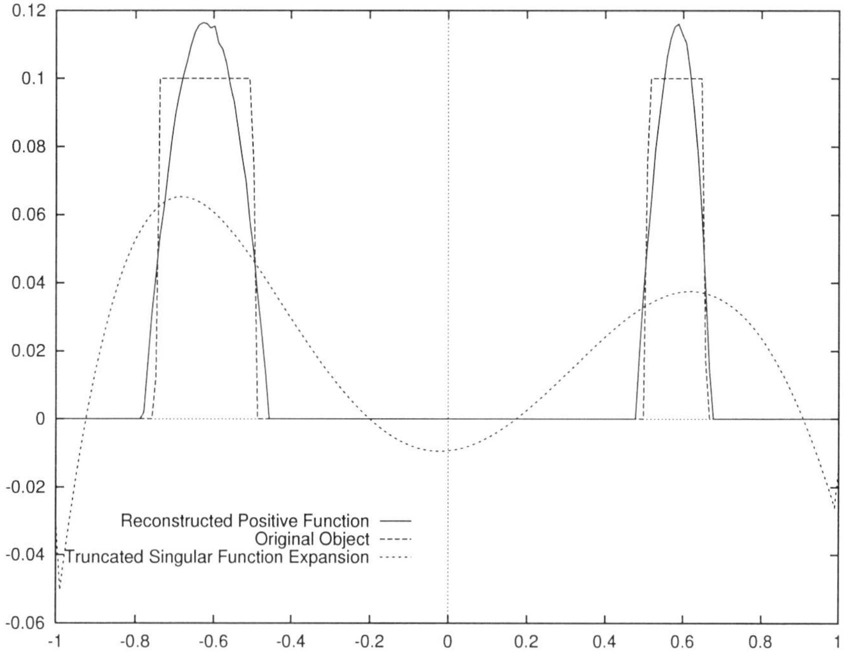

Figure 1. The object, TSVD reconstruction, and $L^{(2)}$ positive minimum norm solution corresponding to a one dimensional coherent imaging problem

Figure 1 shows a reconstruction that has been made non-negative by quadratic programming. The example corresponds to the one dimensional coherent imaging operator. Namely,

$$g(y) = \int_{-\infty}^{\infty} \frac{\sin[c(y-x)]}{\pi(y-x)} f(x)dx \tag{1.4}$$

with $c = 1.0$ and $f \in [-1, 1]$. This value of c has been chosen to match exactly a reconstruction performed by the dual method [5].

2 Noise effects

The only noise present in Figure 1 is that of round-off, and the TSVD reconstruction was truncated to six terms (this is still quite optimistic, as $\sigma_6 \approx 7 * 10^{-7}$). Real-world experiments will, however, by subject to much worse conditions than this. Different experiments will be subject to different types of noise - for example, photon counting is subject to the Poisson probability distribution in the signal, which for high count rates can be simplified to a Gaussian distribution.

In work carried out so far the authors have used a simple white noise model to degrade the data. This was done for reasons of computational simplicity and generality.

One transform that appears in calculation of the spectral weight function for the half-filled Hubbard model [4] is defined as

$$G(\tau) = \int_{-\infty}^{\infty} \frac{e^{-\omega\tau}}{1+e^{-\omega\beta}} A(\omega) d\omega \qquad (0 < \tau \leq \beta). \tag{2.1}$$

$G(\tau)$ is the measured (computed) data and $A(\omega)$ is the unknown (desired) object - the spectral wave function. An SVD approach to this problem has been performed in [4] where the TSVD reconstruction was formed. The rest of this paper concerns this problem, albeit on simulated rather than real data. This was done so that comparisons could be made between the reconstructions and the known object. Examples of $G(\tau)$ before and after noise is added can be seen in Figure 2. Before noise addition, the original object had a χ^2 value ($\chi^2 = ||\mathbf{g} - \mathbf{Kf}||^2$) of $3.2 * 10^{-29}$ and the noiseless TSVD solution had one of $3.8 * 10^{-4}$. For comparison with further χ^2 values that use the noisy data, the original object had a χ^2 value of 0.32 with this data.

Although white noise affects all image-space function weights equally (on a statistical average), it has a far greater *relative* effect on the high order weights. This is why a TSVD reconstruction only uses the weights $k \leq R$. Explicitly, the TSVD model assumes there is zero noise up to index R, and beyond this index the noise is infinite. Clearly this is a great simplification on what actually occurs. Still, a TSVD technique can still be said to cope with noise in the respect that high order perturbations are orthogonalised out of the reconstruction. Figure 3 shows the TSVD reconstruction from the noisy data presented in

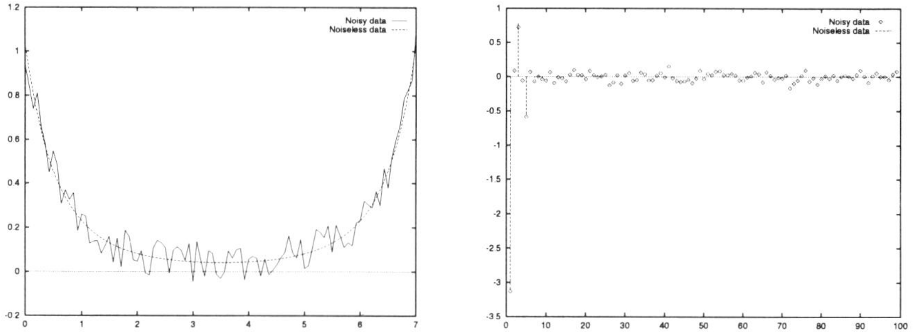

Noisy and noiseless data Projections onto $\mathbf{v}_k$

Figure 2. A comparison of data before and after white noise has been added. Note that (on average) white noise affects all singular function weights to the same extent

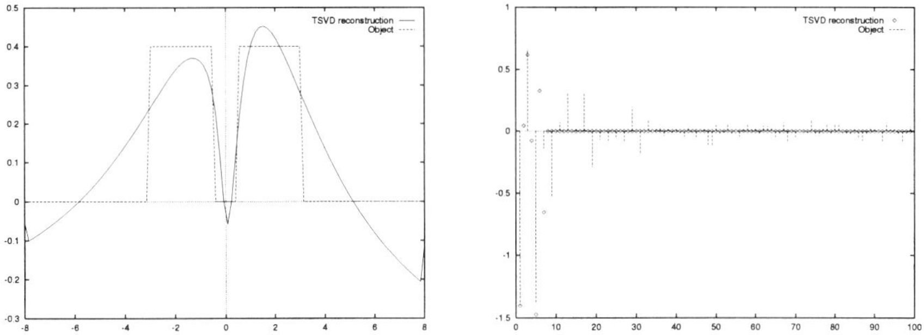

TSVD　　　　Decomposition weights

Figure 3. TSVD reconstruction of the spectral wave density made from the data presented in Figure 2. $\chi^2 = 0.30$. No positive reconstruction was possible with this noise realisation and the first R components defined by their TSVD values

Figure 2. Unfortunately this realisation of noise violated the existence criterion - that a positive function exists with its first R weights equal to those of the TSVD reconstruction. Hence, while a positive reconstruction could have been found from using the noiseless data, the experimentally calculated noisy data has no such reconstruction. The problem of existence is very hard to compute before a reconstruction is attempted, so this is usually the first point at which an existence violation is noticed.

3 Dealing with noise

However, it is possible to treat the effects of noise in a fairly intuative way. Since the first R weights are actually perturbed in some way by noise, it is unlikely that their "true" value is the one calculated from experiment. Setting

$$g = g\prime + n \tag{3.1}$$

where $\mathbf{g}$ is the recorded data, $\mathbf{g}\prime$ is the noiseless data and $\mathbf{n}$ is the noise shows that a reconstruction from $\mathbf{g}$ is actually

$$\begin{aligned} \tilde{\mathbf{f}} &= \sum_{k=1} \frac{\langle \mathbf{g}, \mathbf{v}_k \rangle}{\sigma_k} \mathbf{u}_k \\ \tilde{\mathbf{f}} &= \sum_{k=1} \left\{ \frac{\langle \mathbf{g}\prime, \mathbf{v}_k \rangle}{\sigma_k} + \frac{\langle \mathbf{n}, \mathbf{v}_k \rangle}{\sigma_k} \right\} \mathbf{u}_k . \end{aligned} \tag{3.2}$$

In effect, the reconstruction is made up from a sum of "true" weights $(\langle \mathbf{g}\prime, \mathbf{v}_k \rangle / \sigma_k)$ and a sum of "noise" weights $(\langle \mathbf{n}, \mathbf{v}_k \rangle / \sigma_k)$. This suggests one possible method for dealing with noise. In summary, the steps are as follows

1. Form a data vector containing a multiple, α, of the expected amplitude of the noise at each position. α, should be chosen to represent the maximum noise that is likely in the experiment. Call this vector η.
2. Make the reconstruction from *all* the $\mathbf{u}$ basis functions - not just the first R as in TSVD - but constrain the calculated weights to vary within upper and lower bounds at each index.
3. These bounds are given by

$$\begin{aligned} Upper\ bound &= \frac{\langle \mathbf{g}\prime, \mathbf{v}_k \rangle}{\sigma_k} + \frac{|\langle \eta, \mathbf{v}_k \rangle|}{\sigma_k} \\ Lower\ bound &= \frac{\langle \mathbf{g}\prime, \mathbf{v}_k \rangle}{\sigma_k} - \frac{|\langle \eta, \mathbf{v}_k \rangle|}{\sigma_k}. \end{aligned}$$

4. Employ a QP routine to form a positive reconstruction that has weights within these bounds and some aspect optimised - such as a given norm.

In practice, though, these steps are not optimal. The actual procedure employed by the authors was to apply the above method to only the first R weights, and to treat the rest as completely unknown. In effect this is a trade off between existence and data consistency. Applying tight error bounds to the first R weights keeps the reconstruction as consistent as possible with the data, while

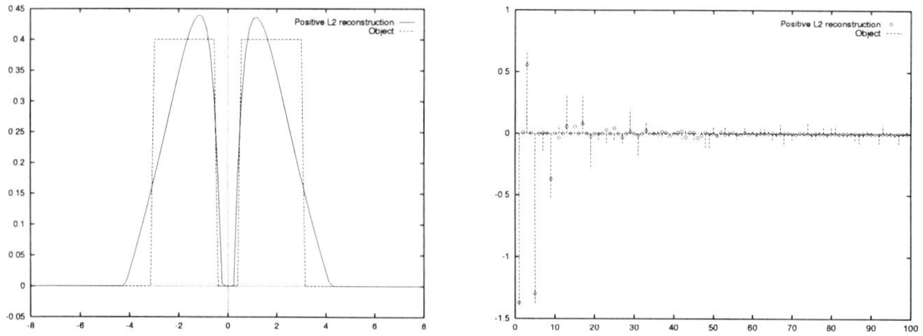

Positive Reconstruction Decomposition

Figure 4. An $L^{(2)}$ positive reconstruction with $\alpha = 1.5$ from the noisy data in Figure 2. This gives a χ^2 value of 0.33

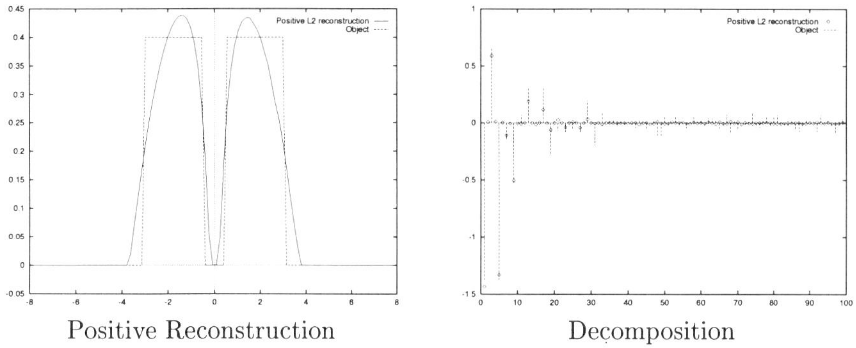

Figure 5. An $L^{(2)}$ positive reconstruction with $\alpha = 1.5$ from the noisy data in Figure 2 and the first three moments constrained to their analytic values. This gives a χ^2 value of 0.32

allowing the rest of the weights to assume any value maximises the chance that a positive solution exists. It is up to the individual to decide an appropriate value for α in this balance. Taking the same example as before, Figure 4 is a positive reconstruction from the noisy data with $\alpha = 1.5$.

3.1 Problems with noise estimation

Many real-world noise distributions (such as Poisson and Gaussian) have no upper limit on their noise. A noise level ten or a thousand times the mean is quite possible, although being increasingly unlikely. Therefore, α may have to be chosen sufficiently large that the weight bounds will bracket the true weight of the original object. Unfortunately, if α is chosen too large the reconstruction will no longer be consistent with the data.

Fortunately, in the case of the Hubbard model there is extra a priori information available in the form of low-order moments. The zeroth, first and second moments have been analytically calculated in [8], and these can be used as extra constraints in the reconstruction process. Figure 5 shows the reconstruction from the noisy data with the first three moments constrained to their analytic value.

4 Problems with uniqueness

So far the optimisation criterion for the quadratic programming module has always been to minimise the $L^{(2)}$ norm of the reconstruction. Micchelli *et.al.* [7] have proved that a unique minimum $L^{(2)}$ norm exists for the first R weights fixed. This avoids problems with degeneracy of solutions that may exist for other optimisation criteria - such as the minimum $L^{(1)}$ norm.

One might think that since a unique $L^{(2)}$ norm exists that it would always be best to use this as the optimisation criterion. This is not so. There is absolutely nothing to suggest that the unknown object is the positive, minimum $L^{(2)}$ norm realisation for the first R weights fixed. In fact, it would be surprising if the unknown object posessed any simple property like that.

The only justification for using the $L^{(2)}$ norm over other optimisation choices is that it is the choice that adds in the *least extra information* to make the reconstruction positive. A corollary of this is that if the TSVD is already non-negative then the $L^{(2)}$ minimised Q.P. enhanced reconstruction will be identical to the TSVD solution. In this situation other minimisation schemes will add in extra spurious information.

Still, it is important to remember that other reconstruction forms with the first R weights correctly chosen are just as valid as the $L^{(2)}$ minimised solution. An example of using different optimisation choices is presented in Figure 6. These

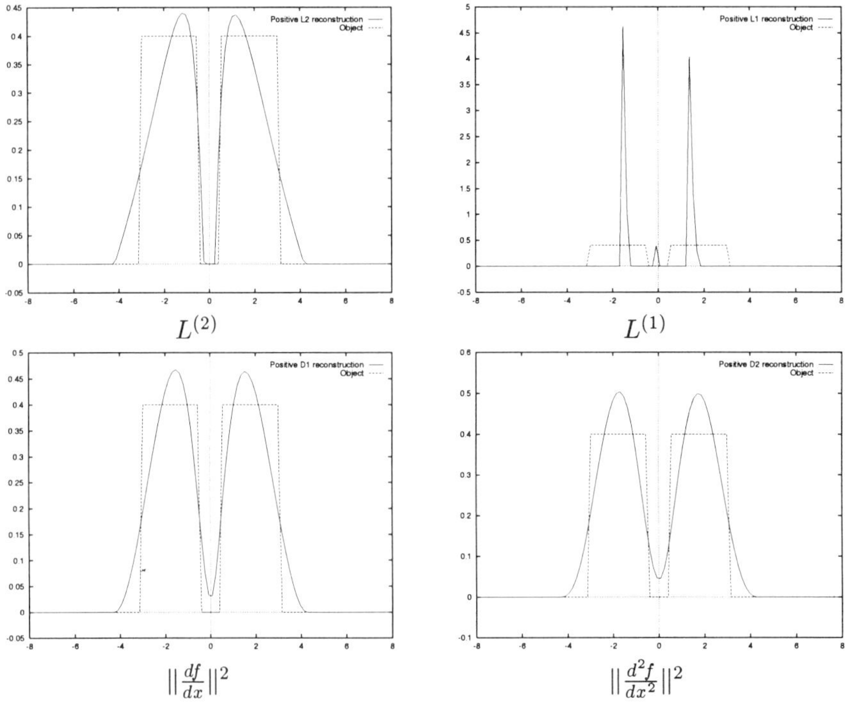

Figure 6. Four different and equally valid possible reconstructions. Moment constraints have not been imposed

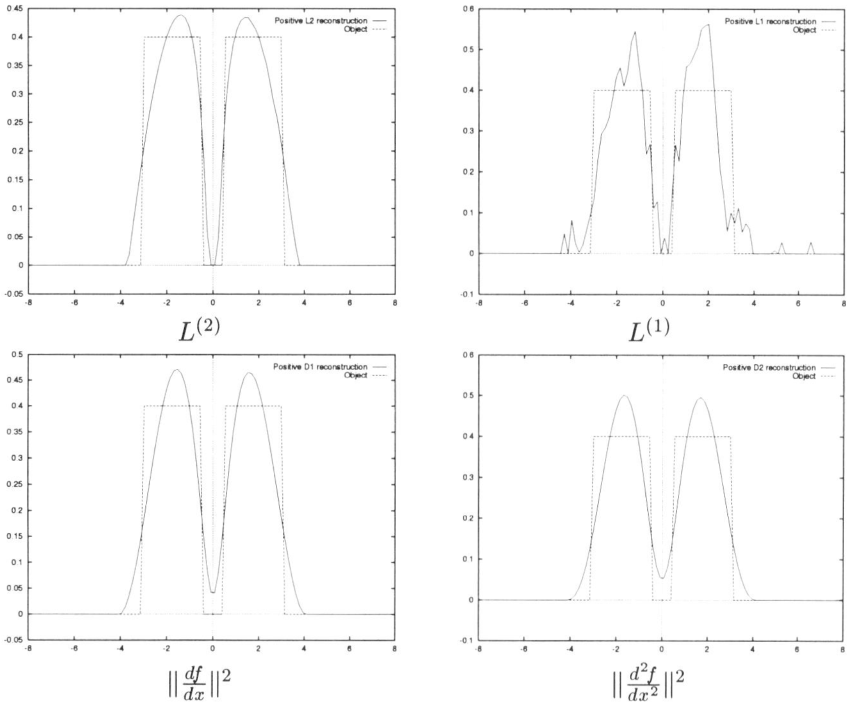

Figure 7. Four different and equally valid possible reconstructions that have had their first 3 moments constrained to that of the unknown object. The norm minimised is shown in each case

are reconstructions from the noisy data in Figure 2. The quantities minimised are shown. Moment constraints have not been imposed for the reason that they are particular to the Hubbard model and are not usually available for other linear inverse problems. Figure 7 shows the convergence to a common solution that the imposition of known moments produces.

5 Conclusion

Several graphical examples have been presented showing the non-uniqueness of positive reconstructions that still fit the recorded data. Noise is present to varying degrees in all real applications, and any inversion technique must be able to cope with it in some manner. The approach taken in this paper is to form upper and lower bounds on the first R (reliable) reconstruction function weights. These bounds are calculated from the singular value spectrum and an estimation of the noise in the problem. Quadratic programming is then used to form a positive reconstruction with an optimal attribute (such as a given norm) by choosing suitable values for the first R and remaining weights. Theoretically, there is no maximum level to many noise distributions, which leads to a compromise between data consistency and existence of a positive solution. However, the χ^2 values produced from the various reconstructions all turned out to be very similar to that produced by the original object - suggesting that data consistency has been maintained.

Acknowledgements

This material is based upon work supported in part by the US Army Research Office under grant No. DAAH04-95-1-0280.

The authors would also like to thank G. de Villiers who has provided frequent support and discussions for this work. E. Klepfish and P. Kornilovitch of KCL physics department have also provided assistance on the Hubbard model problem.

Bibliography

1. B McNally and E R Pike, (1997), *Lecture Notes in Physics, (Springer Verlag, Heidelberg)*, pp. 18–26, Mathematical Programming for Positive Solutions of Ill-conditioned Inverse Problems, G Chavent and P C Sabatier, (eds.).

2. E. G. Steward, (1983), *Fourier Optics*, Ellis Horwood Limited.

3. H Z Cummins & E R Pike, (eds.), (1974), *NATO Advanced Study Institute Series B3: Physics*, Plenum Press.

4. C Creffield, E Klepfish, P Kornilovitch, E R Pike and S. Sarkar, (1995), *Phys. Rev. Letts.*, Spectral Weight Function for the Half-Filled Hubbard Model: A Singular Value Decomposition Approach **75**, 517.

5. G D de Villiers, (1997), On the use of positivity in reconstructing images from noisy data, *IMA Conference Proceedings on "Mathematics in Signal Processing"*.

6. R T Rockafellar, (1970), *Convex Analysis*, Princeton University Press.

7. C.A. Micchelli, P.W. Smith, J. Swetits and J.D. Ward, (1985), *Constructive Approximation*, **1**, 93-102.

8. S R White, (1991), *Phys. Rev. B* **44**, 5670 ; M Vekić and S R White, (1993), *op.cit.* **47**, 1160.

Ill Conditioning and Gradient Based Optimization of Multi-layer Perceptrons

Michael Kirby

Department of Mathematics, Colorado State University, Fort Collins, USA

Abstract

This paper addresses the issue of the ill-conditioning of the Jacobian matrix required for gradient based optimization of sigmoidal feedforward neural networks used for function approximation. The condition number of the Jacobian is computed via the singular value decomposition for several training problems and architectures and is shown to be typically ill-conditioned. In addition, a new *batch* training algorithm, referred to as alternating principal direction descent (APDD), is proposed which appropriate when highly accurate approximations are desired. APDD escapes local minima in a manner analogous to *on-line* training, but exploits information from the entire data set at each iteration. The performance of several optimization procedures is examined and compared including steepest descent, conjugate gradient, memoryless Broyden-Fletcher-Goldfarb-Shanno and APDD.

1 Introduction

The performance of gradient based nonlinear least squares optimization procedures depends on the condition number of the the Jacobian and Hessian matrices of the error function (computed w.r.t. the unknown weight parameters). The fact that feedforward neural networks used for classification are often ill-conditioned was pointed out in [1] within the context of training indicator functions in the plane. In [1] several possible causes for ill-conditioning are enumerated; these include overdetermined architectures and the possibility that the sigmoidal transfer functions are operating in a range where they are saturated.

One consequence of the ill-conditioning is that for some problems traditionally slow methods such as simple steepest descent perform better than more elaborate procedures with generally faster convergence rates, such as conjugate gradient or Newton's method. Thus it appears that the degree of ill-conditioning may play an important role in the selection of the optimization procedure for a given problem. In addition, once ill-conditioning has been identified transformations of the data may be sought to improve the condition number [2]. Furthermore, the condition number of the Jacobian may provide a powerful guide for designing and evaluating dynamical and statistical constraints on feedforward networks such as those considered in [5].

An issue closely related to the ill-conditioning of the optimization problem is the existence of local minima in the error function. It is common practice in training neural networks to avoid local minima by overparametrizing the network, i.e., by using more than the minimal number of weights required. Additionally *on-line* learning methods are employed which compute the gradient direction and update based on single training patterns. While the on-line approach is sufficient for many problems, it is subject to *unlearning*, i.e., there is a non-monotonic approach to the error. As a result it is inadequate for some training architectures such as the bottleneck architecture for nonlinear auto-association where the unlearning problem can be severe [4, 5, 6].

Batch training, the alternative to on-line training, consists of using all the available data to compute an averaged descent direction each iteration. This approach can be problematic due to the natural tendency for the method to discover the local minima. In this paper we present a new batch optimization algorithm which avoids local minima while still using the entire data set at each iteration. The method will be referred to as Alternating Principal Direction Descent (APDD) and can be used in combination with other methods or on its own to produce approximations with higher accuracy. APDD works by first computing the singular value decomposition (SVD) of the Jacobian; it then sweeps through the column space of the gradient matrix using each direction as a descent vector.

In this paper we examine the conditioning of the Jacobian for the problem of smooth function approximation for a variety of training sets and architectures. In Section 2 the matrix formulation of the backpropagation is reviewed. In Section 3 we examine the condition number of a matrix as determined by the SVD. In Section 4 several computations are provided to demonstrate the ubiquitousness of ill-conditioning. In Section 5 the APDD method is outlined and some sample training comparisons are presented. In Section 6 the main results of the paper are summarized with some remarks concerning the future direction of the research.

2 Matrix formulation of backpropagation

In this section we briefly derive the backpropagation equations [8, 9] for computing the derivative emphasizing the matrix form. To avoid any confusion, it should be stressed that backpropagation is only an efficient method for computing the derivatives which in no way contributes to the ill-conditioning under consideration.

A feedforward neural network may be viewed as a composition of L mappings

$$\mathbf{s}(\mathbf{x}) = (\mathbf{s}^{(L)} \circ \cdots \circ \mathbf{s}^{(1)})(\mathbf{x})$$

where each individual mapping is of the form

$$\mathbf{s}^{(i+1)} = \mathbf{s}^{(i+1)}(\mathbf{s}^{(i)}; \mathbf{w}^{(i)})$$

with $\mathbf{s}^{(i)} : X^{(i-1)} \rightarrow X^{(i)}$ and $X^{(i)} \subset R^{l_i}$. In addition let $\mathbf{x} \equiv \mathbf{s}^{(0)}$. Each mapping $\mathbf{s}^{(i)}$ is defined by its functional representation along with a vector of parameters $\mathbf{w}^{(i)}$. Taken together, the composition of mappings depends on the concatenated vector $\mathbf{w} = (\mathbf{w}^{(0)^T}, \ldots, \mathbf{w}^{(L-1)^T})^T$. In what follows we will refer to network architectures by the number of variables in each layer, i.e., $l_0 - l_1 - \ldots - l_L$.

We seek to determine compositions of mappings such that for a given collection of points in the domain, or inputs, $\{\mathbf{x}^{(\mu)}\}_{\mu=1}^{P}$, the associated images of the composition $\mathbf{s}(\mathbf{x}^{(\mu)})$ match desired (or target) values $\mathbf{f}^{(\mu)}$. The difference between the desired value and the output of the mapping $\mathbf{r}|_\mu = \mathbf{s}^{(L)}(\mathbf{x}^{(\mu)}) - \mathbf{f}^{(\mu)}$ (a column vector) is referred to as the residual. In the typical least squares formulation the composition of functions is chosen such that mean square error

$$E_{ms} = 1/2 \sum_{\mu} \sum_{i} r_i^2|_\mu \tag{2.1}$$

is as small as possible. (In the derivation below and in network training we take E to be E_{ms} but the errors tabulated in this paper are root-mean square, i.e., $E_{rms} = \sqrt{\frac{1}{2Pl_L} \sum_\mu \sum_i r_i^2|_\mu}$.)

Gradient based techniques iteratively compute the optimal free parameters along a direction of descent which depends on the gradient. This gradient is obtained by differentiating the error w.r.t. $\mathbf{w}$ via the chain rule which, after transposing, gives

$$\frac{\partial E}{\partial \mathbf{w}} = \mathbf{J}_\mathbf{w}^T \mathbf{r} \tag{2.2}$$

where the Jacobian $\mathbf{J}_\mathbf{w} = \frac{D\mathbf{s}^{(L)}}{D\mathbf{w}}$ and the column vectors are defined

$$\frac{\partial E}{\partial \mathbf{w}} = (\frac{\partial E}{\partial w_1}, \ldots, \frac{\partial E}{\partial w_M})^T \quad \text{and} \quad \mathbf{r} = \frac{\partial E}{\partial \mathbf{s}^{(L)}} = (\frac{\partial E}{\partial s_0^{(L)}}, \ldots, \frac{\partial E}{\partial s_{l_L-1}^{(L)}})^T$$

For batch learning the gradient is averaged over all the points in the domain

$$\langle \frac{\partial E}{\partial \mathbf{w}} \rangle = \sum_{\mu=0}^{P-1} \mathbf{J}_\mathbf{w}^T \mathbf{r}|_\mu \tag{2.3}$$

or in block matrix notation

$$\langle \frac{\partial E}{\partial \mathbf{w}} \rangle = \mathbf{G}\bar{\mathbf{r}} \tag{2.4}$$

where the *gradient matrix* $\mathbf{G} = [\mathbf{J}_\mathbf{w}^T|_0 \cdots \mathbf{J}_\mathbf{w}^T|_{P-1}]$ is a block matrix of size $M \times l_L P$ while $\bar{\mathbf{r}} = (\mathbf{r}_0^T, \ldots, \mathbf{r}_{P-1}^T)^T$ is the column vector of length M consisting of all the residuals.

Now define the Jacobian of the map *between* layers i and $i+1$ as

$$\mathbf{J}^{(i)} = \frac{D\mathbf{s}^{(i+1)}}{D\mathbf{s}^{(i)}}$$

as well the matrix

$$\mathbf{A}^{(i)} = \frac{D\mathbf{s}^{(i+1)}}{D\mathbf{w}^{(i)}}.$$

Given the mapping between layers is the application of an affine transformation followed by the action of a standard sigmoidal function the matrices $\mathbf{J}^{(i)}$ and $\mathbf{A}^{(i)}$ may be readily calculated.

Now the chain rule may be written as

$$\frac{\partial E}{\partial \mathbf{w}^{(i)}} = \mathbf{A}^{(i)T}\mathbf{J}^{(i+1)T}\ldots\mathbf{J}^{(L-1)T}\mathbf{r}. \tag{2.5}$$

Hence the computation of the matrix

$$\mathbf{D}^{(i)} = \mathbf{J}^{(i+1)T}\ldots\mathbf{J}^{(L-1)T} \tag{2.6}$$

may implemented efficiently via the recursion, the main element of the *backpropagation* algorithm, given by

$$\mathbf{D}^{(i)} = \mathbf{J}^{(i+1)T}\mathbf{D}^{(i+1)} \tag{2.7}$$

with the initialization

$$\mathbf{D}^{(L-1)} = \mathbf{I}_{l_{L-1}\times l_L}. \tag{2.8}$$

Writing the gradient matrix in terms of its component block matrices

$$\mathbf{G} = \begin{pmatrix} \mathbf{A}_0^{(0)}\mathbf{D}^{(0)}|_0 & \cdots & \mathbf{A}_0^{(0)}\mathbf{D}^{(0)}|_{P-1} \\ \mathbf{A}_{l_0-1}^{(0)}\mathbf{D}^{(0)}|_0 & & \mathbf{A}_{l_0-1}^{(0)}\mathbf{D}^{(0)}|_{P-1} \\ & & \\ \mathbf{A}_0^{(L-1)}\mathbf{D}^{(L-1)}|_0 & & \mathbf{A}_0^{(L-1)}\mathbf{D}^{(L-1)}|_{P-1} \\ \mathbf{A}_{l_{L-1}-1}^{(L-1)}\mathbf{D}^{(L-1)}|_0 & \cdots & \mathbf{A}_{l_{L-1}-1}^{(L-1)}\mathbf{D}^{(L-1)}|_{P-1} \end{pmatrix}. \tag{2.9}$$

If $M > l_L P$ then the optimization problem is underdetermined and if $M < l_L P$ then the problem is overdetermined.

3 The conditioning of G

The rank of any matrix is defined as its number of non-zero singular values which may be obtained via the SVD. The decomposition of the matrix $\mathbf{G}$ in terms of its singular values and vectors is

$$\mathbf{G} = \mathbf{USV}^T$$

(note rank $\mathbf{G}$ = rank $\mathbf{J_w}$) where

$$\mathbf{U}^T\mathbf{U} = \mathbf{I}_M \quad \text{and} \quad \mathbf{V}^T\mathbf{V} = \mathbf{I}_N$$

and $N = l_L P$. The $M \times N$ diagonal matrix $\mathbf{S} = (\sigma_1, \ldots, \sigma_n)$ consists of the ordered, non-negative singular values $\sigma_1 \geq \cdots \geq \sigma_n \geq 0$. Formally, if there are $r \leq \min(M, N)$ non-zero singular values, i.e., $\sigma_r \neq 0$ and $\sigma_{r+1} = \cdots = \sigma_q = 0$ then the rank of the matrix is r. Furthermore, the *condition number* of $\mathbf{G}$ is defined as $\kappa(\mathbf{G}) = \sigma_1/\sigma_r$. The matrix $\mathbf{G}$ is said to be ill-conditioned if $\kappa(\mathbf{G})$ is large.

In practice determining the condition number from the singular spectrum is difficult since it is standard that many eigenvalues may be very close to zero, therefore estimating the rank is not easy. Hence, as in [1], we include plots of the full spectrum of singular values. More details concerning the SVD may be found in [7].

4 Computations

In the following sections we train several feedforward neural networks using a variety of optimization methods including simple steepest descent, steepest descent with a method of lines (MOL) routine, conjugate gradient, memoryless Broyden-Fletcher-Goldfarb-Shanno (m-BFGS) and alternating principal direction descent (APDD). APDD is described in Section 5 while the other methods are well described in the optimization literature, see [2] and [3].

4.1 The training algorithm

All of the networks were trained using the batch-mode approach in the standard fashion which we briefly summarize here. To begin, the points in each data set are cyclicly presented to the network. For each point the output of the network is computed, as well as the resulting residual. The matrices $\mathbf{D}^{(i)}$ and $\mathbf{A}^{(i)}$ are then evaluated at this point and the associated block entry of $\mathbf{G}$ is computed. After the cycle is completed the P blocks of the gradient matrix $\mathbf{G}$ are known. The weights are then updated using $\mathbf{w} := \mathbf{w} + a\mathbf{d}$ where $\mathbf{d}$ is a function of the gradient $\mathbf{g} = \mathbf{G}\bar{\mathbf{r}}$ according to the method of optimization being used and $a \in R$. Note that the method of lines is employed to determine a in all algorithms except for simple gradient descent where the fixed step size, or *learning rate*, is taken to be $a = 0.01$. This is of course not guaranteed to be the optimal value.

4.2 Function approximation: the parabola

The parabola training data consists of 100 evenly sampled points from the function $y(x) = (x^2 - 1)/2$ on the closed interval $[-1, 1]$. Several training runs of 500 iterations using the m-BFGS method were executed. The initial and final spectra of the gradient matrix $\mathbf{G}$ are shown in Figure 1 for an overparametrized 1-6-1 network. The corresponding spectra for a parsimonious 1-2-1 network are

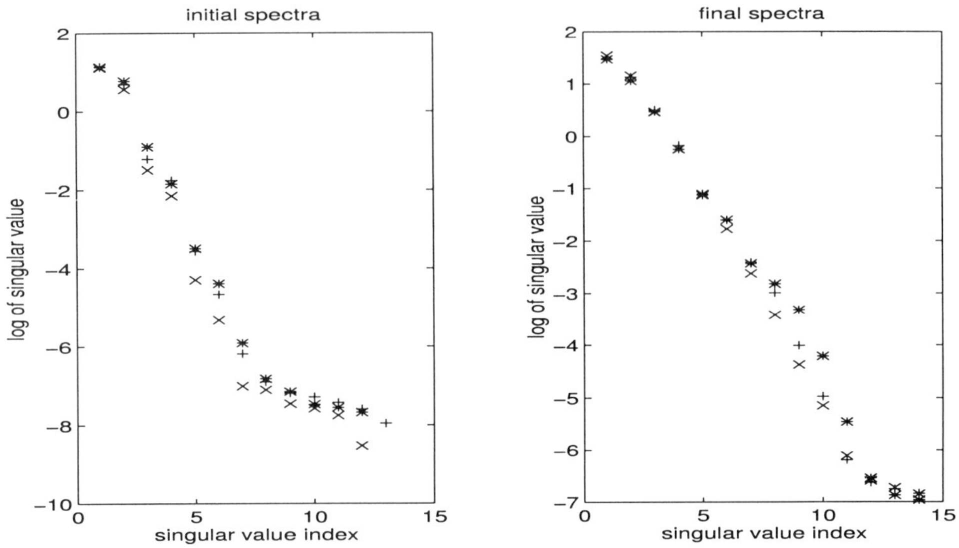

Figure 1. Initial and final singular value spectra (log base 10) for the 1-6-1 network using the 100 evenly sampled points of the parabola as the training set. The final spectra were recorded after 500 iterations and the associated mean square error for each network was small, i.e., $E < .001$. The training method used was the m-BFGS algorithm. Note that $\mathbf{G}$ is a 19×100 matrix and singular values not displayed are zero

shown in Figure 2. Note the local minima found in the 1-2-1 network and the associated extreme ill-conditioning of $\mathbf{G}$. In practice, reducing the size of an overdetermined network may improve the condition number, but may also lead to an increased number of local minima. In Section 5 it will be seen that the APDD approach is suitable for improving solutions for parsimonious networks.

By examining Figure 3 we see that the ill-conditioning persists throughout learning. Note that a 3-dimensional subspace of the columns of $\mathbf{G}$ contains over 98% of the variance.

4.3 Nonlinear dimensionality reduction: the Lorenz attractor

A *sub-composition* of $k-1$ mappings

$$(\mathbf{s}^{(i+k)} \circ \ldots \circ \mathbf{s}^{(i+1)})(\mathbf{s}^{(i)}) : X^{(i)} \rightarrow X^{(i+k)}$$

is a dimensionality reducing mapping, or *reduction* of $X^{(i)}$, if

$$l_{i+1} < l_i.$$

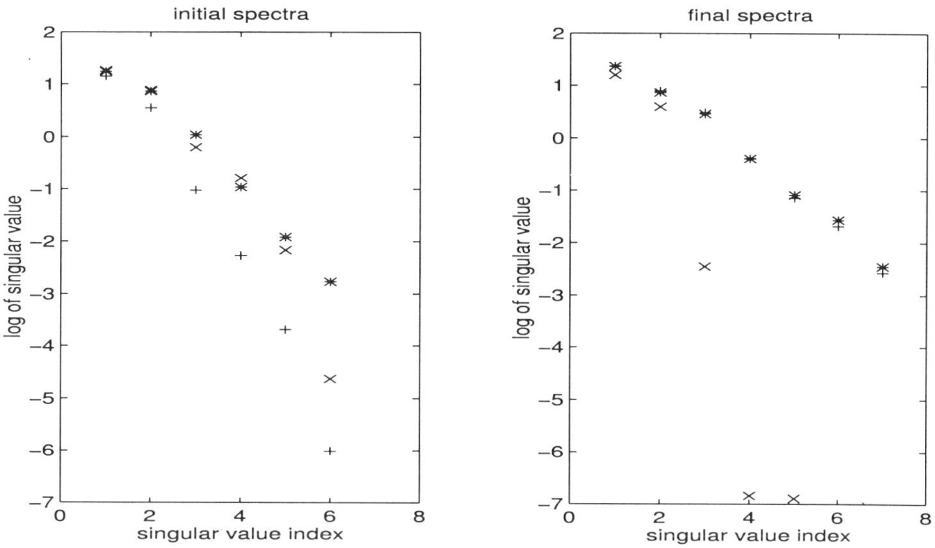

Figure 2. Initial and final singular value distributions for the 1-2-1 network using the 100 evenly sampled points of the parabola. The final spectrum of singular values denoted by crosses "x" corresponds to a local minima reached after 38 iterations. The other two networks both converged after 500 iterations to $E < .001$. The training method was the m-BFGS algorithm. Now $\mathbf{G}$ is a 7×100 matrix

If $\mathbf{s}(\mathbf{x}) = \mathbf{x},\ \ \forall \mathbf{x} \in X^{(0)}$ and $l_i < l_0$ for some i, then $\mathbf{s}^{(i)} = \mathbf{s}^{(i)}(\mathbf{x})$ are reduced coordinate functions for $\mathbf{x}$. Note that it will only be possible to train the identity mapping if the data in the ambient space $X^{(0)}$ reside on a manifold which is homeomorphic to R^{l_i}. If this is not true than a *lossy* nonlinear compression of the data will result.

Also, the mapping

$$\mathbf{s}_\theta : X^{(i)} \rightarrow X^{(L)}$$

$$\mathbf{s}_\theta = \mathbf{s}(\mathbf{s}^{(i)})$$

provides a *parametrization* of the ambient space $X^{(0)}$.

We computed the nonlinear reduction of test data on the Lorenz attractor consisting of 30 well spaced points in R^3 using architectures with dimensions 3-8-1-8-3 and 3-4-4-1-8-3. The gradient matrix $\mathbf{G}$ has size 84×90 for both networks. These architectures correspond to a best 1-dimensional nonlinear fit to the data. The may be viewed either as a 1-dimensional parametrization, or as the first step in the sequential reduction procedure proposed in [6]. (Note that a zero

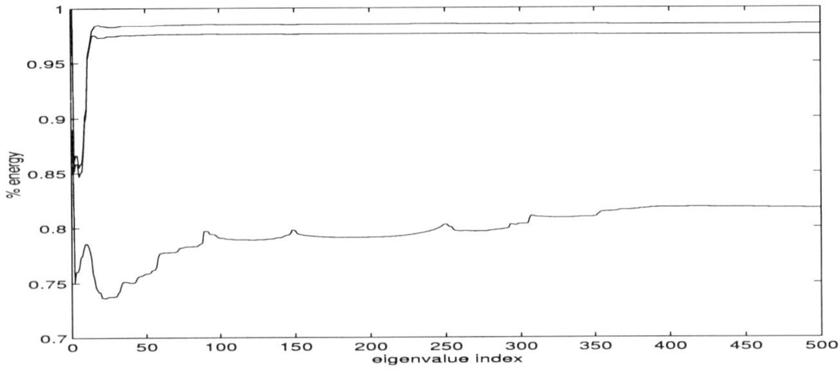

Figure 3. The 3 lines correspond to the variance captured by partial sums of the first 3 eigenvalues (singular values squared) of the covariance matrix. The remaining partial sums look similar establishing that the ill-conditioning is not exceptional, but persists throughout training which has converged to a very small error after 500 iterations on the parabola data using the 1-6-1 network. The training method was again m-BFGS

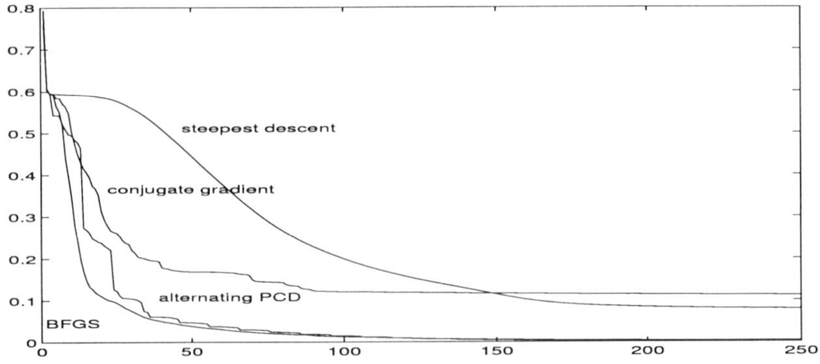

Figure 4. *Average Error vs Iteration.* Errors averaged over 25 different runs are displayed for several different optimization methods. The data is sampled from the parabola and the network has the 1-6-1 architecture

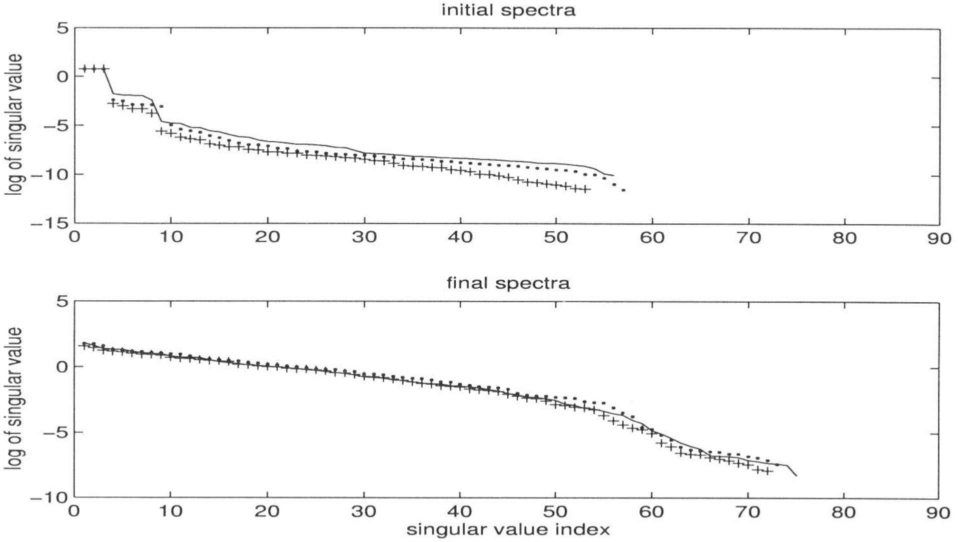

Figure 5. The initial (above) and final (below) singular value spectra for the bottleneck architectures 3-8-1-8-3 and 3-4-4-1-8-3. The latter architecture is trained with both APDD ($E_{rms} = 0.100$, 5000 iterations) and m-BFGS ($E_{rms} = 0.145$, 702 iterations). The corresponding spectra are denoted by the symbols "+" and ".", respectively. The solid lines correspond to the initial and final spectra of the 3-8-1-8-3 network training with m-BFGS, ($E_{rms} = 0.148$, 856 iters)

error for this problem would amount to constructing one of the $P!$ lines through the data points.) More details from the results of these experiments are given in Section 5.0.2. It is clear from Figure 5 that these training problems are ill-conditioned.

5 Alternating principal direction descent

When $\mathbf{G}$ is ill-conditioned, gradient descent algorithms may perform very poorly. In this section we present an idea for exploiting the information provided by the SVD of the gradient matrix $\mathbf{G}$ which is especially useful for ill-conditioned matrices.

A rank 1 approximation to the gradient matrix $\mathbf{G}$ may be defined as

$$\mathbf{G}_k = \mathbf{U}\mathbf{S}_k\mathbf{V}^T$$

where $\mathbf{S}_k = diag(0, \ldots, \sigma_k, \ldots, 0)$, $k = 1, \ldots, r$ and $r = \text{rank } \mathbf{G}$. Once $\mathbf{G}$ has been computed by the training algorithm, APDD executes a sweep of r updates using the r rank-1 approximations $\mathbf{G} \approx \mathbf{G}_k$.

Individually these approximations to $\mathbf{G}$ will degrade as k increases. However, the advantage of defining approximations in this way is that it guarantees that each direction in the subspace spanned by the columns of $\mathbf{G}$ will be updated using a scale factor a_k optimized separately (via MOL).

Note that this procedure requires the computation of the SVD every r iterations. This can be expensive so the method may be best recommended as being used in conjunction with methods susceptible to local minima, which APDD is excellent at escaping as demonstrated in the following runs.

Given the number of variables governing the performance of an optimization procedure, it is very difficult to benchmark optimization routines in any absolute sense. In this paper the emphasis is on determining poor properties of the optimization procedures which which additional CPU time can't cure.

5.0.1 The parabola

It is interesting to compare the training statistics for the 1-2-1 network shown in Table 1 with the 1-6-1 network statistics displayed in Table 2. The average error was computed for 100 individual runs limited to a maximum of 500 iterations using several different optimization methods. Again, it is not the goal of this paper to assess which technique requires the minimum CPU time for a given error tolerance, but rather to infer properties concerning the optimization methods from their behaviors on the different networks. For instance, note the existence of local minima is revealed by a large standard deviation in the average error. In general we see that the standard deviations are higher for the 1-2-1 net as expected since there are fewer degrees of freedom which may be used to escape a local minimum. A notable exception is the conjugate gradient which apparently is more negatively affected by the increase in ill-conditioning.

Table 1. Comparison of parabola training statistics for various descent procedures. Training is based on 500 iterations and the statistics compiled over 100 randomly initialized training runs using a 1-2-1 network

Method *parabola data*	**Average error**	**Minimum error**	**Standard deviation**
Steepest descent	0.112261	0.034907	0.105023
Steepest descent (MOL)	0.086500	0.009718	0.101432
Conjugate Gradient	0.090789	0.014892	0.106893
m-BFGS	0.032042	0.000329	0.090698
APDD	0.008164	0.000248	0.001317

Table 2. Comparison of parabola training statistics for various descent procedures using the 1-6-1 network. Training is based on 500 iterations and the statistics compiled over 100 randomly initialized training runs

Method *parabola data*	**Average error**	**Minimum error**	**Standard deviation**
Steepest descent	0.073055	0.049461	0.057090
Steepest descent (MOL)	0.039751	0.027891	0.040153
Conjugate Gradient	0.095026	0.025400	0.110901
m-BFGS	0.000927	0.000242	0.000943
APDD	0.001290	0.000153	0.001357

The most accurate solutions are seen to be obtained by the APDD. Although this also comes at greater expense, the APDD is far better at escaping local minima (see the low standard deviations) and hence is suitable when highly accurate solutions are required. As an additional example, the APDD approach was able to determine a solution to the XOR classification problem to within an error of 10e-10 while the next best solution was 10e-3 using m-BFGS in 100 trials on a parsimonious 2-1-1 net.

The m-BFGS performs excellently on the very ill-conditioned 1-6-1 network and degrades for the better conditioned 1-2-1 problem given its tendency to compute local minima. This technique has not been widely used in the neural network literature which is surprising given its suitability for problems with a large number of parameters (it is *memoryless*, see [2]).

Steepest descent performs surprisingly well compared with significantly more expensive methods suggesting that its coarse approach is less susceptible to the adverse effects of ill-conditioning. For further comparisons see Figure 4.

5.0.2 The Lorenz data

One of the main disadvantages of the 1-dimensional reduction architectures 3-8-1-8-3 and 3-4-4-1-8-1 is the enormous number of local minima which plague batch learning procedures and the poor performance of on-line optimization due to *unlearning*. Note that the problem itself is ill-posed, but nonetheless it is useful to determine (non-unique) solutions. We allowed the APDD method to train for 5000 iterations during which the error steadily decreased to $E_{rms} = 0.10$. The other sample runs using m-BFGS stopped due to local minima after less than 1000 iterations. The singular value spectra for these runs are shown in Figure 5. It is interesting to note that the Jacobian of these networks span essentially a 3-dimensional subspace at initialization, a fact which is clearly independent of the training method.

6 Conclusions

This paper demonstrates that the approximation of smooth functions using multi-layer perceptrons by means of gradient based methods generally involves using ill-conditioned Jacobians. As a result the standard optimization procedures may behave poorly, such as conjugate gradient, or better than expected, such as gradient descent. The memoryless BFGS method appeared to perform exceedingly well on ill-posed problems.

The alternating principal direction descent (APDD) method was proposed as a means to escape local minima using a batch method. It obtained the most accurate solutions and might be used efficiently in conjunction with other methods, such as the m-BFGS.

Further work is required to exploit the knowledge of the condition number of the Jacobian matrix for optimizing learning. Ill-conditioned problems might be improved by appropriate scaling of the data. In addition, it may be possible to design useful constraints which improve the conditioning of the learning problem.

Acknowledgements

The author would like to thank the EPSRC and the National Science Foundation for their support.

Bibliography

1. Saarinen, S., Bramley, R. and Cybenko, G. (1991). Ill-conditioning in neural network training problems. *Tech. Report, Center for Supercomputing Research and Development*, University of Illinois.

2. Luenberger, David G. (1984). *Linear and Nonlinear Programming* , Addison Wesley, Reading, MA.

3. Webb, A.R., Lowe, David, and Bedworth, M.D. (1988). A comparison of nonlinear optimisation strategies for feed-forward adaptive layered networks. *Tech. Report 4157, Royal Signals and Radar Establishment*, Malvern, U.K.

4. Kirby, M. and Miranda, R. (1994). *Nonlinear reduction of high-dimensional dynamical systems via neural networks*, Phys. Rev. Letters, Vol. 72, No. 12, pp1822.

5. Kirby, M. and Miranda, R. (1996). *Empirical Dynamical System Reduction I: Global Nonlinear Transformations*, to appear.

6. Kramer, M. (1991). *Nonlinear Principal Component Analysis Using Autoassociative Neural Networks*, AIChE Journal, Vol. 37, No. 2, pp. 233-243.

7. Golub, Gene H. and Van Loan, Charles F. (1983). *Matrix Computations*, North Oxford Academic, Oxford.

8. Rumelhart, D.E., McClelland, J.L. and the PDP Research Group (1986), *Parallel Distributed Processing: Explorations in the Microstructure of Cognition*, Vol. 1, MIT Press, Cambridge.

9. Werbos, P.J, (1974) *Beyond Regression: New Tools for Prediction and Analysis in the Behavioral Sciences*, Harvard University, PhD Dissertation.

Online Modelling of Time Series with Resource Allocating Neural Networks

A. McLachlan

Neural Computing Research Group, Department of Computer Science and Applied Mathematics, Aston University, Birmingham

Abstract

Online model order complexity estimation remains one of the key problems in neural network research. The problem is further exacerbated in situations where the underlying system generator is non-stationary. In this paper, we introduce a novelty criterion for resource allocating networks (RANs) which is capable of being applied to both stationary and slowly varying non-stationary problems. The deficiencies of existing novelty criteria are discussed and the relative performances are demonstrated on two real-world problems : electricity load forecasting and exchange rate prediction.

1 The RAN network

The resource allocating network (RAN) was introduced by Platt [1] as a means of constructing a network of adequate complexity online. It is a Gaussian radial basis function network, whose training requires a novelty criterion (which determines whether a new basis function is to be added) and an online training algorithm (which updates the existing weights).

Considerable work has been performed regarding training algorithms for both stationary and non-stationary data [2, 3, 4], but little has been done regarding novelty criteria, and it is this problem which we shall be addressing in this paper. It should be emphasised that we will be concerning ourselves with improvements to the RAN prescription rather than with comparisons of RAN performance to other techniques. Such analysis can be found elsewhere in the literature [1, 2, 5].

2 The extended Kalman filter

As the networks in this paper will be trained using the extended Kalman filter, we shall briefly describe the algorithm [6] before proceeding with the discussion of novelty criteria.

We are interested obtaining an online estimate of the vector of network weights $\boldsymbol{w}$ from the data sequence $Y_t = \{y_t, y_{t-1}, ...\}$. The state space equations specify both the relation between the measurements and the weights as well as the evolution of the weights between timesteps. In the following, we denote the network output $\boldsymbol{f}_t$ as a function of the weights rather than the input vector. (The dependence on the input vector is implicit in the t suffix.) The state equations for our network are thus

$$\begin{aligned} \boldsymbol{y}_t &= \boldsymbol{f}_t(\boldsymbol{w}_t) + \boldsymbol{\nu}_t \\ \boldsymbol{w}_t &= \boldsymbol{w}_{t-1} + \boldsymbol{\mu}_{t-1} \end{aligned} \tag{2.1}$$

where $\boldsymbol{\nu}$ and $\boldsymbol{\mu}$ are noise processes. Our estimate of the weights on receipt of the datum $\boldsymbol{y}_t$ is obtained by maximising the posterior

$$p(\boldsymbol{w}_t|Y_t) = \frac{p(\boldsymbol{y}_t|\boldsymbol{w}_t)p(\boldsymbol{w}_t|Y_{t-1})}{p(\boldsymbol{y}_t|Y_{t-1})} \tag{2.2}$$

where the likelihood $p(\boldsymbol{y}_t|\boldsymbol{w}_t)$ contains the information from the new datum and the prior $p(\boldsymbol{w}_t|Y_{t-1})$ contains information gathered from previous data Y_{t-1}. The prior is obtained from the posterior at the previous timestep thus

$$p(\boldsymbol{w}_t|Y_{t-1}) = \int d\boldsymbol{w}_{t-1} p(\boldsymbol{w}_t|\boldsymbol{w}_{t-1})p(\boldsymbol{w}_{t-1}|Y_{t-1}). \tag{2.3}$$

We shall now take the noise processes $\boldsymbol{\nu}$ and $\boldsymbol{\mu}$ in (2.1) to be zero mean with covariances R and Q respectively,

$$\begin{aligned} p(\boldsymbol{w}_t|\boldsymbol{w}_{t-1}) &= N(\hat{\boldsymbol{w}}_{t-1|t-1}, Q_{t-1}) \\ p(\boldsymbol{y}_t|\boldsymbol{w}_t) &= N(\boldsymbol{f}_t(\boldsymbol{w}_t), R_t) \end{aligned} \tag{2.4}$$

where the notation $\hat{\boldsymbol{w}}_{t|t}$ signifies our estimate of the weights $\boldsymbol{w}_t$ given the information available up to and including time t. We further assume that the weights are drawn from a normal distribution, centered on our current estimate $\hat{\boldsymbol{w}}$ and with covariance $\hat{P}$. This gives us the following distributions for the prior and posterior

$$\begin{aligned} p(\boldsymbol{w}_t|Y_t) &= N(\hat{\boldsymbol{w}}_{t|t}, \hat{P}_{t|t}) \\ p(\boldsymbol{w}_t|Y_{t-1}) &= N(\hat{\boldsymbol{w}}_{t|t-1}, \hat{P}_{t|t-1}). \end{aligned} \tag{2.5}$$

The only distribution still to be specified is the evidence in (2.2). The evidence specifies the distribution of the new datum conditioned on the data history, and we shall model this as a Gaussian centred on our prediction $\boldsymbol{f}_t(\hat{\boldsymbol{w}}_{t|t-1})$, with covariance (error bar matrix) S.

$$p(\boldsymbol{y}_t|Y_{t-1}) = N(\boldsymbol{f}_t(\hat{\boldsymbol{w}}_{t|t-1}), S_t). \tag{2.6}$$

Linearising the network output around our current estimate,

$$\boldsymbol{f}_t(\boldsymbol{w}_t) = \boldsymbol{f}_t(\hat{\boldsymbol{w}}_{t|t-1}) + \nabla_w \boldsymbol{f}_t(\hat{\boldsymbol{w}}_{t|t-1})(\boldsymbol{w}_t - \hat{\boldsymbol{w}}_{t|t-1}) \tag{2.7}$$

leads to the extended Kalman filter algorithm. Between timesteps, our estimates of the weights and covariance are updated to take account of the system noise in (2.1).

$$\begin{aligned}\hat{\boldsymbol{w}}_{t|t-1} &= \hat{\boldsymbol{w}}_{t-1|t-1} \\ \hat{P}_{t|t-1} &= \hat{P}_{t-1|t-1} + Q_{t-1}.\end{aligned} \tag{2.8}$$

On the receipt of a new datum, the prediction error $\boldsymbol{e}$ is calculated, and the information contained therein is used to update the prediction error covariance S and thence $\hat{\boldsymbol{w}}$ and $\hat{P}$.

$$\begin{aligned}\hat{\boldsymbol{y}}_{t|t-1} &= \boldsymbol{f}_t(\hat{\boldsymbol{w}}_{t|t-1}) \\ \boldsymbol{e}_t &= \boldsymbol{y}_t - \hat{\boldsymbol{y}}_{t|t-1} \\ S_t &= R_t + \nabla_w \boldsymbol{f}_t(\hat{\boldsymbol{w}}_{t|t-1})\hat{P}_{t|t-1}\left(\nabla_w \boldsymbol{f}_t(\hat{\boldsymbol{w}}_{t|t-1})\right)^T \\ K_t &= \hat{P}_{t|t-1}\left(\nabla_w \boldsymbol{f}_t(\hat{\boldsymbol{w}}_{t|t-1})\right)^T S_t^{-1} \\ \hat{\boldsymbol{w}}_{t|t} &= \hat{\boldsymbol{w}}_{t|t-1} + K_t \boldsymbol{e}_t \\ \hat{P}_{t|t} &= (I - K_t \nabla_w \boldsymbol{f}_t(\hat{\boldsymbol{w}}_{t|t-1}))\hat{P}_{t|t-1}.\end{aligned} \tag{2.9}$$

3 The original novelty criterion

Platt's novelty criterion is formulated for a single output for simplicity and consists of two distinct parts. Firstly, the prediction error e_t is compared to a critical value e^c : if $e_t < e^c$, then the existing network is taken to be performing adequately and no unit is added. If $e_t > e^c$, then the network may be attempting to extrapolate, and a new unit is added if the Euclidean distance d from the input vector to the nearest basis function centre exceeds a critical distance d^c. The critical distance is allowed to decrease exponentially with time in order to inhibit excessive addition of units early in the training while still permitting network growth at a later stage if neccessary.

In summary, for a new unit to be added we must have

$$\begin{aligned}e_t &> e^c \text{ and } d_t^{\text{nearest unit}} > d_t^c \ , \quad \text{where} \\ d_t^c &= \max(\gamma^t d^c_{\max}, d^c_{\min}) \quad , \quad \gamma \in (0,1).\end{aligned} \tag{3.1}$$

When a new unit is added, the extra terms in the weight vector $\hat{\boldsymbol{w}}$ and the corresponding elements in the augmented matrix $\hat{P}$ require initialisation. As the basis function is Gaussian, the logical choice is to centre it on the input vector of the novel pattern. The weighting of this Gaussian in the network is chosen such that the network output matches the target pattern exactly. The covariance matrix P is augmented by a multiple of the identity matrix.

4 Prediction error and network error-bars

The obvious drawback of the above criterion lies in the arbitrariness of the choice of critical parameters $e^c, d^c_{\max}, d^c_{\min}$ and γ. (If we have more than one output, then there will be a critical error for each.) Minor changes to these parameters can lead to major changes in network growth, in many cases with no significant improvement in performance [3, 4, 5].

With this in mind, Kadirkamanathan introduced IncNet [7]. This RAN trains using the extended Kalman filter and has a novelty criterion which tests the compatibility of the prediction error, e_t (again dealing with one output for simplicity) and the Kalman filter's online estimation of its variance S_t, as shown in Equation (2.9). The statistic

$$e_t/\sqrt{S_t} \tag{4.1}$$

follows a t distribution, and a new unit is added only if the prediction error exceeds the 95% confidence limit at that timestep. This removes the need for critical parameters, and has been shown to lead to the construction of more compact networks than the original RAN *for stationary data sets*. It can be extended to multiple outputs by testing each component of $\boldsymbol{e}$ against the diagonal elements of S.

5 A new novelty criterion

An undesirable feature of both the above novelty criteria is that, by basing the decision on the instantaneous error, the RAN becomes sensitive to outliers. Similarly, as the new elements in the Kalman filter's weight covariance matrix $\hat{P}$ require initialisation after unit addition, the filter should be given time to compensate before further assessments of performance are made. The above criteria do not take this into account, and hence on occasion many unnecessary units are added in consecutive timesteps. A further problem with IncNet comes when RAN networks are applied to non-stationary problems. In order to prevent the Kalman filter from converging, and thus allow the network to be more adaptable without having always to rely on unit addition, zero mean Gaussian system noise is added to widen the prior in Equation (2.8) As this artificially extends the error bars obtained from the matrix S, the IncNet criterion will only be satisfied when the prediction error is large. This leads to the construction of inadequate networks whose prediction performance is unacceptably poor, and can even stop unit addition altogether.

We therefore propose a new novelty criterion with the following rationale. If the model at time t produces an adequate fit to the underlying data generator, then the subsequent output sequence should be distributed as a white noise process. As we do not have access to an ensemble at time t, the best we can do is to estimate the statistics using the outputs at previous timesteps. (It is clear that a point estimate procedure, such as used in IncNet, is inadequate.) As we may be dealing with non-stationary systems, then the size of the window we use must be less that the characteristic timescale of the non-stationarity. Our novelty criterion therefore involves testing that the prediction error sequence does indeed correspond to a zero mean Gaussian at the 95% confidence level.

To test for a zero mean sequence we simply construct the t statistic

$$< \hat{e} >_t \sqrt{N/\hat{S}_t} \tag{5.1}$$

where N is the window length over which we have obtained the sample estimates of S to replace the artificially noise-corrupted filter estimate. (As with IncNet, this can be extended to multiple outputs.) The normality test can be performed with the *weighted-sum-squared-residual* (WSSR) statistic [6] which integrates information across all outputs

$$\sum_{k=t-N+1}^{t} e_k^T \hat{S}_k^{-1} e_k \tag{5.2}$$

where we are continually updating our estimate S from a travelling window of length N at each timestep k. For a network with p outputs, this statistic is distributed as $\chi^2(Np)$. N should be large enough to allow reliable statistics to be obtained, but so large that necessary growth is inhibited or the property of approximate ergodicity is lost.

As the network requires time to re-adjust after the addition of a new unit, we do not permit further unit addition for N timesteps after a previous increment. This gives us time to aggregate prediction error information over a window which does not contain error information obtained from the previous architecture, and also allows the filter time to adapt to a potentially unsuitable initialisation of the new components of the weight covariance matrix.

6 Electricity load forecasting

This data set represents averaged daily electricity load demand over a period of 298 days. As there is a weekly cycle superimposed on the seasonality, a window of the previous seven days' load was used as input during training, giving a data set of 291 patterns.

Figure 1 and Table 1 illustrate the effects of altering just two critical parameters in the original criterion and the window size in the new criterion.

The new criterion leads to the construction of smaller networks, some of whose performances are comparable to the much larger nets constructed via the original RAN criterion. The sensitivity of the original RAN to the settings of the critical parameters can clearly be seen in Figure 1, as can the bursts of unit addition mentioned previously. The slightly lower errors obtained with the original criterion are a feature of the seasonal nature of this data set : some of the excess units added early in the training can come back into play as the data generator returns to its original form.

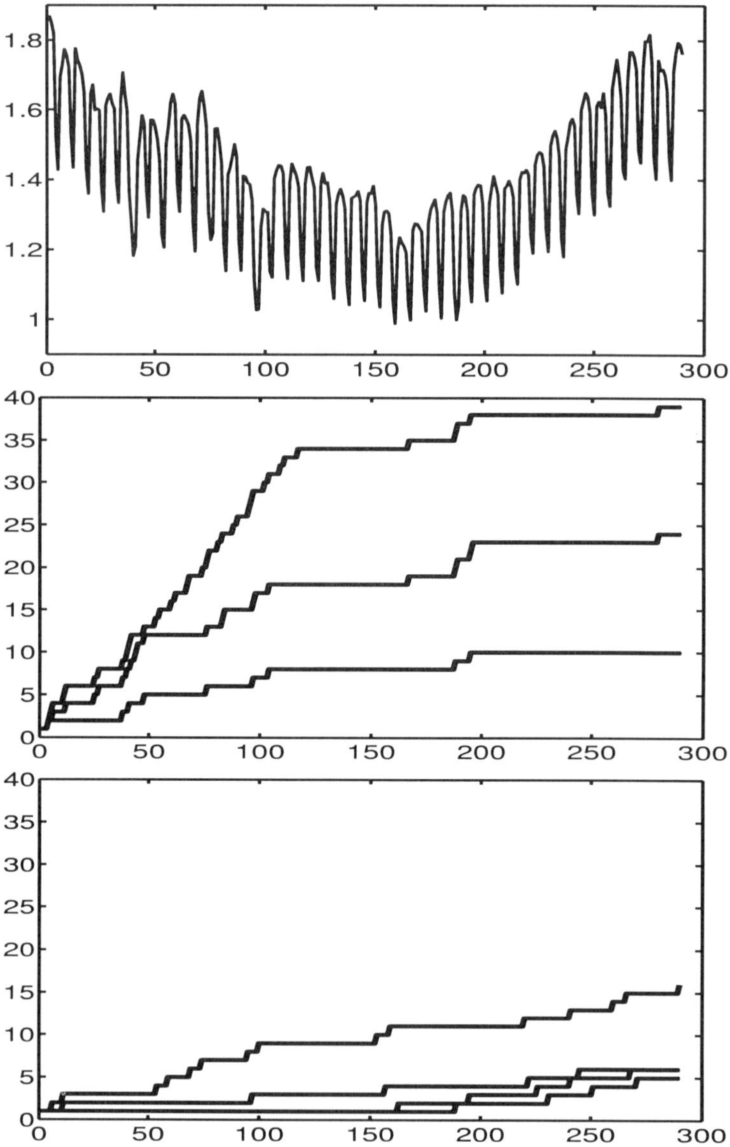

Figure 1. Electricity load (top), growth of RAN using original criterion with varying critical parameters (centre), growth of RAN using new criterion with varying window lengths (bottom)

d^c_{max}	e^c	Network Size	NPE
0.5	0.1	24	0.376
0.25	0.1	39	0.395
0.5	0.2	10	0.370

window size	Network Size	NPE
5	16	0.391
10	6	0.383
15	6	0.402
20	5	0.400

Table 1. Comparison of old and new criteria for load data. Naive predictor gives NPE = 0.668

While the larger windows in this case make the network slow to react and thus increase the normalised prediction error (NPE), it is easier to optimise the window length than it is to optimise the hyperparameters of the original criterion.

7 Exchange rate forecasting

The data in Figure 2 shows daily prices for the Deutsche Mark/French Franc market over 701 days. This dataset has irregular nonstationary components due in the main to government intervention in the European Exchange Rate mechanism. As there can be "day of week" effects in such data, a window of 5 previous values was used as input, giving a data set of 696 patterns.

Figure 2 and Table 2 reinforce our conclusions from the load forecasting case. Using the original criterion leads to rapid bursts of unit addition between steps 250 and 300, indicating a failure to give the Kalman filter sufficient time to adapt. The new procedure consistently gives smaller networks of comparative predictive power, and shows considerably less variability in the size of networks constructed.

8 Conclusions

In this paper, we have demonstrated the deficiencies of previous novelty criteria for resource allocating networks, and have introduced a more robust criterion derived from standard signal processing theory.

d^c_{max}	e^c	Network Size	NPE
0.5	0.0075	30	0.332
0.25	0.0075	28	0.332
0.5	0.015	19	0.333

window size	Network Size	NPE
10	11	0.336
15	13	0.334
20	11	0.332
25	10	0.329

Table 2. Comparison of old and new criteria for exchange rate data. Naive predictor gives NPE = 0.357

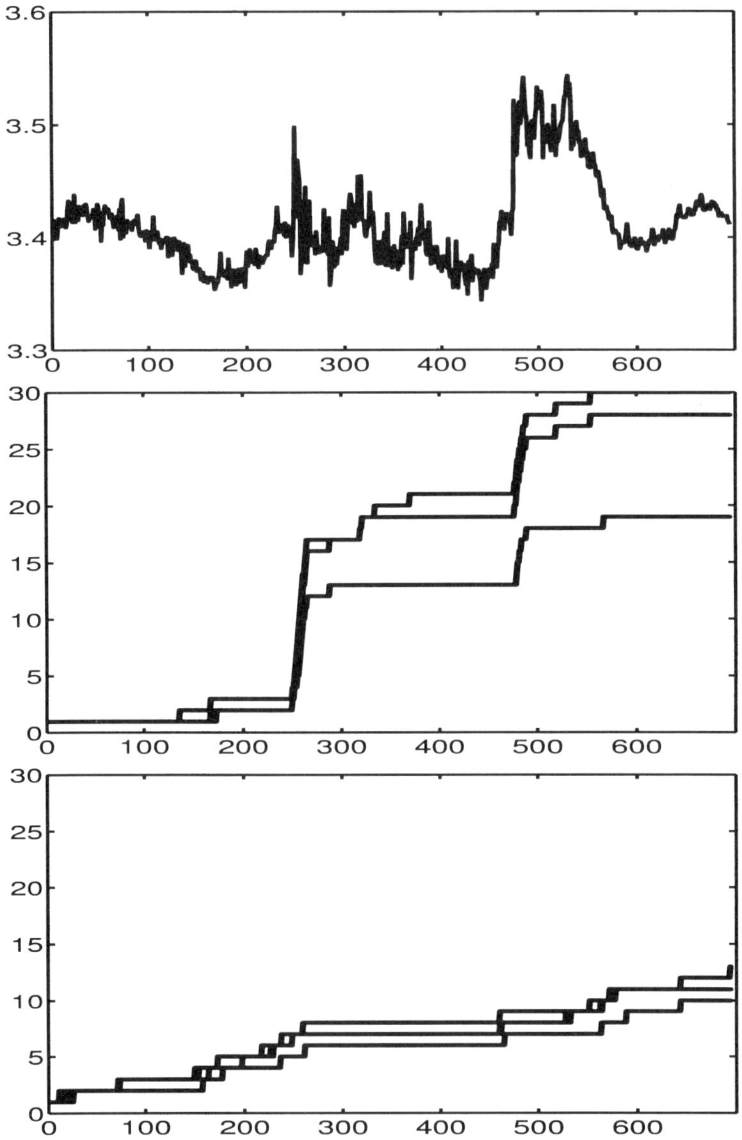

Figure 2. DM/Fr exchange rate (top), growth of RAN using original criterion with varying critical parameters (centre), growth of RAN using new criterion with varying window lengths (bottom)

This new criterion

- calculates statistics over a travelling window to reduce the effects of outliers and to allow time for filter adaptation,
- uses sample estimates to remove any dependence on the Kalman filter,
- contains only a single hyperparameter - the window width,
- is relatively insensitive to the choice of window width, and
- can be applied in slowly varying non-stationary environments.

Acknowledgements

This work was supported by EPSRC grant GR/J75425 "Novel Developments in Learning Theory for Neural Networks". I would like to thank Midland Electricity plc, UK and Pareto Partners, London, UK for the provision of the data, and David Lowe for helpful comments.

Bibliography

1. Platt, J. C. (1991). A resource allocating network for function interpolation. *Neural Computation*, **3**, 213–225.
2. Kadirkamanathan, K. and Niranjan M. (1993). A function estimation approach to sequential learning with neural networks. *Neural Computation*, **5**, 954–975.
3. Lowe, D. and McLachlan, A. (1995). Modelling of nonstationary processes using radial basis function networks. *Fourth IEE International Conference on Artificial Neural Networks*, IEE Conference Proceedings No. 409, 300–305.
4. McLachlan, A. and Lowe D. (1996). tracking of nonstationary time series using resource allocating RBF networks. *Cybernetics and Systems '96*, Editor : R. Trappl, Austrian Society for Cybernetic Studies, 1066–1071.
5. Nabney, I. T., McLachlan, A. and Lowe D. (1996). Practical methods of tracking nonstationary time series applied to real world data. *AeroSense '96 - Applications and Science of Artificial Neural Networks II*, Editors : S. K. Rogers and D. W. Ruck, SPIE Publications Vol. 2760, 152–163.
6. Candy, J. V. (1986). *Signal Processing : The Model-Based Approach.* McGraw-Hill.
7. Kadirkamanathan, K. (1994). A statistical inference based growth criterion for the RBF network. *Proceedings of the IEEE Workshop on Neural Networks for Signal Processing IV*, 12–21.

Bayesian Model Selection for Linear and Non-linear Time Series using the Gibbs Sampler

Paul T. Troughton and Simon J. Godsill

Department of Engineering, University of Cambridge

Abstract

We present a stochastic simulation technique for model selection in time series, based on the use of indicator variables with the Gibbs sampler within a hierarchical Bayesian framework. As an example, the method is applied to the selection of subset AR models, in which only significant lags are included. The same approach is then used to identify the structure of a non-linear time series. We discuss the possibility of model mixing where the model is not well determined by the data.

1 Introduction

Until recently, research into time series modelling has concentrated on those models which are analytically convenient, without necessarily justifying the underlying assumptions, such as linearity. With the rapid increase in available computing power, it is now possible to consider a much wider range of models, including hybrids containing terms from several non-linear model families. The problem becomes one of *subset selection* — we wish to select the best subset of terms from the pool available.

We will take a Bayesian approach, since this leads to consistent model selection criteria and avoids the need to introduce explicit penalisation of complex models. By selecting models on the basis of posterior probabilities, we also have the opportunity to incorporate any prior knowledge.

We wish to fit to a time series $\mathbf{x}$ a model which consists of a number of terms with (possibly vector) parameters $\boldsymbol{\theta}_1, \ldots \boldsymbol{\theta}_P$. We associate a binary indicator γ_i with each term such that if $\gamma_j = 1$ then the term with parameter $\boldsymbol{\theta}_j$ is included in the model; otherwise it is excluded. We gather any parameters which are common to all models into $\boldsymbol{\phi}$. The most probable combination of terms, and hence the model we wish to select, as represented by $\boldsymbol{\gamma}$, is then:

$$\operatorname*{argmax}_{\boldsymbol{\gamma}}\Big(p(\gamma_1, \gamma_2, \ldots \gamma_P \mid \mathbf{x}) = \int_{\boldsymbol{\theta}_1, \boldsymbol{\theta}_2, \ldots \boldsymbol{\theta}_P, \boldsymbol{\phi}} \cdots \int p(\boldsymbol{\theta}_1, \gamma_1, \boldsymbol{\theta}_2, \gamma_2, \ldots, \boldsymbol{\phi} \mid \mathbf{x})\, d\boldsymbol{\theta}_1 \cdots d\boldsymbol{\theta}_P\, d\boldsymbol{\phi}\Big). \quad (1.1)$$

If we have a pool of P candidate terms, there are 2^P possible combinations. For sizeable P, it becomes impractical to evaluate the probability of *all* subsets. To avoid this, there is a variety of sub-optimal search algorithms (see for example [1, 2]). It is possible that our posterior distributions will be multimodal. This can cause problems with deterministic search algorithms, as they tend to stop at local maxima. Hence we will concentrate on stochastic methods.

Hastings [3] generalises the stochastic simulation methods of [4] to produce, for any given distribution π, an ergodic Markov chain which has π as a limiting distribution. Geman & Geman [5] take a special case of this Metropolis-Hastings algorithm, in which each variable is sampled in rotation from its full conditional density, and call it the Gibbs sampler:

$$\begin{aligned}
\psi_1^{(n+1)} &\sim p(\psi_1 \mid \psi_2^{(n)}, \psi_3^{(n)}, \ldots \psi_d^{(n)}, \mathbf{x}) \\
\psi_2^{(n+1)} &\sim p(\psi_2 \mid \psi_1^{(n+1)}, \psi_3^{(n)}, \ldots \psi_d^{(n)}, \mathbf{x}) \\
&\vdots \\
\psi_d^{(n+1)} &\sim p(\psi_d \mid \psi_1^{(n+1)}, \psi_2^{(n+1)}, \ldots \psi_{d-1}^{(n+1)}, \mathbf{x}).
\end{aligned}$$

If we allow the ψ_i to be multivariate, we have a multi-move Gibbs sampler.

Returning to Equation (1.1), a Markov chain can be constructed which moves around the model space by sampling both the indicators and the other model parameters to produce a sequence of states $\gamma^{(1)}, \gamma^{(2)}, \ldots$, which converges in the limit to produce (dependent) samples from the posterior $p(\gamma \mid \mathbf{x})$, thereby performing numerically the integration of Equation (1.1) [6]. From these sampled states, we can obtain Monte Carlo estimates of the marginal posterior density of the indicators.

Both George & McCulloch [7] and Carlin & Chib [8] use variations on this approach. George & McCulloch's approach is to use a Gibbs sampler, as above, and sample each parameter and indicator separately, from its full conditional. Carlin & Chib effectively sample the indicators as a block, conditional on the parameters.

In George & McCulloch [7], disabled terms are not completely switched out but rather their parameters are given a narrow prior density so that they remain sufficiently close to zero not to affect the modelling error. However, we also require the parameter to be able to take on sufficiently large values for the term to be switched in again. Carlin & Chib avoid this compromise by reposing the model in terms of the product of each parameter with its indicator. The model is now completely unaffected by the value of the parameters for disabled terms, and hence an arbitrary *pseudoprior* can be used.

Chen [9] applies George & McCulloch's method to subset autoregressive models. By contrast, our approach completely removes disabled terms, as this gives computational advantages.

George & McCulloch [7] argue that, as we are only interested in the subsets with highest posterior probability, rather than evaluation of the full posterior, a run of length $\ll 2^P$ should suffice; for all but the most degenerate multimodal posteriors, this seems reasonable.

If variables are strongly dependent, the Gibbs sampler will tend to converge slowly [10]. Since there is likely to be strong interdependence between the indicator and parameter(s) of each term, we speed convergence by sampling *jointly* from the indicators and their associated parameters, in a similar manner to that already used for impulse detection [11, 12, 13].

There will also be interdependence between the parameters and indicators of different terms. We can address this by multivariate sampling of the indicators, in blocks of size Q. Each iteration then requires the evaluation of the conditional for 2^Q combinations of terms. Varying Q allows a trade-off between the number of iterations required for convergence and the computational complexity of each iteration.

Following from Stark [14], we sample the indicators in random order, but sample the different types of component in a fixed sequence.

2 Example

We now illustrate this method with a simple linear model.

2.1 Subset AR model

The subset autoregressive model [15] with maximum lag P can be represented in terms of parameters a_i:

$$x_t = e_t + \sum_{i=1}^{P} x_{t-i}\, a_i\, \gamma_i$$

where e_t is an i.i.d. (independent, identically distributed) Gaussian excitation sequence with constant variance. With appropriate matrix and vector definitions [16], the approximate conditional likelihood can be expressed as:

$$p(\mathbf{x} \mid \mathbf{a}, \gamma, \sigma_e) \approx (2\pi\sigma_e^2)^{-\frac{N-P}{2}} \exp\left(-\tfrac{1}{2}\sigma_e^{-2}\, ||\mathbf{x}_1 - \mathbf{X}(\mathbf{a}\cdot\gamma)||^2\right) \tag{2.1}$$

where $\mathbf{x}_0$ contains the first P elements of $\mathbf{x}$, and $\mathbf{x}_1$ the remainder.

2.2 Priors

We use a Bernoulli prior for the indicators, $p(\gamma_i = 1) = \alpha$. For the AR parameter values, we use a convenient prior: independent zero-mean univariate Gaussians, all of variance ${\sigma_p}^2$. Since the noise variance is a scale parameter, we use a Jeffreys' prior, $p({\sigma_e}^2) \propto \frac{1}{{\sigma_e}^2}$. With suitable bounds, this can be made proper. Alternatively, an inverse gamma prior could be used on ${\sigma_e}^2$.

2.3 Conditional distributions

Our implementation of the Gibbs sampler involves two types of sampling step:

$$\mathbf{a}_u, \gamma_u \sim p(\mathbf{a}_u, \gamma_u \mid \mathbf{x}, \mathbf{a}_k, \gamma_k, \sigma_e) \tag{2.2}$$
$$\sigma_e \sim p(\sigma_e \mid \mathbf{x}, \mathbf{a}, \gamma) \tag{2.3}$$

where the subscripts $(\cdot)_u$ and $(\cdot)_k$ denote partitioning into, respectively, those elements corresponding to terms whose indicators are being sampled, and those which are currently being regarded as fixed.

The joint sampling operation of step (2.2) can be performed in two steps using the method of composition [17]:

$$\gamma_u \sim p(\gamma_u \mid \mathbf{x}, \mathbf{a}_k, \gamma_k, \sigma_e) \tag{2.4}$$
$$\mathbf{a}_u \sim p(\mathbf{a}_u \mid \mathbf{x}, \mathbf{a}_k, \gamma, \sigma_e). \tag{2.5}$$

Note that step (2.4) is *not* conditional on $\mathbf{a}_u$.

We derive the (discrete) distribution for γ_u from the conditional likelihood as follows:

$$p(\mathbf{x}, \mathbf{a}_u \mid \mathbf{a}_k, \gamma, \sigma_e) = \overbrace{p(\mathbf{x} \mid \mathbf{a}, \gamma, \sigma_e)}^{\text{likelihood}} \cdot \overbrace{p(\mathbf{a}_u)}^{\text{prior}}$$
$$p(\mathbf{x} \mid \mathbf{a}_k, \gamma, \sigma_e) = \int p(\mathbf{x}, \mathbf{a}_u \mid \mathbf{a}_k, \gamma, \sigma_e)\, d\mathbf{a}_u$$
$$p(\gamma_u \mid \mathbf{x}, \mathbf{a}_k, \gamma_k, \sigma_e) \propto p(\mathbf{x} \mid \mathbf{a}_k, \gamma, \sigma_e) \cdot \underbrace{p(\gamma_u)}_{\text{prior}}$$

giving (we omit the details for the sake of brevity):

$$p(\gamma_u \mid \mathbf{x}, \mathbf{a}_k, \gamma_k, \sigma_e) \propto \alpha^{n_1}(1-\alpha)^{(l-n_1)} \cdot (2\pi\sigma_e^2)^{-\frac{N-P}{2}} \cdot \exp\left(-\tfrac{1}{2}(\boldsymbol{\mu}_p^T \mathbf{C}_p^{-1} \boldsymbol{\mu}_p - \boldsymbol{\mu}_s^T \mathbf{C}_s^{-1} \boldsymbol{\mu}_s)\right) \cdot \sqrt{\frac{|\mathbf{C}_s|}{|\mathbf{C}_p|}}$$

where l is the dimension of γ_u, n_1 is the number of components of γ_u which are switched in, $\boldsymbol{\mu}_p$ and $\mathbf{C}_p$ are the mean vector and covariance matrix of the Gaussian prior $p(\mathbf{a}_u)$, and:

$$\mathbf{C}_s = (\sigma_e^{-2}\mathbf{X}_u{}^T\mathbf{X}_u + \mathbf{C}_p^{-1})^{-1}$$
$$\boldsymbol{\mu}_s = \mathbf{C}_s(\sigma_e^{-2}\mathbf{X}_u{}^T(\mathbf{x} - \mathbf{X}_k\mathbf{a}_k) + \mathbf{C}_p^{-1}\boldsymbol{\mu}_p).$$

In these terms, the distribution required for step (2.5) is simply a multivariate Gaussian:

$$p(\mathbf{a}_u \mid \mathbf{x}, \mathbf{a}_k, \gamma, \sigma_e) \propto \mathrm{N}(\mathbf{a}_u \mid \boldsymbol{\mu}_s, \mathbf{C}_s). \tag{2.6}$$

Sampling all of $\mathbf{a}$ is a simple operation, based on Equation (2.6) with $\mathbf{a}_u = \mathbf{a}$ and $\mathbf{a}_k$ empty. Occasionally including this step:

$$\mathbf{a} \sim p(\mathbf{a} \mid \mathbf{x}, \gamma, \sigma_e)$$

can further reduce the effect of interdependence between AR parameters.

Finally, the full conditional distribution of the noise variance (Equation 2.3) is found to be an inverse square-root gamma distribution, for which well-known sampling methods exist.

2.4 Sampling scheme

The sampling steps are carried out as follows:

for $i = \{1 \ldots I\}$
    $\sigma_e^{(i)} \sim p(\sigma_e \mid \mathbf{x}^{(i-1)}, \mathbf{a}^{(i-1)}, \gamma^{(i-1)})$
    for $u = \{\text{random subsets of candidate terms}\}$
        $\gamma_u{}^{(i)} \sim p(\gamma_u \mid \mathbf{x}^{(i-1)}, \mathbf{a}_k{}^{(i-1)}, \gamma_k{}^{(i-1)}, \sigma_e^{(i-1)})$
        $\mathbf{a}_u{}^{(i)} \sim p(\mathbf{a}_u \mid \mathbf{x}^{(i-1)}, \mathbf{a}_k{}^{(i-1)}, \gamma^{(i-1)}, \sigma_e^{(i-1)})$
    end
    $\mathbf{a}^{(i)} \sim p(\mathbf{a} \mid \mathbf{x}^{(i-1)}, \gamma^{(i-1)}, \sigma_e^{(i-1)})$
end

where I is the number of iterations and $\mathbf{a}^{(0)}$, $\gamma^{(0)}$ and $\sigma_e^{(0)}$ are arbitrary initial values.

3 Results

The above sampler was implemented, and experiments were performed using both synthetic and real data.

3.1 Synthetic data

800 samples of synthetic data were generated from a subset AR model containing terms of lag $\{1, 2, 3, 5, 7\}$. The sampler was run with candidate terms up to lag 9. The initial values of the indicators, parameters and noise variance were zero. Indicators were sampled in triples.

Figure 1 shows the results of a typical run of 150 iterations. The top plot shows the mean value of each of the indicators over the final 100 iterations, which can be interpreted as an estimate of the marginal posterior probability of inclusion of each of the model terms. It can be seen that the correct terms come out with clearly higher probability.

An alternative method for choosing a model from this data is to find the *combination* of terms which appears most frequently. This frequency should be an estimate of the subset's posterior probability. The middle plot shows (again

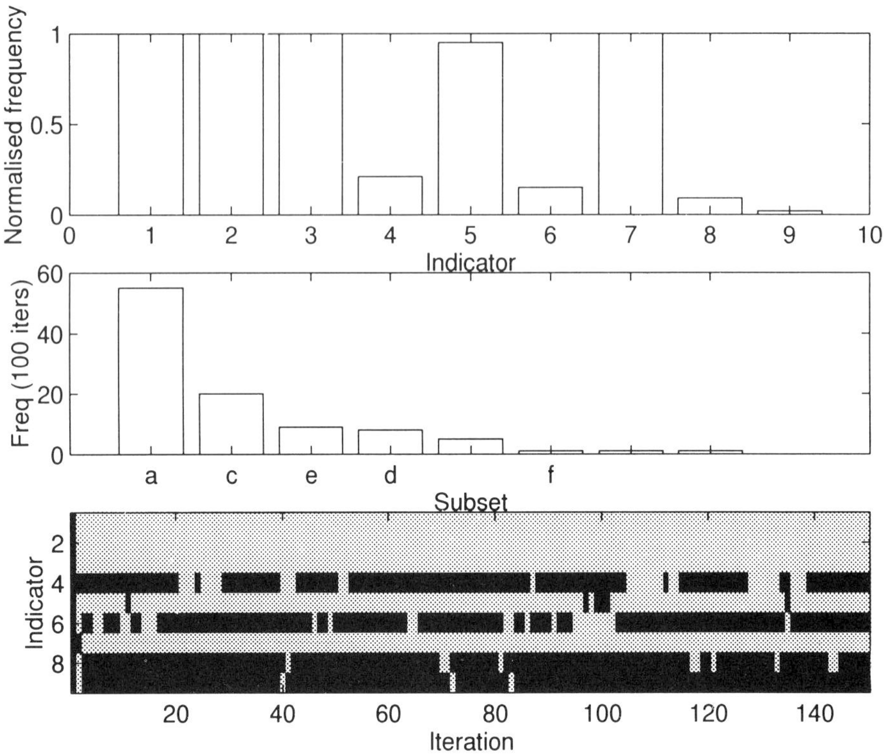

Figure 1. Synthetic data: Simulation results after 150 iterations (including 50 iterations burn-in): (top) Normalised indicator frequencies; (middle) Indicator subset frequencies, labelled according to Table 1; (bottom) Raw indicator values

using only the final 100 iterations) that the correct model comes out as the most popular one. This approach avoids the need to set a threshold when choosing terms.

The bottom plot shows the values of the indicators in each iteration — those indicators which are switched on are shown as white pixels.

It was found that the above results were insensitive to variations of σ_p and α over a wide range when the model was well determined by the data.

3.2 Analytic results

To verify that the sequence of states being produced is correct for subset selection, the posterior model probability was evaluated analytically for each of the possible subsets. The same model was used as for Section 3.1, but this time

Table 1. Composition of subsets

Subset	Terms included								
a	1	2	3		5		7		
b	1	2	3	4		6			
c	1	2	3	4	5		7		
d	1	2	3		5		7	8	
e	1	2	3		5	6	7		
f	1	2	3		5		7		9
g	1	2	3	4		6	7		
h	1	2	3	4	5	6			
i	1	2	3	4		6		8	

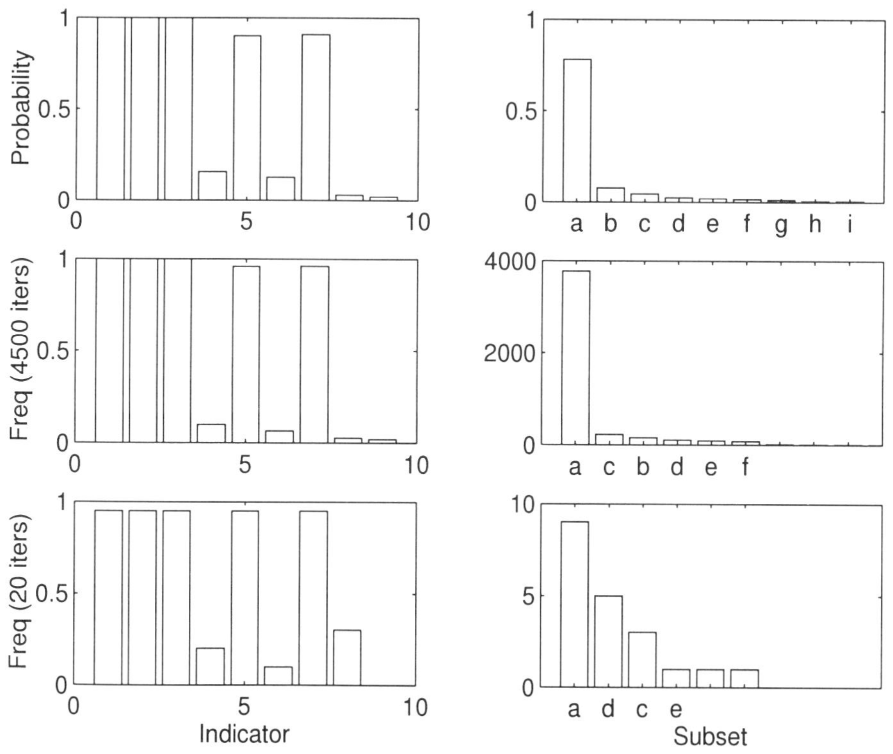

Figure 2. Comparison of (top) analytically calculated posterior model probabilities with (middle) simulation results from 5000 iterations (including 500 iterations burn-in) and (bottom) from 30 iterations (including 10 iterations burn-in). See Table 1 for a key to the subsets

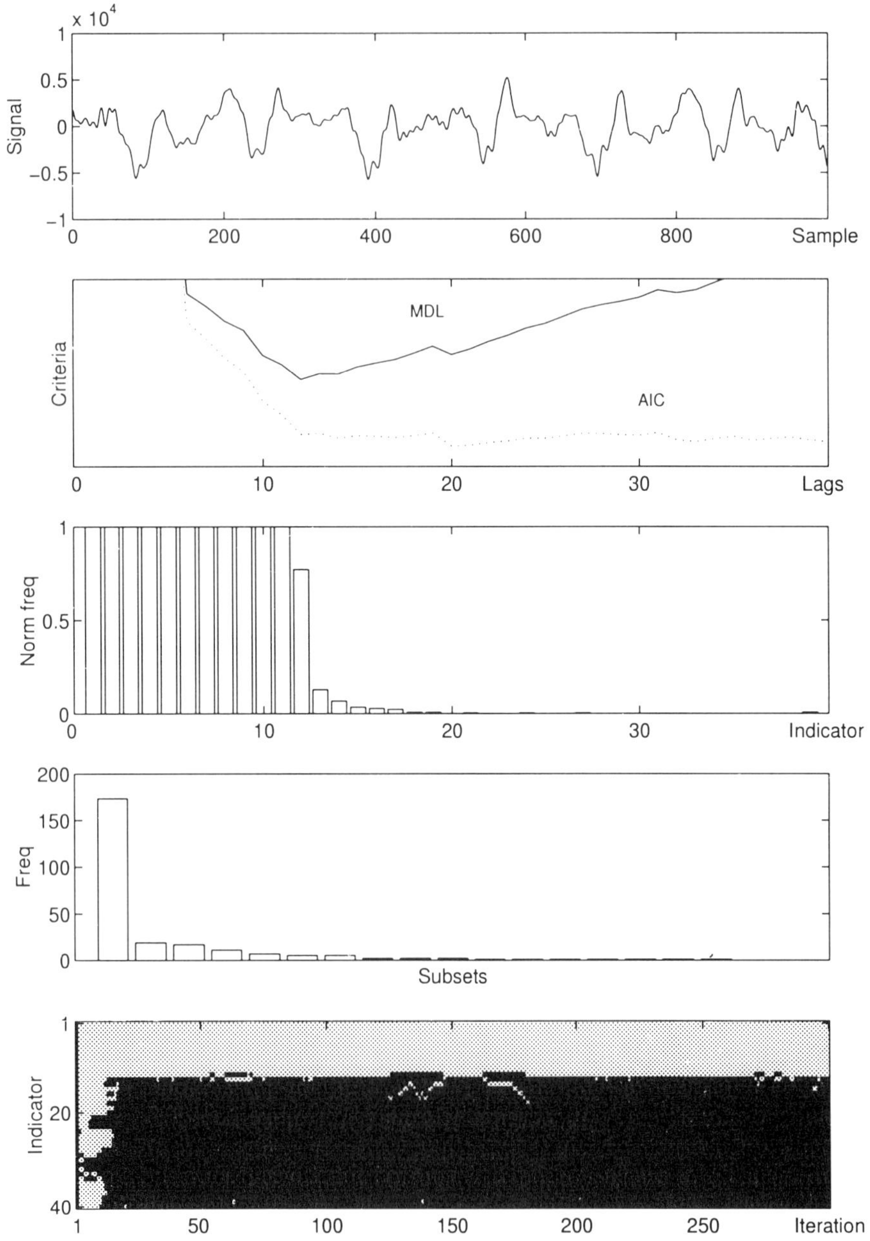

Figure 3. Orchestral recording: (from top) (a) Signal; (b) MDL (solid) and AIC (dotted) values for non-subset AR models; (c) Normalised indicator frequencies; (d) Indicator subset frequencies; and (e) Raw indicator values

only 400 samples were generated. This exhaustive calculation is feasible only for small P, and requires knowledge of the correct value of σ_e.

Figure 2 shows histograms generated from the calculated probabilities, together with the simulation results (with fixed σ_e) for both a large and a small number of iterations. It can be seen that the long run agrees closely with the calculated probabilities, and the short run, although more coarse, would lead to the selection of the same model.

3.3 Audio data

Figure 3 shows 1000 samples from an orchestral recording, together with the values of the AIC and MDL criteria for different lengths of non-subset AR models. The AIC would lead to a choice of an AR(20) model, whereas the MDL favours an AR(12) model. The AIC is known to tend to overfit.

The figure also gives the results of running the sampler with candidate terms to lag 40 for 300 iterations, discarding the first 50 as burn-in. The highest frequency subset, a plain AR(12) model, accounts for some 85% of the post burn-in iterations. This agrees with the MDL criterion.

4 Non-linear models

It is straightforward to extend these methods to include polynomial terms, forming a truncated Volterra series [18]. In the second-order case, the modelling equations take the form:

$$x_t = e_t + \sum_{i=1}^{P} a_i \, \gamma_i \, x_{t-i} + \sum_{i=1}^{P} \sum_{j=1}^{i} b_{ij} \, \xi_{ij} \, x_{t-i} \, x_{t-j}$$

where redundant terms have been eliminated by making the kernels triangular [19].

This can be converted into matrix-vector notation as:

$$\mathbf{e} = \mathbf{x}_1 - \mathbf{X}'(\mathbf{a}' \cdot \gamma')$$

where $\mathbf{a}'$ contains both $\{a_i\}$ and $\{b_{ij}\}$, γ' contains both $\{\gamma_i\}$ and $\{\xi_{ij}\}$, and $\mathbf{X}'$ is a block diagonal matrix. Note that this system is still *linear in the parameters*, and of the same form as Equation (2.1).

As an example, 5000 samples were generated from the following polynomial AR system:

$$x_t = e_t + 0.5x_{t-1} - 0.2x_{t-2} - 0.02x_{t-1}^3 - 0.07x_{t-2}^2 x_{t-3}.$$

The sampler was run for 100 iterations with all 34 possible candidate terms up to third-order with 4 lags. The results are shown in Figure 4. The four terms which appear with clearly highest frequency are indeed the correct ones.

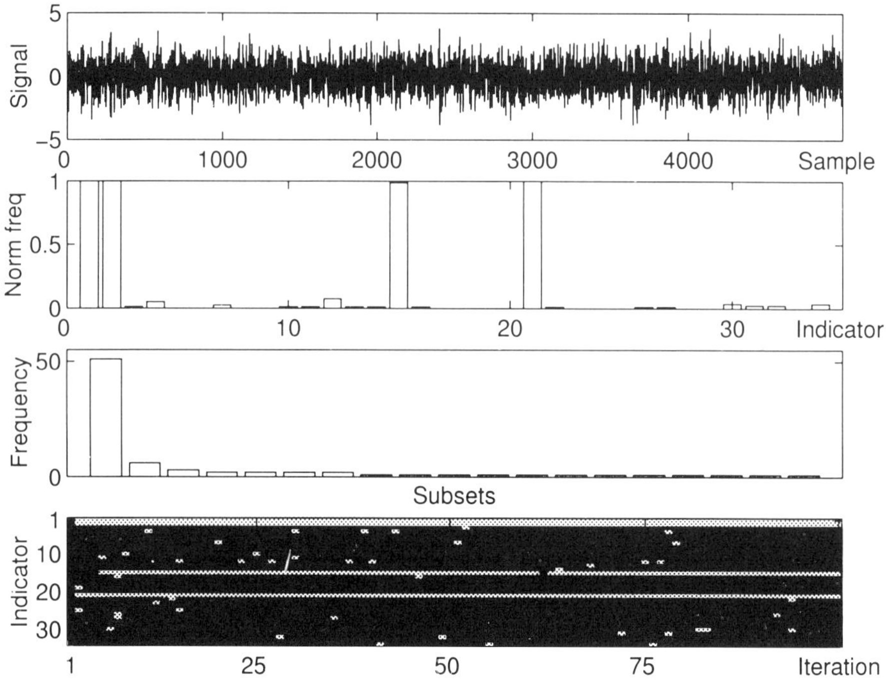

Figure 4. Non-linear data: Simulation results after 100 iterations (including 20 iterations burn-in): (a) Signal; (b) Normalised indicator frequencies; (c) Indicator subset frequencies; and (d) Raw indicator values

5 Applications

Having developed a sampler for model selection, we can incorporate extra steps to use the model to produce any required output, such as forecasts or a reconstruction of missing data.

In terms of Equation (1.1), the required output can be included in ϕ.

In addition to its simplicity, this approach has the advantage that, in the event of model uncertainty, the output will be based on processing using *all* the probable models, rather than just the one with highest posterior probability, *i.e.*

$$p(\phi \mid \mathbf{x}) = \int_{\boldsymbol{\theta}_1,\boldsymbol{\theta}_2,\ldots\boldsymbol{\theta}_P} \cdots \int \sum_{\gamma_1,\gamma_2,\ldots\gamma_P} \cdots \sum p(\boldsymbol{\theta}_1, \gamma_1, \ldots \mid \mathbf{x})\, d\boldsymbol{\theta}_1 \cdots d\boldsymbol{\theta}_P.$$

This approach has been used with a linear model for signal reconstruction in the presence of impulsive and continuous noise [20, 21]; the ability to incorporate non-linear model terms should make it possible to reconstruct audio which has suffered distortion by a poor recording chain [22].

6 Conclusions

We have shown, using the example of a simple linear model, that this method provides a means of avoiding the 2^P combinatorial explosion associated with subset selection. The MCMC framework has the advantage that it can be applied to models which are not analytically tractable. Further, it allows much flexibility in producing output, and copes elegantly with model uncertainty. Since this is a Bayesian method, we have a clear means for introducing prior knowledge, and also for expressing ignorance.

Acknowledgment

The work of the first author is supported by the Engineering & Physical Sciences Research Council.

Bibliography

1. Pope, K. J. and Rayner, P. J. W. (1994). Non-linear system identification using Bayesian inference. *Proceedings of IEEE ICASSP-94*, **IV**, 457–460.

2. Furnival, G. M. and Wilson, Jr., R. W. (1974). Regressions by leaps and bounds. *Technometrics*, **16**, 499–511.

3. Hastings, W. K. (1970). Monte Carlo sampling methods using Markov chains and their applications. *Biometrika*, **57**, 97–109.

4. Metropolis, N., Rosenbluth, A. W., Rosenbluth, M. N., Teller, A. H. and Teller, E. (1953). Equation of state calculations by fast computing machines. *Journal of Chemical Physics*, **21**, 1087–1092.

5. Geman, S. and Geman, D. (1984). Stochastic relaxation, Gibbs distributions, and the Bayesian restoration of images. *IEEE Transactions on Pattern Analysis and Machine Intelligence*, **PAMI-6**, 721–741.

6. Gelfand, A. E. and Smith, A. F. M. (1990). Sampling-based approaches to calculating marginal densities. *Journal of the American Statistical Association*, **85**, 398–409.

7. George, E. I. and McCulloch, R. E. (1996). Stochastic search variable selection. In W. R. Gilks, S. Richardson and D. J. Spiegelhalter (eds.), *Markov Chain Monte Carlo in Practice: Interdisciplinary Statistics*, pp. 203–214. Chapman & Hall.

8. Carlin, B. P. and Chib, S. (1995). Bayesian model choice via Markov chain Monte Carlo. *Journal of the Royal Statistical Society B*, **57**, 473–484.

9. Chen, C. W. S. (1996). On the selection of best subset autoregressive time series models. Tech. Rep., Department of Statistics, Feng-Chia University, Taiwan.

10. Tierney, L. (1994). Markov chains for exploring posterior distributions. *Annals of Statistics*, **22**, 1701–1762. With discussion.

11. Godsill, S. J. and Rayner, P. J. W. (1995). Robust reconstruction and analysis of autoregressive signals in impulsive noise using the Gibbs sampler. *Proceedings of IEEE Tr. Speech and Audio - 98*, (to appear).

12. Godsill, S. J. and Rayner, P. J. W. (1996). Robust treatment of impulsive noise in speech and audio signals. In *Bayesian Robustness*, IMS Lecture Notes – Monograph Series (1996) **29**, 331–342.

13. Barnett, G., Kohn, R. and Sheather, S. (1995). Robust Bayesian estimation of autoregressive-moving average models. Tech. Rep. WP95-012, Australian Graduate School of Management, University of New South Wales, Australia.

14. Stark, J. A. (1995). *Variable Selection in Data and Signal Modelling*. Ph.D. thesis, University of Cambridge.

15. McClave, J. T. (1978). Estimating the order of autoregressive models; the max χ^2 method. *Journal of the American Statistical Association*, **73**, 122–128.

16. Box, G. E. P. and Jenkins, G. M. (1976). *Time Series Analysis: Forecasting and Control*. Holden-Day Series in Time Series Analysis. Holden-Day, revised edition.

17. Tanner, M. A. (1993). *Tools for Statistical Inference: Methods for the Exploration of Posterior Distributions*. Springer series in statistics. Springer-Verlag, second edition.

18. Tong, H. (1990). *Non-linear Time Series: A Dynamical System Approach*. Oxford Statistical Science Series. Oxford University Press.

19. Morrison, I. J. (1990). *The Application of Volterra Series to Signal Detection and Estimation*. Ph.D. thesis, University of Cambridge.

20. Godsill, S. J. and Rayner, P. J. W. (1996). Robust noise reduction for speech and audio signals. *Proceedings of IEEE ICASSP-96*.

21. Godsill, S. J. (1997). Bayesian enhancement of speech and audio signals which can be modelled as ARMA processes. *International Statistical Review*, **65(1)**, 1–21.

22. Mercer, K. J. (1993). *Identification of Distortion Models*. Ph.D. thesis, University of Cambridge.

Bayesian Image Estimation from Sparse Raw MRI Data

Gert Jan Marseille*, Coen van Meijeren*, Iraklis M. Spiliotis, Dirk van Ormondt* and Basil G. Mertzios****

**Department of Applied Physics, Delft University of Technology, The Netherlands, and **Department of Electrical and Computer Engineering, Democritus University of Thrace, Hellas, Greece*

Abstract

A perennial challenge in MRI is reduction of the scan time. An obvious way to achieve this goal is to simply acquire fewer samples in the k-space. This strategy poses a problem to estimation (reconstruction) of the image from the k-space data because the attendant system of equations has become underdetermined. We solve the problem by imposing general prior knowledge in a Bayesian setting.

1 Introduction

This work concerns reduction of the MRI scan time. A full scan usually involves acquisition of 256×256 samples, arranged in a "raw" data matrix $S(k_x, k_y)$. The independent k space variables k_x and k_y are integers in the range $-128, -127, \ldots, +127$. Reduction of scan time can be achieved by omitting a number of k_y-values (rows). The omitted k_y's can be chosen such that the loss of information is minimized [1]. See example in Figure 1. We call data matrices with empty rows, sparse.

If no samples at all are omitted, *i.e.* if the scan is full, mere 2D FFT of $S(k_x, k_y)$ suffices to generate an MR image $I(x, y)$, x and y being integers in the range $-128, -127, \ldots, +127$. However, in the case of sparse sampling such as in Figure 1, the attendant inverse problem is strongly underdetermined. Zero-filling of omitted samples and subsequent application of FFT produces strong artefacts, rendering the image useless. To overcome this problem, one may invoke prior knowledge about the image. However, such prior knowledge ought to be of general nature, so as to avoid bias. In the following, we describe our choice of prior knowledge and its imposition by way of an iterative Bayesian procedure that shuttles back and forth between measurement domain and image domain using FFT and inverse FFT. As already mentioned in the caption of Figure 1, k_x values are *not* omitted because this yields no scan time reduction [2]. Hence, estimation of missing data is not necessary in the k_x-space, and 1D FFT of all rows of $S(k_x, k_y)$ can be tacitly carried out prior to all processing described

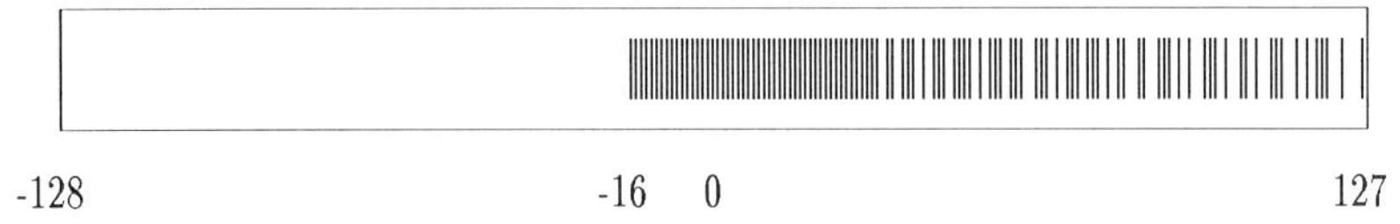

Figure 1. k_y values (bars) of a sparse 2D sample distribution yielding 57% scan time reduction. The positions of the used sample positions k_y is such that the acquired information is maximized [1]. The densely sampled central region $|k_y| \leq n_{\text{central}} = 16$ or 32 is used for estimating a low-resolution version of the image. The omitted samples are estimated with a Bayesian procedure that invokes prior knowledge. Omission of sample positions in the k_x space does not yield scan time reduction and is therefore not done [2]

below. In keeping with this, the data matrix is henceforth written as $S(x, k_y)$, where x is an integer in the range $-128, -127, \ldots, +127$.

2 Prior knowledge of MR images

2.1 Preliminaries

In MR, both the data $S(x, k_y)$ and the image $I(x, y)$ are complex-valued. In absence of noise and phase errors, the real part of the image, $I'(x, y) \stackrel{\text{def}}{=} \operatorname{Re} I(x, y)$, is nonnegative within the perimeter of the object O, and zero elsewhere. The imaginary part of the image, $I''(x, y) \stackrel{\text{def}}{=} \operatorname{Im} I(x, y)$, is zero everywhere. In actual practice, both $I'(x, y)$ and $I''(x, y)$ contain white Gaussian measurement noise at all coordinates x, y. The 256×256 discrete pairs of these coordinates are called *pixels.* In addition, phase errors appear unavoidable and result in mixing of $I'(x, y)$ and $I''(x, y)$. These phase errors need to be estimated and taken into account. The remainder of this section treats the various kinds of prior knowledge used in the Bayesian estimation of the image from sparse (*i.e.*, incomplete) data, including the object perimeter and the phase correction.

2.2 The object prior

An important piece of general information is contained in the probability distribution of *differences* of neighbouring pixel intensities within O: Fuderer found empirically that this distribution possesses Lorentzian (Cauchy) shape [3], and used it to reduce MR image estimation artefacts. This property was applied later by Marseille *et al.* [4, 5], Lettington and Hong [6], and McNally [7].
Defining

$$\delta_y(x, y) \stackrel{\text{def}}{=} I'(x, y) - I'(x, y-1), \tag{2.1}$$

one can express the probability distribution of the δ's as

$$\mathrm{p}(\delta_y(x,y)) = \frac{1}{a\pi(1+\delta_y(x,y)^2/a^2)}, \tag{2.2}$$

where $(x,y) \in O$, and $2a$ is the width of the distribution at half height. Because no k_x samples were omitted, we do not consider intensity differences in the x direction.

For later use, note that one can write for the case of an image column I'_x comprising for example two separated pieces of object, labeled 1 and 2,

$$\delta_y(x) = \mathbf{D}I'_x, \tag{2.3}$$

where $\delta_y(x)$ is the column vector of all $\delta_y(x,y)$ with common x, and $\mathbf{D}$ is defined by

$$\mathbf{D} = \begin{pmatrix} 0 & \mathbf{D}'(1) & 0 & 0 & 0 \\ 0 & 0 & 0 & \mathbf{D}'(2) & 0 \end{pmatrix}, \ \mathbf{D}'(.) = \begin{pmatrix} 1 & & & & \\ -1 & 1 & & \mathcal{O} & \\ & \ddots & \ddots & & \\ \mathcal{O} & & -1 & 1 & \\ & & & -1 & 1 \end{pmatrix}. \tag{2.4}$$

In Equation (2.4), 0 and $\mathcal{O}$ represent rectangular and triangular zero matrices of various sizes, respectively.

The size of a can be found by least squares fitting the model of Equation (2.2) to the experimental distribution. However, this distribution is strongly affected by the intentional omission of samples. Rather than fitting the model, our approach is as follows [5]. First we produce an image by zero-filling the omitted samples and applying mere 2D FFT and phase correction. Note that this image is distorted by Gibbs ringing. Next, a is estimated from

$$a = \frac{1}{2}\sqrt{\frac{\sum_{(x,y)\in O} \delta_y(x,y)^2}{N_O - 1}}, \tag{2.5}$$

where N_O is the number of pixels belonging to the object. In words, a is determined by the standard deviation of the intensity differences of the Gibbs ring-distorted image. The size of a is not altered in any stage of the remaining reconstruction process, experiments indicating that such alterations have only marginal effect.

For the sake of simplicity, we assume that the Lorentz distribution applies also to individual columns I'_x of I, and that a is the same for each column.

2.3 The background prior

In the real part of a phase-corrected image, one distinguishes the object O and the background B. Both O and B contain white Gaussian measurement noise with zero mean and standard deviation σ. Thus the probability distribution for a background pixel can be expressed as

$$\mathrm{p}(I'(x,y)) = \frac{1}{\sigma\sqrt{2\pi}}\exp(-\frac{I'(x,y)^2}{2\sigma^2}), \quad \text{with } (x,y) \in B. \tag{2.6}$$

The imaginary part of a phase-corrected image contains only white Gaussian measurement noise, with zero mean and the same standard deviation σ and can therefore be called background everywhere. Hence, the probability distribution of a pixel at any position in the field of view (FOV) can be expressed as

$$\mathrm{p}(I''(x,y)) = \frac{1}{\sigma\sqrt{2\pi}}\exp(-\frac{I''(x,y)^2}{2\sigma^2}), \quad \text{with } (x,y) \in \mathrm{FOV}. \tag{2.7}$$

2.4 The object perimeter

The perimeter of an object can be estimated from a histogram of pixel intensities [8]. Figure 2 shows an example for a full (*i.e.* 256 × 256) scan of a slice of a human head. The histogram shows two distinct peaks, the left-hand one originating from the noise, the right-hand one from the object. In the transition region between the two peaks it is difficult to classify the pixel values. One way to classify is to assign all pixel values between zero and the left-most minimum in the transition region as belonging to the noise. In Figure 2 this criterion appears easily applicable. However, for other objects, such as a spine with lungs, the classification can be problematic [5].

As in Section 2.2, an additional complication is that our scans are sparse. In order to avoid hampering the pixel classification by Gibbs ringing, we estimate the object perimeter from only the fully sampled central part of the data matrix $S(x,k_y)$, $|k_y| \leq n_{\mathrm{central}}$ [5, 8], n_{central} usually being 16 or 32. The procedure is as follows.

1. Zero-filling of the samples $k_y > n_{\mathrm{central}}$. This removes ringing due to the irregular sampling shown in the right half of Figure 1.
2. Row-wise weighting of the central samples ($|k_y| \leq n_{\mathrm{central}}$) by a Hanning window. This removes ringing due the fact that the signal has not nearly died out at $|k_y| = n_{\mathrm{central}}$.
3. FFT of each column.

The resulting image has been freed from ringing, but now the resolution is low so that classification of pixels remains problematic.

To cope with these complications, we devised an alternative classification method based on the probability density function for measurement noise in a background pixel of the *magnitude* image $|I(x,y)|$ [9]

$$\mathrm{p}(|I(x,y)|,\sigma) = \frac{|I(x,y)|}{\sigma^2}\exp(\frac{-|I(x,y)|^2}{2\sigma^2}). \tag{2.8}$$

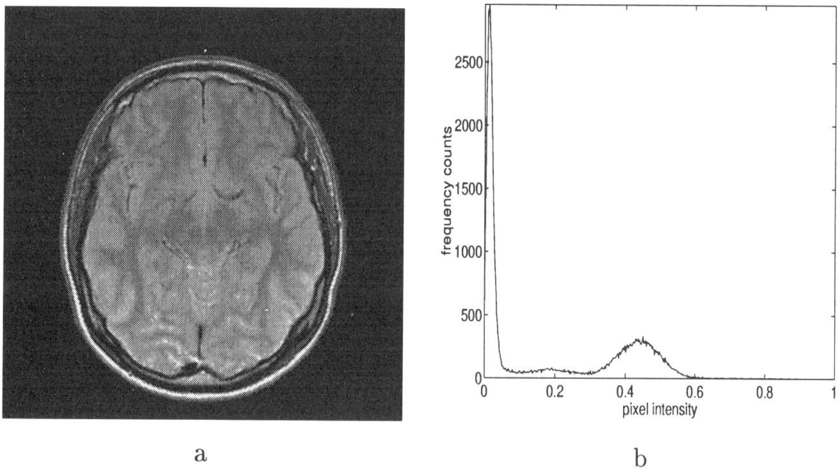

Figure 2. (a) 256 × 256 real-world MR image of an axial slice of a human brain, and (b) histogram of pixel intensities $I'(x, y)$ corresponding to (a). Pixel values are scaled between 0 and 1. Two peaks can be distinguished in (b). The left-hand peak, near zero, originates from the noise in the backgrond; the right-hand peak, around 0.45, originates from the object

Note that the distribution of Equation (2.8) peaks at $|I(x, y)| = \sigma$. Hence, fitting Equation (2.8) to the left-hand peak of the pixel intensity histogram derived from the low-resolution magnitude image, immediately yields σ. Figure 3 shows the result of this fit for the same object as in Figure 2 and $n_{\mathrm{central}} = 32$. Empirically, assigning all pixels with $|I(x, y)| \geq 5\sigma$ to the object appeared a good criterion for automatic perimeter estimation. It turned out that the remaining classification errors have no significant consequences so long as they amount to assigning a background pixel to the object. However, the reverse error, *i.e.* assigning an object pixel to the background, can lead to unacceptable image distortion.

2.5 The phase correction

The phase correction is estimated from the low-resolution image described above, prior to taking the absolute value. The phase of each pixel follows simply from the arctan of the ratio of the real and imaginary parts. A typical result for a slice of a human head scanned with the low phase-distortion "spin-echo" technique [5] is shown in Figure 4. It can be seen that the phase varies smoothly over the low-resolution image. There appeared no need to improve the phase estimate at a later stage in the iterative Bayesian reconstruction in which omitted samples are approaching their true values and need not be zero-filled per se to avoid strong Gibbs ringing. However, when using the high phase-distortion (but faster)

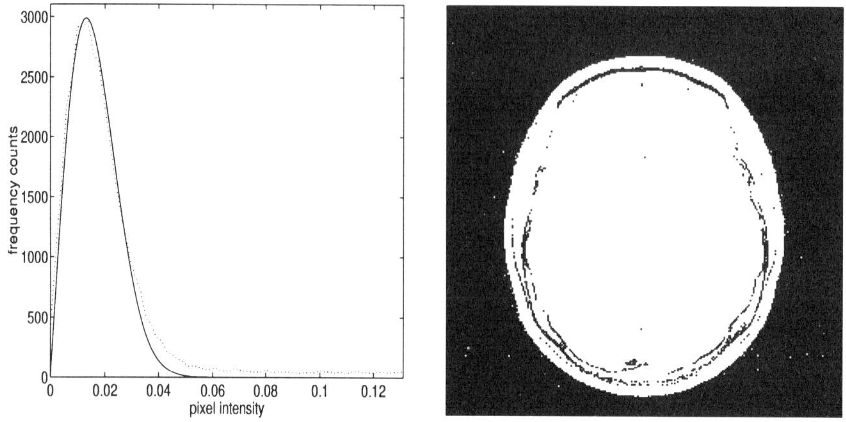

a b

Figure 3. (a) Fit of the model function Equation (2.8) to the noise-peak of the histogram of the low-resolution magnitude image of the object shown in Figure 2(a). The standard deviation of the noise σ is equated to the position of the top of the model function, and (b) resulting object perimeter estimate by setting the threshold at 5σ

"gradient-echo" technique, it appeared advantageous to update the phase after each iteration of the reconstruction procedure.

2.6 The prior knowledge combined

The prior knowledge about pixel intensities and pixel intensity differences will now be combined. We do this for each column I'_x of the phase-corrected image I' separately, ignoring correlation between columns. According to [10], the probability that events $\{A_1, A_2, \ldots, A_n\}$ occur simultaneously can be written as

$$\mathrm{p}(A_1, A_2, \ldots, A_n) = \mathrm{p}(A_1)\,\mathrm{p}(A_2|A_1)\,\mathrm{p}(A_3|A_2, A_1), \ldots, \mathrm{p}(A_n|A_{n-1}, \ldots, A_1). \tag{2.9}$$

If event A_m is independent, then $\mathrm{p}(A_m|A_{m-1}, \ldots, A_1) = \mathrm{p}(A_m)$. However, if event A_m depends on event A_{m-1} but is independent of all others, then $\mathrm{p}(A_m|A_{m-1}, \ldots, A_1) = \mathrm{p}(A_m|A_{m-1})$. Application of these results to the real part of a phase-corrected image column I'_x with elements $\{I'_x(1), I'_x(2), \ldots, I'_x(n)\}$, and comprising for example a single object with perimeters l and m, yields

$$\begin{aligned}\mathrm{p}(I'_x) \;=\;& \mathrm{p}(I'_x(1))\,\mathrm{p}(I'_x(2)) \ldots \mathrm{p}(I'_x(l)) \times \\ & \mathrm{p}(I'_x(l+1)|I'_x(l))\,\mathrm{p}(I'_x(l+2)|I'_x(l+1)) \ldots \mathrm{p}(I'_x(m-1)|I'_x(m-2)) \times \\ & \mathrm{p}(I'_x(m))\,\mathrm{p}(I'_x(m+1)) \ldots \mathrm{p}(I'_x(n)).\end{aligned} \tag{2.10}$$

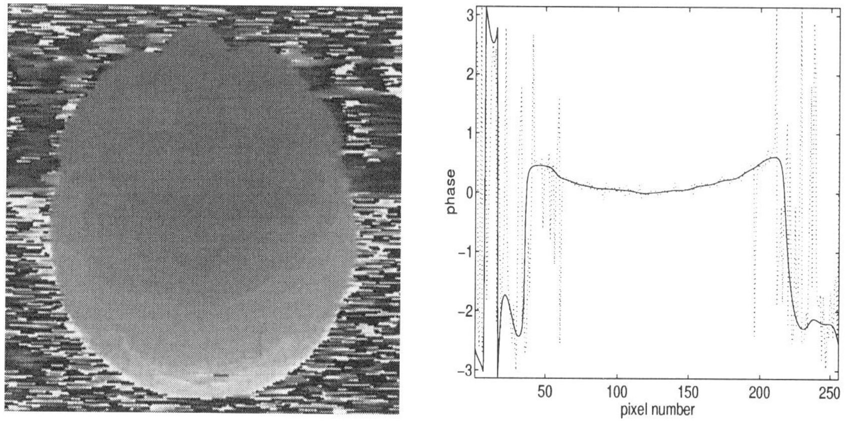

a b

Figure 4. (a) Low-resolution ($n_{\text{central}} = 32$) phase estimate (proportional to grey value) of a slice of a human brain scanned with the "spin-echo" technique. Within the object, the phase varies gradually between $-\pi$ (black) and π (white), and (b) phase (in radians) of one image row. The dotted and solid line are the phase of the full-resolution image and low-resolution image respectively

Noting that terms of the form $\mathrm{p}(I_x'(l+r)|I_x'(l+r-1))$ are governed by the Lorentzian probability distribution treated in Section 2.2, one can finally write

$$\begin{aligned}
\mathrm{p}(I_x) &= \frac{1}{(\sigma\sqrt{2\pi})^{b_x}} \exp[-\frac{1}{2\sigma^2} \sum_{y \notin O_x} I'(x,y)^2] \\
&\times \frac{1}{(\sigma\sqrt{2\pi})^{n}} \exp[-\frac{1}{2\sigma^2} \sum_{\forall y} I''(x,y)^2] \\
&\times \prod_{y \in O_x} \frac{1}{a\pi(1+\delta_y(x,y)^2/a^2)} \qquad (2.11)
\end{aligned}$$

with $x = -128, -127, \ldots, 127$, b_x is the number of background pixels (*i.e.*, outside O_x), $n = 256$.

We mention that additional knowledge is often available when measuring a series of MR scans of an object that varies with time. This pertains to the fact that each scan of the series can benefit from information gained from previous scans [11]. The latter is beyond the scope of this contribution.

3 Bayesian image reconstruction

3.1 Strategy for underdetermined systems

Bayesian estimation lends itself well to accommodation of prior knowledge [12, 13, 14]. For the present problem, the well-known Bayes formula can be written as

$$\mathrm{p}(I_x|S_x) = \frac{\mathrm{p}(S_x|I_x)\,\mathrm{p}(I_x)}{\mathrm{p}(S_x)}, \tag{3.1}$$

in which I_x is a column of the image I and S_x is the related column of the sparse raw data matrix $S(x, k_y)$. Furthermore, the posterior $\mathrm{p}(I_x|S_x)$ is the probability density function of I_x after collecting the data, the likelihood $\mathrm{p}(S_x|I_x)$ is the probability density function of the noise superimposed on the data, the prior knowledge $\mathrm{p}(I_x)$ has been treated above, and the evidence $\mathrm{p}(S_x)$ is just a scaling factor. The task is to find for each x the image column I_x that maximizes the posterior $\mathrm{p}(I_x|S_x)$. In the standard Bayesian procedure, the resulting image is a trade-off between the measured data and the invoked prior knowledge. The present approach differs from this in the following sense. Since S_x is sparse, our system is underdetermined. We maximize the posterior by adjusting the omitted samples subject to the available prior knowledge and leaving the measured samples untouched. Note that not touching the data amounts to treating them as ideal which in turn implies that the likelihood $\mathrm{p}(S_x|I_x)$ becomes a constant factor.

3.2 Maximization of the posterior

Maximizing the posterior is equivalent to minimizing minus its natural logarithm. Dropping constant terms, the natural logarithm of the posterior becomes

$$\begin{aligned} &\ell_x \stackrel{\text{def}}{=} -\ln \mathrm{p}(I_x|S_x) \Longrightarrow \\ &\tfrac{1}{2\sigma^2} I_x'^{\mathrm{T}} \mathbf{B} I_x' + \sum_{y \in O} \ln[1 + \delta(y)^2/a^2] + \tfrac{1}{2\sigma^2} I_x''^{\mathrm{T}} I_x'', \end{aligned} \tag{3.2}$$

where $\mathbf{B}$ is an n $\times$ n diagonal matrix with diagonal entries

$$\mathbf{B}(y,y) = \begin{cases} 1 & \text{if } I_x(y) \notin O_x, \\ 0 & \text{if } I_x(y) \in O_x. \end{cases} \tag{3.3}$$

The quantity ℓ_x defined in Equation (3.2) is to be minimized as a function of the omitted samples, subject to the conditions

$$\begin{aligned} &1)\ I_x = \mathbf{W}^{-1} S_x, \\ &2)\ \text{measured data remain unchanged,} \end{aligned} \tag{3.4}$$

where $\mathbf{W}$ is the DFT matrix

$$\mathbf{W}(u,v) = w^{uv} \text{ with } \begin{cases} w = \exp(2\pi i/n), \\ u = -n/2, \ldots, n/2-1, \\ v = -n/2, \ldots, n/2-1. \end{cases} \tag{3.5}$$

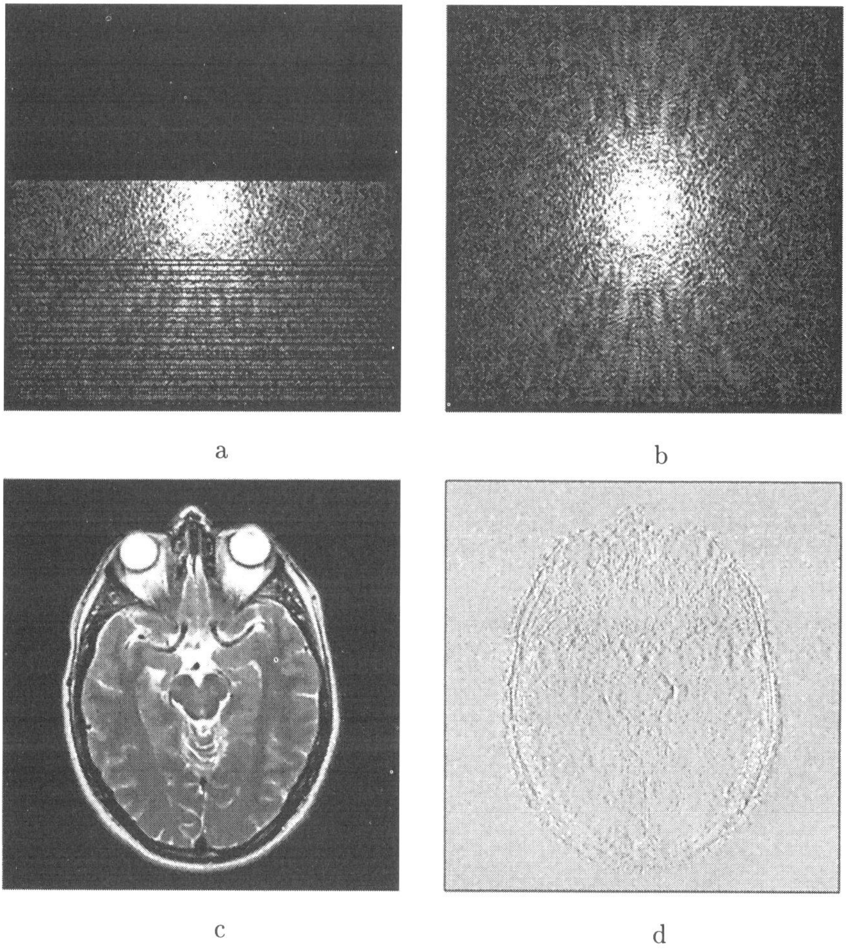

Figure 5. Spin-echo scan time reduction by 57% for a slice of a human head by sparse sampling. (a) Magnitude of the sparse data matrix, $|S(k_x, k_y)|$, on a logarithmic scale. The black rectangle and black lines represent omitted data, (b) same as (a), but now the omitted samples have been estimated with the iterative Bayesian procedure described in this paper, (c) 2D FFT of the phase-corrected reconstructed data matrix $S(k_x, k_y)$, and (d) difference between (c) and the 2D FFT of the full scan

Although ℓ_x is not convex we found empirically that a gradient search for its minimum does not critically depend on the choice of starting values of the omitted samples [5]. Hence, starting with zeros is adequate. For minimization, we use the iterative conjugate gradients method [15]. The search direction, $\mathrm{d}S_x^{(j)}$, in iteration j is given by

$$\mathrm{d}S_x^{(j)} = -\nabla \ell_x(S_x^{(j)}) + \frac{\|\nabla \ell_x(S_x^{(j)})\|^2}{\|\nabla \ell_x(S_x^{(j-1)})\|^2}\,\mathrm{d}S_x^{(j-1)}. \tag{3.6}$$

The columns S_x comprise measured data and omitted data. As mentioned earlier, the former are left untouched whereas the latter are updated in each iteration. The gradient of ℓ_x with respect to the omitted data is [5]

$$\nabla \ell_x(S_x) = \mathbf{MW}(\frac{\partial \ell_x}{\partial I_x})^{\mathrm{T}}, \tag{3.7}$$

in which $\mathbf{M}$ is a diagonal matrix with diagonal entries

$$\mathbf{M}(k_y, k_y) = \begin{cases} 0 & \text{if } S_x(k_y) \text{ has been measured,} \\ 1 & \text{if } S_x(k_y) \text{ has been omitted,} \end{cases} \tag{3.8}$$

and

$$\frac{\partial \ell_x}{\partial I_x} = \frac{\partial \ell_x}{\partial I'_x} + i\frac{\partial \ell_x}{\partial I''_x} \tag{3.9}$$

$$\frac{\partial \ell_x}{\partial I'_x} = \frac{\partial \ell_x}{\partial I'_x} + \frac{\partial \ell_x}{\partial \delta_y(x)}\frac{\partial \delta_y(x)}{\partial I'_x} = (\frac{1}{\sigma^2}\mathbf{B}I'_x + \mathbf{D}^{\mathrm{T}}(\frac{\partial \ell_x}{\partial \delta_y(x)})^{\mathrm{T}})^{\mathrm{T}} \tag{3.10}$$

$$\frac{\partial \ell_x}{\partial I''_x} = (\frac{1}{\sigma^2}I''_x)^{\mathrm{T}} \tag{3.11}$$

$$\frac{\partial \ell_x}{\partial \delta_y(x,y)} = \frac{2\delta_y(x,y)}{a^2 + \delta_y(x,y)^2}, \tag{3.12}$$

where $i = \sqrt{-1}$. Finally, the linear search parameter λ_j of the omitted data update in iteration j is the smallest possible positive number that minimizes $\ell_x(S_x^{(j)} + \lambda_j \mathrm{d}S_x^{(j)})$ [5]. The iterations are pursued until the changes of the omitted samples become insignificant. Note that in each iteration the agreement between the inverse FFT of the image and the measured samples is exact.

Summarizing, the iterative Bayesian image estimation from sparse raw data runs as follows.

1. FFT in k_x space of measured rows of $S(k_x, k_y)$, resulting in $S(x, k_y)$.
2. Zero-filling of omitted rows of $S(x, k_y)$.
3. Estimation of a starting image by FFT in k_y space.
4. Estimation of a low-resolution image from the fully sampled region $|k_y| \leq n_{\mathrm{central}}$ (see Figure 1), n_{central} =16 or 32, $-128 \leq k_x \leq +127$.

5. Estimation of the object perimeter and phase error map from the low-resolution image, to be used as *prior knowledge.*

6. Correction of phase errors in the current image.

7. Computation of an image update from prior knowledge. This can be done for each column separately.

8. Undoing of the phase correction.

9. Conversion of the image update to the k_y space by IFFT. Measured data are left intact.

10. If the changes of omitted samples are sufficiently small, then the current image becomes the final image. Else, go to 6 preceded by a better phase error estimate in the case of gradient-echo scanning.

Reconstruction (estimation) of an image column from real-world raw data usually converges in ten to fifteen iterations. Using a SUN SPARCstation 5 and Fortran77, this takes 0.2 seconds. For a complete image, this is to be repeated for up to 256 columns. Figure 5 shows a successful application to a spin-echo scan of a slice of a human head yielding 57% scan time reduction. With gradient echoes, the scan time reduction is less because of the phase errors incurred.

4 Conclusions

- Sparse irregular sampling combined with Baysian image estimation and prior knowledge yields substantial scan time reduction.
- Additonal prior knowledge is sought, especially for "dynamic" scans.
- The computation time is presently too long for on-line use.

Acknowledgment

This work is supported by the EU programme HCM/Networks, the Dutch Research Foundation STW, Philips Medical Systems, and the Advanced School for Computing and Imaging (ASCI).

Bibliography

1. Marseille, G.J., Beer, R. de, Fuderer, M., Mehlkopf, A.F., Ormondt, D. van (1996), Nonuniform Phase-Encode Distributions for MRI Scan Time Reduction, *J. Magn. Reson. B*, **111**, 70-75.

2. Liang, Z.P., Boada, F.E., Constable, R.T., Haacke, E.M., Lauterbur, P.C, Smith, M.R. (1992), Constrained Reconstruction Methods in MR Imaging, *Reviews of Magnetic Resonance in Medicine*, **4**, 67-185.

3. Fuderer, M. (1989), Ringing Artefact Reduction by an Efficient Likelihood Improvement Method, *Proc. SPIE*, **1137**, 84-90.

4. Marseille, G.J., Fuderer, M., Beer, R. de, Mehlkopf, A.F., and Ormondt, D. van (1994), Reduction of MRI scan time through nonuniform sampling and edge-distribution modeling, *J. Magn. Reson. B*, **103**, 292-295.

5. Marseille (1997), MRI scan time reduction through nonuniform sampling, *Ph D Thesis*, Delft University of Technology. http://dutnsic.tn.tudelft.nl:8080/main/main.html

6. Lettington, A.H. and Quong, Q.H. (1995), Image restoration using a Lorentzian probability model, *J. Modern Optics* , **42**, 1367-1376.

7. McNally, B. (1996), Lorentzian Probability Model, *Internal Report*, King's College London, London, UK, 16-17.

8. Spiliotis, I.M., Ormondt, D. van, Mertzios, B.G. (1995), Prior Knowledge for Improved Image estimation from Raw MRI Data, *Conf. Proc. 2nd International Workshop on Image and Signal Processing: Theory, Methodology and Applications*, Budapest, Hungary, 8-10 November 1995, 274-283.

9. Bernstein, M.A., Thomasson, D.M., and Perman, H. (1989), Improved detectability in low signal-to-noise ratio magnetic resonance images by means of a phase-corrected real reconstruction, *Med. Physics* **16** (5), 813-817.

10. Blum, J.R., Rosenblatt, J.I. (1972), Probability and Statistics, Saunders, Philadelphia.

11. Chandra, S., Liang, Z.P., Webb, A., Lee, H., Morris, H.D., Lauterbur, P.C. (1996), Application of Reduced-Encoding Imaging with Generalized-Series Reconstruction (RIGR) in Dynamic MR Imaging, *JMRI*, **6**, 783-797.

12. Norton, J.P. (1986), An Introduction to Identification, Academic Press, London.

13. Gelman, A., Carlin, J.B., Stern, H.S., Rubin, D.B. (1996), Bayesian Data Analysis, Chapman & Hall, London.

14. Sivia, D.S. (1996), Data Analysis. A Bayesian Tutorial, Clarendon, Oxford.

15. Fletcher, R., Reeves, C.M. (1964), Function minimization by conjugate gradients, *The Comp. J.* **7**, 149-154.

Polynomial Models for 3D Motion Parameters and Depth Estimation in Monocular Computer Vision

J.M. Menéndez, N. García, L. Salgado and E. Rendón

Grupo de Tratamiento de Imágenes, E.T.S. Ingenieros de Telecomunicación, Universidad Politécnica de Madrid, Spain

Abstract

Egomotion requires the determination of the position and trajectory of the moving sensor just by analysing the sequence of images acquired along its movement. Here, we present a new analytical procedure for the estimation of the three-dimensional kinematic parameters of the sensor and the depth map that describes the three-dimensional structure of the world that surrounds the acquisition camera (only monocular gray-level images are considered). The procedure is simple but robust, providing a closed-form solution to the egomotion problem.

1 Introduction

Image motion analysis stems as one of the key elements for image analysis and interpretation. One of its main goals is the detection and tracking of the moving objects in the scene, as well as the global movement estimation due to camera motion. It is always possible to consider the camera as the center of the analysed space (the coordinate system is located in the *static* camera) and so the images are grabbed from a moving world surrounding the camera. This approach simplifies the analysis in a large number of computer vision applications and it will be the one followed here. Moreover, only monochrome information (gray level images) will be considered.

Motion analysis allows not only for image interpretation. It provides the possibility to derive information about the three-dimensional motion parameters of the moving camera, as well as to obtain a parametric estimation of the world *depth map*, that is, a map of distances between each visualized object point and the image plane. Activities such as automatic robot guidance or autonomous navigation deeply rely on the success of the correct estimation of the 3D structure of the environment in which the camera is moving.

2 Previous work

Egomotion requires the determination of the position and trajectory of the moving sensor just by analysing the sequence of images acquired along its movement.

An extension of the statement of such a problem is to obtain an estimation of the three-dimensional structure of the surrounding world, required, for example, to avoid from crashing with any other vehicle or object in the scene [5].

During the last two decades, several important works have analysed the feasibility of the problem. Longuet–Higgins and Prazny [4] already proved the possibility to estimate the motion parameters and the depth map from the analysis of the *vector field*. Later, Waxman, Kamgar-Parsi and Subbarao [15] suggested to use first and second order approaches to the inverse function of the depth map to obtain estimation of the motion parameters, providing the transformation that makes null one of the translational components, and therefore extending the usability of the first order approach. Recently, some very interesting works try to emphasize on different mechanisms to estimate the motion parameters through the analysis of the separability of the translational and rotational components [3] [14], through the search of the Focus of Expansion (FOE) [1] [2] [7], or through the analysis of the ambiguities of the global mathematical approach [5] [9].

Here, we provide a new analytical procedure for the estimation of the three-dimensional kinematic parameters and the depth map, relying on the assumption of second order curved surfaces on the visualized objects, and patch analysis of the optic flow field. This proposal relies on the assumption of the knowledge of the translational component in the direction of the optical axis, as this component is usually known and controlled in most of the autonomous navigation applications. It is simpler and more robust than the one previously mentioned [15], therefore offering an alternative approach.

3 The Euclidean coordinate system

We will consider the Euclidean coordinate system and the motion parameters shown in Figure 1. The three-dimensional coordinate system (X, Y, Z) is centered at the camera's center of projection, and the two-dimensional projected image surface (x, y) is located on the plane $Z = f$, where f is the focal length of the acquisition camera. So, the origin of the image coordinates (x, y) will be located at the center of the image. Assuming an ideal pin-hole camera model (perspective projection), the projected image location $p(x, y)$ corresponding to the 3-D point $P(X, Y, Z)$ is given by

$$(x, y, f)^T = \frac{f}{Z}(X, Y, Z)^T. \tag{3.1}$$

Assuming rigid motion in the Euclidean three-dimensional coordinate system, any movement can be described in terms of its three-dimensional translational $\vec{V} = (V_x, V_y, V_z)^T$ and rotational $\vec{W} = (W_x, W_y, W_z)^T$ velocities. So, the instantaneous velocity $\vec{V_I}$ of point P with respect to the three-dimensional coordinate system can be expressed as

$$\vec{V_I} = (\dot{X}, \dot{Y}, \dot{Z}) = -\vec{V} - \vec{W} \times \vec{R} \tag{3.2}$$

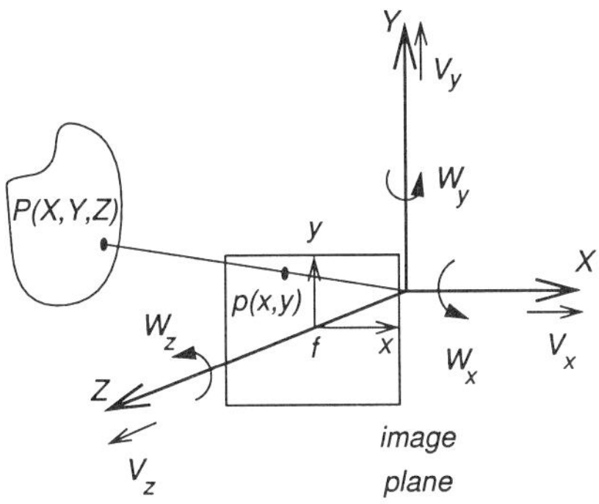

Figure 1. The Euclidean coordinate system

where $\vec{R} = (X, Y, Z)^T$ is the position vector of point P. Therefore, the component form of Equation (3.2) is

$$\begin{array}{lllll} V_{Ix} & = & \dot{X} & = & -V_x - W_y Z + W_z Y \\ V_{Iy} & = & \dot{Y} & = & -V_y - W_z X + W_x Z \\ V_{Iz} & = & \dot{Z} & = & -V_z - W_x Y + W_y X \end{array} . \tag{3.3}$$

We are interested in the two-dimensional projection $\vec{v} = (v_x, v_y) = (\dot{x}, \dot{y})$ of the velocity $\vec{V_I}$, as it represents the instantaneous velocity of the brightness pattern centered at that point, the so called *optic flow* [13]. From Equation (3.1) we obtain

$$x = \frac{f}{Z}X \quad \text{and} \quad y = \frac{f}{Z}Y. \tag{3.4}$$

The optic flow at point (x, y) is obtained by differentiating Equations (3.4) with respect to time

$$\begin{array}{lllll} v_x & = & \dot{x} & = & \frac{f}{Z}\dot{X} - \frac{f}{Z^2}X\dot{Z} \\ v_y & = & \dot{y} & = & \frac{f}{Z}\dot{Y} - \frac{f}{Z^2}Y\dot{Z} \end{array} . \tag{3.5}$$

Using Equations (3.2) (3.4) and (3.5) we obtain the expression of the projected two-dimensional velocity at point (x, y) in terms of the three-dimensional velocity components

$$\begin{array}{lll} v_x & = & \frac{1}{Z}(-fV_x + xV_z) + W_x\frac{xy}{f} - W_y\left(f + \frac{x^2}{f}\right) + W_z y \\ v_y & = & \frac{1}{Z}(-fV_y + yV_z) + W_x\left(f + \frac{y^2}{f}\right) - W_y\frac{xy}{f} - W_z x \end{array} . \tag{3.6}$$

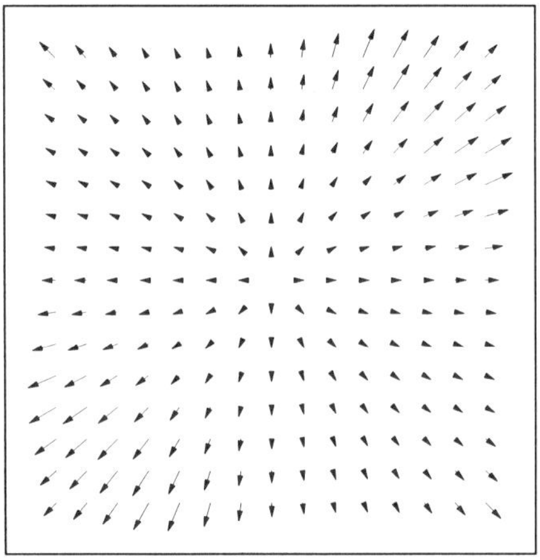

Figure 2. Example of typical vector field

So far, we have derived all the expressions for a single point P. To extend the applicability for every point of the image, we will assume that P lies on a surface defined by a projected function $Z(x, y) = 1/\delta(x, y)$, usually known as the *depth map*. Considering this, we can state that the optic flow was generated by the relative motion of all the surface points with respect to the camera. Figure 2 shows an example of a typical optic flow vector field, that denotes the motion (or variation) of intensity points between two consecutive images of a sequence.

4 Optic flow model

Several algorithms [8] [13] have been proposed for the estimation of the generated velocity field. All these methods usually provide large number of estimated image velocity vectors $\hat{\vec{v}} = (\hat{v}_x, \hat{v}_y)$. The quality of the estimation varies from one method to another, depending on the assumed set of constraints imposed in the vector estimation step. Although noisy, the large number of vectors provide a robust estimation of the velocity field when considered globally. Therefore, it seems quite reasonable to use the whole set of vectors to infer the analytical modeled expression of the vector field, also called *phase portrait* [12].

Several authors have made use of the optic flow for different sorts of applications: texture analysis [12], camera motion parameters estimation [14], FOE location [7], moving object segmentation [6], etc. Most of them make use of a first or second order polynomial model to approximate the global (and noisy) vector field by an analytical expression like

$$\begin{aligned} \hat{v}_x &= f_1 + f_2 x + f_3 y + f_4 x^2 + f_5 xy + f_6 y^2 + \mathcal{R}_x \\ \hat{v}_y &= g_1 + g_2 x + g_3 y + g_4 x^2 + g_5 xy + g_6 y^2 + \mathcal{R}_y \end{aligned} \tag{4.1}$$

where $\mathcal{R}_x$ and $\mathcal{R}_y$ are the remainders of each polynomial, with orders highers than two. The coefficients f_i and g_i are easily obtained by fitting the dense field of estimated vectors to the selected model, through any classical least-squares procedure.

In our case, we are trying to model Equation (3.6), whose exact order depends on the function $\delta(x,y)$. This inverse function of the depth map is approximated in [5] by the following Taylor series:

$$\delta(x,y) = k_1 + k_2 x + k_3 y + k_4 x^2 + k_5 xy + k_6 y^2 \tag{4.2}$$

where it is assumed that $\delta(x,y)$ is a continuous and differentiable up to second order function. Although restrictive, this assumption provides a first approximation of the inverse of the distance function by a second order smooth curved surface. The validity of this approximation is increased if the analysis is applied to a small image region, corresponding then to a small part of the object surface—the so called *patch* analysis [14]. In other words, the validity of the patch analysis is higher if only a small set of vectors are considered and evaluated. After substituting the expression of $\delta(x,y)$ in (3.6) it is clear that a third order polynomial is required for both $(\hat{v}_x, \hat{v}_y)$. This implies the need to use a polynomial system whose order is higher than the second order model that has been commonly used in the previously published works. Imposing that $v_x \simeq \hat{v}_x$ and $v_y \simeq \hat{v}_y$ the polynomial model that considers the possible three-dimensional movement of the sensor and approximates the global vector field requires the following couple of equations:

$$\begin{array}{rcl} \hat{v}_x & = & f_1 + f_2 x + f_3 y + f_4 x^2 + f_5 xy + f_6 y^2 + f_7 x^3 + f_8 x^2 y + f_9 xy^2 \\ \hat{v}_y & = & g_1 + g_2 x + g_3 y + g_4 x^2 + g_5 xy + g_6 y^2 + g_7 x^2 y + g_8 xy^2 + g_9 y^3 \end{array} . \tag{4.3}$$

5 Motion parameters estimation

Equations (4.3) state that:

1. The optic flow vectors corresponding to a patch are analytically expressed by a couple of finite order polynomial equations on x and y.

2. Each one of the polynomial equations determines the value of each vector component $(\hat{v}_x, \hat{v}_y)$.

3. The origin of each vector is the analysed point $p(x,y)$.

Keeping this in mind, the following equations can be established:

$$\begin{array}{lll|lll} f_1 & = & -k_{00}fV_x - W_yf & g_1 & = & -k_{00}fV_y + W_xf \\ f_2 & = & -k_{10}fV_x + k_{00}V_z & g_2 & = & -k_{10}fV_y - W_z \\ f_3 & = & -k_{01}fV_x + W_z & g_3 & = & -k_{01}fV_y + k_{00}V_z \\ f_4 & = & -k_{20}fV_x + k_{10}V_z - \frac{W_y}{f} & g_4 & = & -k_{20}fV_y \\ f_5 & = & -k_{11}fV_x + k_{01}V_z + \frac{W_x}{f} & g_5 & = & -k_{11}fV_y + k_{10}V_z - \frac{W_y}{f} \\ f_6 & = & -k_{02}fV_x & g_6 & = & -k_{02}fV_y + k_{01}V_z + \frac{W_x}{f} \\ f_7 & = & k_{20}V_z & g_7 & = & k_{20}V_z \\ f_8 & = & k_{11}V_z & g_8 & = & k_{11}V_z \\ f_9 & = & k_{02}V_z & g_9 & = & k_{02}V_z \end{array} . \quad (5.1)$$

As $f_7 = g_7$, $f_8 = g_8$ and $f_9 = g_9$, the actual number of independent coefficients that have to be fitted (through the Levenberg-Marquardt [11] method for example) is 15 instead of 18. With this set of 15 independent coefficients $\{f_i, g_i\}$, it should be possible to recover the whole set of 13 unknowns: $\vec{V} = (V_x, V_y, V_z)$, $\vec{W} = (W_x, W_y, W_z)$, $\{k_{i,j}\}$, and even the focal distance f. However, it can be proved that there is a non-linear dependency that relates four coefficients with the others and the focal distance. Those relations are shown in Equations (5.2).

The relations have been computed for the coefficients f_3, f_5, g_5 and g_6. Nevertheless, they could be obtained for any other four coefficients following similar procedures. Equations (5.2) reduce the number of estimated unknowns from 15 to 11. Therefore, only up to 11 of the 13 motion parameters ($\vec{V}, \vec{W}, \{k_{i,j}\}$, and f) can be recovered, with 2 degrees of freedom.

In order to provide a practical closed-form solution to the non-linear stated problem, it has to be considered that the main application behind this work is egomotion, that is, the determination of the position and trajectory of the moving vehicle just by analysing the sequence of images acquired along its movement. With this in mind, the following has to be considered:

- The focal distance f can be assumed to be known in most of the applications, as the camera is usually selected by the user or designer of the system.
- There is still another restriction to reduce the number of unknowns. This degree of freedom can be assigned to the translational velocity in the direction of the optical axis V_z, as this is usually the magnitude that can be controlled by the user in most of the egomotion or autonomous navigation applications.

Taking the above into account, we can provide the closed-form set of Equations (5.3) that allow for the estimation of the motion parameters and the depth map, which are alternative and simpler to the ones provided by Waxman et al. [15].

Equations (5.3) reflect some independency between $\vec{W}$ and V_z. Although this is not a definite prove, it goes in the line of several authors [3] [10] that state the

$$
\begin{aligned}
f_3 &= \frac{f_9 g_4}{f_7\left(f_6^2+f^2 f_9^2\right)}\left(f_2 f_6-f^2 f_6 f_7-f_1 f_9+f^2 f_4 f_9\right) \\
&- g_2-\frac{f_6 f_7}{f_9 g_4\left(f_6^2+f^2 f_9^2\right)}\left(f^2 f_6^2 f_7+f_1 f_6 f_9\right. \\
&- \left. f^2 f_4 f_6 f_9+f^2 f_2 f_9^2-f_6^2 g_3-f^2 f_9^2 g_3\right) \\
f_5 &= \frac{1}{f^2 f_7\left(f_6^2+f^2 f_9^2\right)}\left(f_6^2 f_7 g_1+f^2 f_7 f_9^2 g_1\right. \\
&- \left. f^2 f_6^2 f_7 g_4-f_1 f_6 f_9 g_4+f^2 f_4 f_6 f_9 g_4-f^2 f_2 f_9^2 g_4\right) \\
&- \frac{f_7}{g_4\left(f_6^2+f^2 f_9^2\right)}\left(f^2 f_6^2 f_7+f_1 f_6 f_9-f^2 f_4 f_6 f_9\right. \\
&+ \left. f^2 f_2 f_9^2-f_6^2 g_3-f^2 f_9^2 g_3\right)+\frac{f_6 f_8}{f_9} \\
g_5 &= \frac{f_9\left(f_2 f_6-f^2 f_6 f_7-f_1 f_9+f^2 f_4 f_9\right)}{f_6^2+f^2 f_9^2}+\frac{f_8 g_4}{f_7} \\
&- \frac{f_6^3 f_7-f_4 f_6^2 f_9+f_2 f_6 f_9^2-f_1 f_9^3}{f_9\left(f_6^2+f^2 f_9^2\right)} \\
g_6 &= \frac{1}{f^2 f_7\left(f_6^2+f^2 f_9^2\right)}\left(f_6^2 f_7 g_1+f^2 f_7 f_9^2 g_1\right. \\
&- \left. f^2 f_6^2 f_7 g_4-f_1 f_6 f_9 g_4+f^2 f_4 f_6 f_9 g_4-f^2 f_2 f_9^2 g_4\right) \\
&+ \frac{f_9 g_4}{f_7}-\frac{1}{g_4\left(f_6^2+f^2 f_9^2\right)}\left(f_7\left(f^2 f_6^2 f_7+f_1 f_6 f_9\right.\right. \\
&- \left.\left. f^2 f_4 f_6 f_9+f^2 f_2 f_9^2-f_6^2 g_3-f^2 f_9^2 g_3\right)\right)
\end{aligned}
\tag{5.2}
$$

possibility to obtain separate estimates for $\vec{V}$ and $\vec{W}$. The above equations can be further simplified in a progressive version, as shown in Equations (5.4).

The presented solution has been tested with a set of synthetic and real images. In all the cases, the procedure has proven to be robust, as the mean squared error (MSE) of the estimated parameters was below the 1% of its actual value for both the translational and rotational parameters, evaluated when the error presented in the fitted coefficients were up to the 20% of their actual value. The MSE of the $\{k_{ij}\}$ parameters was also below the 1.5% of their actual valued under the same conditions, as displayed in the Appendix.

6 Conclusions

A new analytical procedure for the estimation of the kinematic parameters and the surrounding three-dimensional world structure of a monocular moving observer has been presented.

$$
\begin{aligned}
V_x &= -\frac{f_6 V_z}{f f_9} \\
V_y &= -\frac{g_4 V_z}{f f_7} \\
W_x &= \frac{1}{f f_7 \left(f_6^2 + f^2 f_9^2\right)} \left(f_6^2 f_7 g_1 + f^2 f_7 f_9^2 g_1\right. \\
&- \left. f^2 f_6^2 f_7 g_4 - f_1 f_6 f_9 g_4 + f^2 f_4 f_6 f_9 g_4 - f^2 f_2 f_9^2 g_4\right) \\
W_y &= \frac{f_6 \left(f^2 f_6^2 f_7 + f_1 f_6 f_9 - f^2 f_4 f_6 f_9 + f^2 f_2 f_9^2\right)}{f f_9 \left(f_6^2 + f^2 f_9^2\right)} - \frac{f_1}{f} \\
W_z &= \frac{f_9 g_4 \left(f_2 f_6 - f^2 f_6 f_7 - f_1 f_9 + f^2 f_4 f_9\right)}{f_7 \left(f_6^2 + f^2 f_9^2\right)} - g_2 \\
k_{00} &= \frac{f^2 f_6^2 f_7 + f_1 f_6 f_9 - f^2 f_4 f_6 f_9 + f^2 f_2 f_9^2}{V_z \left(f_6^2 + f^2 f_9^2\right)} \\
k_{10} &= \frac{f_9 \left(f_2 f_6 - f^2 f_6 f_7 - f_1 f_9 + f^2 f_4 f_9\right)}{V_z \left(f_6^2 + f^2 f_9^2\right)} \\
k_{01} &= \frac{f_7}{V_z g_4 \left(f_6^2 + f^2 f_9^2\right)} \left(f^2 f_4 f_6 f_9 + f_6^2 g_3 + f^2 f_9^2 g_3\right. \\
&- \left. f^2 f_6^2 f_7 - f_1 f_6 f_9 - f^2 f_2 f_9^2\right) \\
k_{20} &= \frac{f_7}{V_z} \\
k_{11} &= \frac{f_8}{V_z} \\
k_{02} &= \frac{f_9}{V_z}
\end{aligned}
\tag{5.3}
$$

The input data to the method is the vector field that has been generated by the evaluation of the relative movement between the moving sensor and its surrounding environment. The optic flow is fitted to a set of two third degree polynomials, whose 15 coefficients gather all the information concerning the searched parameters. Due to the fact that the number of independent coefficients is actually 11, the fitting process requires the implementation of a non-linear procedure.

This new proposal is alternative and simpler compared to the classic one proposed by Waxman et al. [15], and relies on the assumption of the knowledge of the translational component in the direction of the optical axis, as this component is usually known and controlled in most of the autonomous navigation applications. The test performed on synthetic and real images confirm that the presented solution is very robust against noise in the fitted coefficients.

$$
\begin{aligned}
k_{20} &= \frac{f_7}{V_z} \\
k_{11} &= \frac{f_8}{V_z} \\
k_{02} &= \frac{f_9}{V_z} \\
V_x &= -\frac{V_z f_6}{f f_9} \\
V_y &= -\frac{V_z g_4}{f f_7} \\
k_{10} &= \frac{V_z f^2 f_4 + V_x V_z k_{20} f^3 - V_x f f_2 - V_z f_1}{f^2 \left(V_x^2 + V_z^2\right)} \\
W_z &= -V_y k_{10} f - g_2 \\
k_{00} &= \frac{V_x k_{10} f + f_2}{V_z} \\
W_y &= -V_x k_{00} - \frac{f_1}{f} \\
W_x &= \frac{V_y k_{00} f + g_1}{f} \\
k_{01} &= \frac{f f_5 + f^2 k_{11} V_x - W_x}{f V_z}
\end{aligned}
\tag{5.4}
$$

7 Acknowledgements

This work has been partially supported by the Comisión Interministerial de Ciencia y Tecnología of the Spanish Government.

Bibliography

1. Y. Aloimonos and Z. Duric, *Estimating the heading direction using normal flow.* Int. J. of Computer Vision, 13(1), pp. 33–56, 1994.

2. W. Burger and B. Bhanu, *Qualitative motion understanding.* Kluwer Academic Publishers, 1992.

3. R. Hummel and V. Sundaeswaran, *Motion parameter estimation from global flow field data.* IEEE Trans. on Pattern Analysis and Machine Intelligence, Vol 15, No. 5, pp. 459–476, 1993.

4. H. C. Longuet–Higgins, *The visual ambiguity of a moving plane.* Proc. of the Royal Society of London, B223(2), pp. 165–175, 1984.

5. S. Maybank, *Theory of reconstruction from image motion.* Springer–Verlag, 1993.

6. J. M. Menéndez, E. Rendón, L. Salgado, and N. García, *Differential model-based moving object segmentation in dynamic monocular sequences.* Proc. 5th Int. Conf. of Signal Processing Applications and Technology, pp. 939–944, Dallas, 1994.

7. J. M. Menéndez, N. García, L. Salgado, and E. Rendón, *An algorithm for FOE localization.* Proc. IEEE Int. Conf. on Image Processing, pp. 811–814, Lausanne, 1996.

8. H.-H. Nagel and W. Enkelmann, *An investigation of smoothness constraints for the estimation of displacement vector fields from image sequences.* IEEE Trans. on Pattern Analysis and Machine Intelligence, Vol. 8, No. 5, pp. 565–593, September 1986.

9. S. Negahdaripour, *Multiple interpretations of the shape and motion of objects from two perspective image.* IEEE Trans. on Pattern Analysis and Machine Intelligence, Vol. 12, No. 11, pp. 1025–1039, 1990.

10. K. Prazny, *Determining the instantaneous direction of motion from optical flow generated by a curvilinearly moving observer.* Computer Graphics and Image Processing, Vol. 17, No. 3, pp. 238–248, 1981.

11. W. H. Press, S. A. Teukolsky, W. T. Vetterling and B. P. Flannery, *Numerical Recipes in C: the art of scientific computing (Second Edition).* Cambridge University Press, 1992.

12. A. R. Rao and R. C. Jain, *Computerized flow fields analysis: oriented texture fields.* IEEE Trans. on Pattern Analysis and Machine Intelligence, Vol. 14, No. 7, pp. 693–709, July 1992.

13. A. Singh, *Optic flow computation. A unified perspective.* IEEE Computer Society Press, 1991.

14. V. Sundareswaran, *Global Methods for Image Motion Analysis.* PhD. Thesis, New York University, October, 1992.

15. A. M. Waxman. B. Kamgar-Parsi and M. Subbarao. *Closed-form solutions to image flow equations for 3-d structure and motion.* Int. Journal of Computer Vision, 1, pp. 239-258, 1987.

Appendix

The following figures display the results obtained from the sensibility analysis applied to Equations 5.2 after disturbing the fitted coefficients $\{f_i, g_i\}$ with additive white Gaussian noise.

The figures show the evolution of the Mean Squared Error (MSE) of each estimated motion parameter, computed as a percentage % of its actual value, while the standard deviation of the additive white Gaussian noise increases up to the 30% of the fitted coefficient actual value.

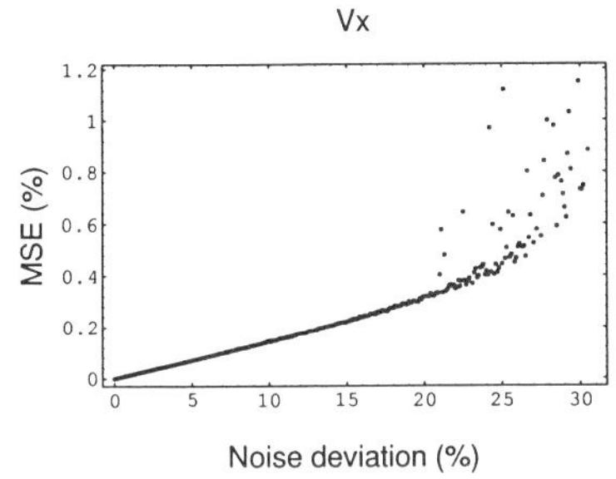

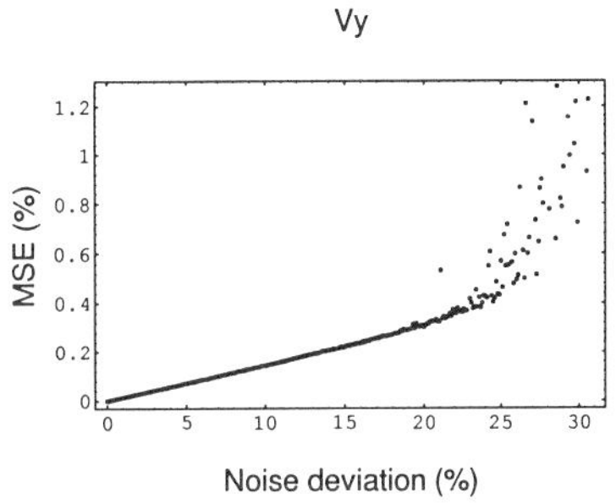

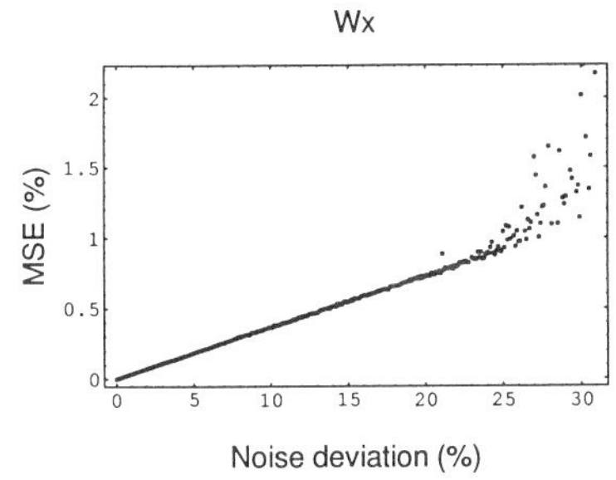

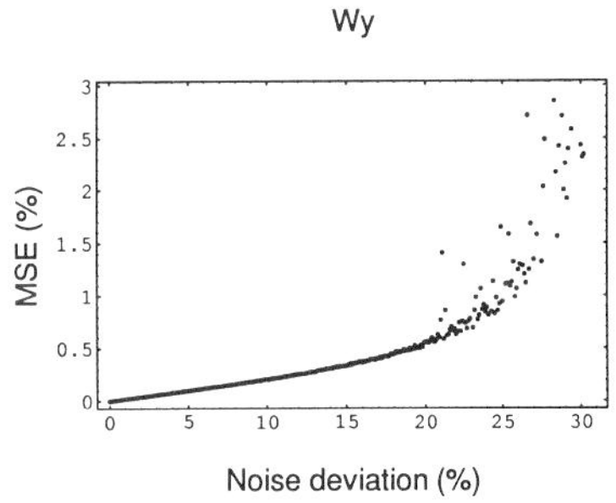

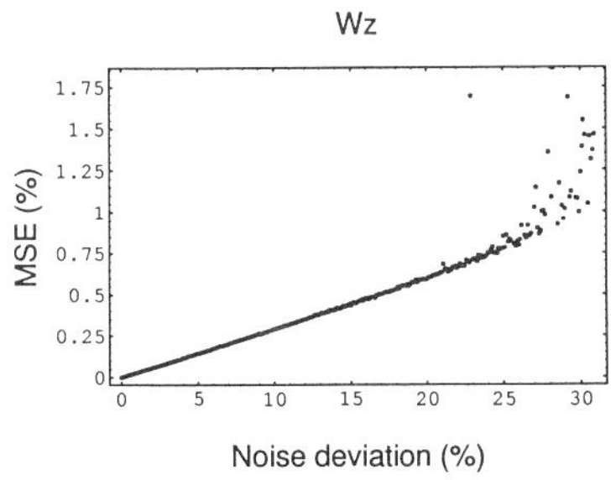

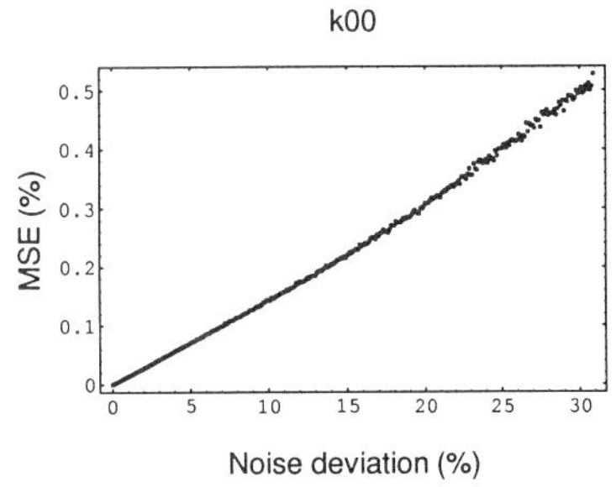

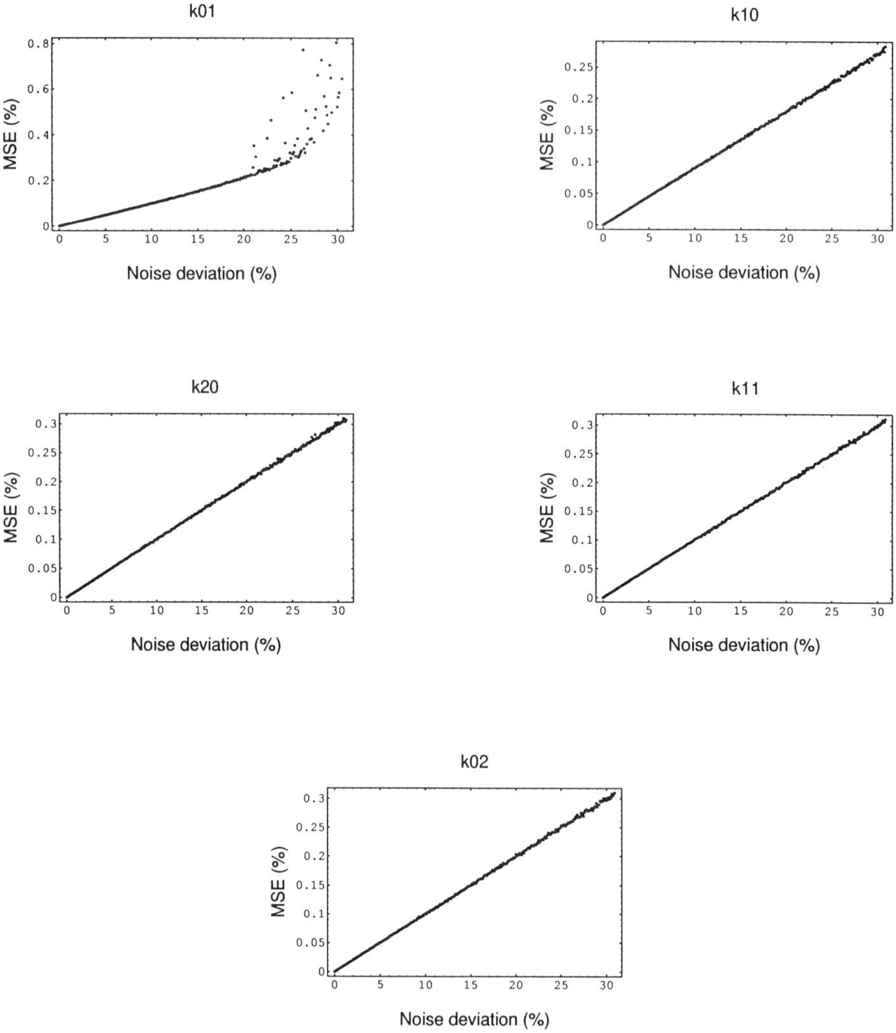
k01
MSE (%)
Noise deviation (%)
k10
MSE (%)
Noise deviation (%)
k20
MSE (%)
Noise deviation (%)
k11
MSE (%)
Noise deviation (%)
k02
MSE (%)
Noise deviation (%)

Symmetry and Wavelet Transforms for Image Data Compression

R. Wilson, I. Levy and P.R. Meulemans

Department of Computer Science, University of Warwick, Coventry

Abstract

This paper describes work in the application of wavelet transforms to the data compression of both still images and image sequences. The main principle underlying the work is the representation of the natural symmetries of image data, which form a subgroup of the 2-d affine group, itself an approximation in 2-d of the motions resulting from perspective projection of the rigid motions in 3-d. It will be shown how the use of a suitable wavelet transform can simplify the representation of these motions, to good effect in two data compression applications. The first is a recasting of fractal compression into a more conventional predictive framework, using an orthonormal wavelet basis. The second makes use of an overcomplete wavelet transform in estimating motions for video coding.

1 Symmetry and image representation

The use of wavelet transforms (WT) in image analysis and data compression is one of the more significant developments in the subject over the last ten years or so. The introduction of orthonormal wavelet bases by Daubechies [1] has led to a wide range of techniques using various types of wavelet representation [2, 3, 4, 11]. One obvious feature of wavelets, which distinguishes them from other image representations, is their symmetry properties, the continuous WT in 1-d representing the coherent states of the affine group, for example [1]. Such symmetries as rotation, translation and dilation, which constitute a subgroup of the 2-d affine group, are clearly important in image analysis and ought to be considered, when selecting an image representation. Indeed, the widespread use of Fourier transforms in signal analysis is directly attributable to their role in the representation of the translation group [8]. Symmetries are reflected in statistical measures such as the autocorrelation tensor, revealing themselves in the eigen-analysis of that tensor, ie. the Karhunen-Loève transform (KLT) of the signal. Over the years, much attention has been paid in image data compression to the statistical aspects of the problem (see for example [7]), but relatively little to symmetry. A notable exception to this can be found in the work of Jacquin on iterated-function data compression [5]. His work draws heavily on the ideas of self-similarity under the affine group, which were expounded by Barnsley [6] and although in its simple form it has its weaknesses, it nevertheless represents a direct way of incorporating symmetries into an image description. More recently, some work on texture description has taken the idea of self-similarity in a different direction, building on a 2-d affine motion model to generalise the simple idea of a texture as a periodic placement of a texture element [10]. In the compression of video sequences, of course, it has long been recognised that motion estimation is a central component [9], but limited computational power has meant that much of the work has focused on simple methods such as block matching to provide the velocity estimates. Inevitably this leads to poor estimates, since the block structure, though it underlies the current data compression algorithms such as JPEG and MPEG [9], is hardly symmetry-adapted. While use of multiresolution processing can mitigate these effects, it is no substitute for a properly designed image representation.

The work presented in this paper is concerned with exploiting the symmetry properties of certain types of wavelet transform, to see how these might be used in applications such as image data compression. After a brief discussion of the underlying principles, two examples will be used to illustrate the importance of symmetry in image representation. The first is based on an orthonormal wavelet transform, which will be shown to lead to a simpler and more elegant approach to fractal image coding, with results approaching those of the best systems currently reported for still image compression. The second uses an *overcomplete* WT, the multiresolution Fourier transform (MFT), in video coding using a combination of featureless and feature-based motion estimation. The key property of the MFT in this application is that it reflects the structure of the 2-d affine group, simplifying estimation of the motion parameters. Results of motion estimation using this technique will be used to illustrate its potential. The paper is concluded with a discussion of the relationship between symmetry and statistics in image representation.

2 Representing symmetry in images

The role of symmetry in image representation is widely misunderstood to be limited to axial or rotational symmetries, such as occur in simple geometric shapes. In the present context, a broader definition is needed - one which reflects the role of 2-d or 3-d motions. To this end, suppose there are two images, or rather two image patches, labelled $f_1(\vec{x}), f_2(\vec{x})$ respectively, where the image co-ordinates are $\vec{x} \in R^2$. They might be parts of the same image or from two images in a temporal sequence, for example. To say that they are related by a symmetry is to assert that there exists a co-ordinate transformation $\boldsymbol{T}$ such that

$$f_2(\vec{x}) = f_1(\boldsymbol{T}^{-1}\vec{x}) \tag{2.1}$$

where the exponent "-1" is used to indicate the inverse, which will be assumed to exist. (2.1) asserts that the two patches are related by a *motion*: the second can be found by simply moving the first appropriately. For example, if $f_1(\vec{x})$

represents a horizontal edge and $f_2(\vec{x})$ a vertical one, then the transformation is just a rotation by $\pi/2$. While ideal image features like edges may exhibit such perfect symmetry, a more realistic model includes an element of approximation, replacing (2.1) by

$$f_2(\vec{x}) = f_1(\boldsymbol{T}^{-1}\vec{x}) + \nu(\vec{x}) \tag{2.2}$$

where the *residual*, $\nu(\vec{x})$, expresses the approximation error. This formulation can also be viewed as a rather general model for image data compression: the patch f_2 is *predicted* from f_1 and the prediction error is the residual. Typically, most of the work in compression algorithms, both intraframe and interframe, is done by the residuals - the predictor structure is fixed or has relatively few degrees of freedom. Notable exceptions to this are the iterated function system (IFS) for still image compression [5] or motion compensation schemes for video compression, both of which typically spend more information on the transformation than on the residuals. This raises the prospect of choosing the transformation, among some admissible set of motions, to minimise the average residual error energy:

$$\boldsymbol{T}_{opt} = \arg\min_{\boldsymbol{T}} \sum_{\vec{x}} (f_2(\vec{x}) - f_1(\boldsymbol{T}^{-1}\vec{x}))^2 \tag{2.3}$$

where the sum is over all pixels within the patch. The main problem with this formulation is the implied search of the parameter space: the full affine group is a 6-parameter continuous group; even the discrete subgroups used below constitute a huge space. For example, in a $N \times N$ pixel patch, there are of the order of N^2 translations. Identifying the minimum error fit using cross-correlation requires N^2 multiplications per position, giving an overall burden of N^4. But if the computation is done in the Fourier domain, (2.3) can be replaced by

$$\vec{\tau}_{opt} = \arg\max_{\vec{\tau}} \int d\vec{\omega} \hat{f}_2(\vec{\omega}) \bar{\hat{f}}_1(\vec{\omega}) e^{\jmath\vec{\omega}.\vec{\tau}} \tag{2.4}$$

where $\hat{f}$ is the FT of f and $\bar{\cdot}$ denotes complex conjugate. This trick reduces the computation to one of order N^2. This results from the diagonalisation of translations by the FT. It is this sort of computational issue which makes the choice of image representation significant.

When translations are the only motion, their representation by cyclic shifts is a convenient simplification, which leads to use of discrete FT methods based on (2.4), but the general affine case is less obliging - the group is not abelian and no representation will provide a full diagonalisation. Moreover, it is seldom the case that motions of the whole image are involved - most relevant motions apply *locally*, rendering simple Fourier methods ineffective. The best which can be done in general is to require that the function set $\mathcal{W} = \{w_{a,b}(\vec{x})\}$ is *invariant* to the group operations, ie.

$$\boldsymbol{T}\mathcal{W} = \{w_{a,b}(\boldsymbol{T}^{-1}\vec{x})\} = \mathcal{W}, \quad \boldsymbol{T} \in \mathcal{T}. \tag{2.5}$$

In other words, the vectors $w_{a,b}(\vec{x})$ are *coherent states* of the group [1]. The significant feature of this invariance is that any motion can be represented by a motion of the parameters

$$w_{a,b}(\boldsymbol{T}^{-1}\vec{x}) = w_{T(a),T(b)}(\vec{x}), \quad \boldsymbol{T} \in \mathcal{T} \tag{2.6}$$

where $T(a)$ is a mapping of the parameter a. Among other things, this implies that the motions and the parameter space have the same cardinality, whether finite or infinite. If such a representation exists, then the problem of identifying motions avoids the computation of $f(\boldsymbol{T}^{-1}\vec{x})$ for each transformation. As an example, the continuous 1-D wavelet transform of a signal is [1]

$$Wf(x,\sigma) = \frac{1}{\sqrt{\sigma}} \int_y dy\; f(y) w(\frac{y-x}{\sigma}) \tag{2.7}$$

where $w(.)$ is the so called "mother wavelet". This two-parameter transform is the natural representation for the 1-d affine group, in the sense of (2.5), since if $(\alpha, \beta) : x \mapsto \beta x + \alpha$ is the general group element, then its action on the WT is just

$$Wf(x,\sigma) \overset{(\alpha,\beta)}{\mapsto} \beta Wf(\frac{x-\alpha}{\beta}, \frac{\sigma}{\beta}) \tag{2.8}$$

which is no more than a "relabelling" of the transform. The commonest form of 2-d WT is the Cartesian product of 1-d transforms in the horizontal and vertical directions

$$Wf(\vec{x},\sigma) = \frac{1}{\sigma} \int d\vec{y}\; f(\vec{y}) w(\frac{\vec{y}-\vec{x}}{\sigma}) \tag{2.9}$$

where the wavelet $w(.)$ is separable. This has rotations by multiples of $\pi/2$ and axial symmetry about the vertical axis, in addition to translation and uniform dilations, as its natural group. Continuous WT's adapted to the full rotation group, translations and dilations have also been described [12].

In the general affine case, a suitable choice of representation is the over-complete wavelet transform called the multiresolution Fourier transform (MFT) [11],which gives estimates of the Fourier spectrum of image patches at a range of scales. The MFT for continuous 2-d signals is defined as

$$Mf(\vec{x},\vec{\omega},\boldsymbol{L}) = |\boldsymbol{L}|^{\frac{1}{2}} \int d\vec{\xi}\; f(\vec{\xi}) w(\boldsymbol{L}^{-1}(\vec{\xi}-\vec{x})) e^{-j\vec{\omega}.\vec{\xi}} \tag{2.10}$$

where $w(.)$ is a window function, $\boldsymbol{L}$ is an invertible linear transformation, $\vec{x}$ is the spatial and $\vec{\omega}$ the frequency co-ordinate. This represents a generalisation of the form given in [11], to accommodate the affine symmetry model. As a signal representation, the MFT is not square-integrable, but it has a form of inverse given by

$$f(\vec{x}) = \frac{1}{4\pi^2} \int d\vec{\omega} d\vec{\xi} d\boldsymbol{L}\; |\boldsymbol{L}|^{\frac{1}{2}} \rho(\boldsymbol{L})\; Mf(\vec{\xi},\vec{\omega},\boldsymbol{L}) w(\boldsymbol{L}^{-1}(\vec{\xi}-\vec{x})) e^{j\vec{\omega}.\vec{x}} \tag{2.11}$$

where $\rho(\boldsymbol{L})$ is a *density* over the linear group

$$\int d\boldsymbol{L}\ \rho(\boldsymbol{L}) = 1 \tag{2.12}$$

and provided $w(.)$ satisfies the weak requirement

$$\int d\vec{x}\ w(\vec{x})w(\vec{x}) = 1. \tag{2.13}$$

This is readily proved by an extension of the arguments in [11]. In the present context, the MFT has two important properties: symmetry and locality.

First, consider the action of an affine group element $(\boldsymbol{A}, \vec{a})$ on the MFT. From (2.10), if $(\boldsymbol{A}, \vec{a}) : \vec{x} \mapsto \boldsymbol{A}\vec{x} + \vec{a}$, then its action on the MFT is

$$Mf(\vec{x}, \vec{\omega}, \boldsymbol{L}) \overset{(\boldsymbol{A},\vec{a})}{\mapsto} |\boldsymbol{A}|\ e^{-\jmath\vec{\omega}.\vec{a}}\ Mf(\boldsymbol{A}^{-1}(\vec{x} - \vec{a}), \boldsymbol{A}^T\vec{\omega}, \boldsymbol{A}^{-1}\boldsymbol{L}), \tag{2.14}$$

which is an obvious generalisation of (2.8). More significantly, for any finite energy image $f(\vec{x})$ and group element $(\boldsymbol{A}, \vec{a})$, there is *some* window scale, σ, for which

$$Mf(\vec{x}, \vec{\omega}, \boldsymbol{L}) \overset{(\boldsymbol{A},\vec{a})}{\mapsto} |\boldsymbol{A}|\ e^{-\jmath\vec{\omega}.\vec{a}}\ Mf(\vec{x}, \boldsymbol{A}^T\vec{\omega}, \boldsymbol{A}^{-1}\boldsymbol{L}) + O(\frac{1}{|\boldsymbol{L}|}), \quad |\boldsymbol{L}| > \sigma. \tag{2.15}$$

In other words, the translation can be determined from the *phase* and the linear transformation from the magnitude of the MFT's of the original and transformed images. This observation underpins the use of the MFT in video coding.

Locality is a key issue in signal modelling, which goes beyond the simplistic argument that because WT's use basis functions of arbitrary size, they avoid the limitations imposed by the uncertainty principle. It is exemplified in (2.15) by the $O(\frac{1}{\sigma})$ term on the right hand side: *any* model of images is valid only over some range of scales. In the above case, the window scale has to be large enough to allow the factoring of the group action into its two components - translation and linear transform. But there is a catch: the affine model itself is an approximation, only valid over a range of scales up to some largest scale; beyond that, multiple motions will be encountered, due for example to the effects of perspective, occlusions and so on. Since the range of usable scales varies across images and even within many images, it requires a representation allowing such variation of scale *without* sacrificing invertibility.

In this respect, the MFT has a significant advantage over the simpler forms of the WT: the window scale can be selected to get the best model fit; it is not tied to the basis functions. This is why the inversion formula contains a density over $\boldsymbol{L}$ - the lack of a conventional inverse, far from being unfortunate side-effect of the definition, is essential in allowing the selection of the optimum window for a given data set. To be more specific, consider again the model of (2.1), expressed in terms of the corresponding MFT's and using (2.15)

$$Mf_2(\vec{x},\vec{\omega},\sigma \boldsymbol{I}) = |\boldsymbol{A}|\, e^{-\jmath\vec{\omega}.\vec{a}}\, Mf_1(\vec{x},\boldsymbol{A}^T\vec{\omega},\sigma\boldsymbol{A}^{-1}) + \nu(\vec{x},\vec{\omega},\sigma\boldsymbol{I}) \tag{2.16}$$

where the patch being predicted $f_2(\vec{x})$ is circular, reducing the general case to the scale σ only. Now choose the scale to maximise the likelihood, ie.

$$\sigma_{opt} = \arg\max_s \left(\max_T prob(f_2(\vec{x}), \|\vec{x}\| < r | f_1(\boldsymbol{T}^{-1}\vec{y}), \|\vec{y}\| < s \leq r)\right), \tag{2.17}$$

where $prob(.|.)$ is the conditional density. A simple model illustrating the idea is that the two signals are related via (2.16), for scales $s \leq \sigma$. Following the standard approach to the problem [14], with appropriate normalisation, for jointly normal, white $f_1(.), f_2(.)$, the log-likelihood is maximised by choosing the scale and affine parameters according to

$$\begin{aligned}\sigma_{opt} \quad = \quad & \arg\max_s \left(\max_{(\boldsymbol{A},\vec{a})} 2|\boldsymbol{A}| \int d\vec{\omega} Mf_2(\vec{x},\vec{\omega},s\boldsymbol{I})\bar{M}f_1(\vec{x},\boldsymbol{A}^T\vec{\omega},s\boldsymbol{A}^{-1})\, e^{\jmath\vec{\omega}.\vec{a}} - \right.\\ & \left. |\boldsymbol{A}|^2 \int d\vec{\omega}|Mf_1(\vec{x},\boldsymbol{A}^T\vec{\omega},s\boldsymbol{A}^{-1})|^2\right),\end{aligned} \tag{2.18}$$

which takes advantage of (2.16) to factor out the translations, which are estimated using cross-correlation, once the linear transformation has been identified. By using the MFT, the scale can be handled as an unknown parameter - the best scale being selected in a data-dependent fashion, without losing invertibility. The combination of symmetry and locality marks the MFT as uniquely fitted to motion estimation.

In practice, signals are sampled, with dilations being restricted to the subgroup $s_i = 2^i$ and the affine group actions are approximated by a linear action on the signal space, so that (2.2) is replaced by the vector form

$$\boldsymbol{f}_2 = \boldsymbol{T}\boldsymbol{f}_1 + \boldsymbol{\nu} \tag{2.19}$$

where $\boldsymbol{T} : R^{N^2} \mapsto R^{N^2}$ is a linear operator. The computational complexity of this operator depends on the structure of the matrix $\boldsymbol{T}$, which again is greatly simplified by the choice of image representation. It is perhaps worth noting that (2.19) is formally identical to the standard predictive models used so widely in image processing [7]. The significant difference from the use here is that in the conventional predictive models the transformations are effectively symmetrised by averaging over the group, leading to a predictor which is lowpass and, by definition, carries no information - it is spatially invariant.

3 Applications

As a first example, the use of affine maps forms the basis of fractal coding, in which 4×4 blocks of pixels, called *range* blocks are represented as "motions" of 8×8 *domain* blocks. The allowable transformations are products of inversion about the $y-$axis, $\boldsymbol{F}$, rotations $\boldsymbol{R}$ by multiples of $\pi/2$ and scaling by a factor of 2 to reduce the block size, which is accomplished by a 2×2 block averaging operator $\boldsymbol{A} : R^{64} \mapsto R^{16}$. To these co-ordinate transformations are added magnitude scaling by $s < 1$ and translation by a constant c

$$\boldsymbol{f}_r = s_{dr}\boldsymbol{R}^i\boldsymbol{F}^j\boldsymbol{A}\boldsymbol{f}_d + c_{dr}\mathbf{1}, \quad 0 \leq i < 4,\ 0 \leq j \leq 1 \tag{3.1}$$

where $\boldsymbol{f}_r$ is a 4×4 range block and $\boldsymbol{f}_d$ an 8×8 domain block. With s, c quantized appropriately, most of the computation and the bit rate are consumed by locating the best domain block for a given range block. The constraint on s guarantees that the mapping thus defined on the image is a contraction, whose fixed point is an approximation to the original image [5]. This is essential because the image is reconstructed from an arbitrary initial image by iteration of the image-image mapping defined by the domain-range block transformations:

$$\boldsymbol{f}_n = \boldsymbol{L}\boldsymbol{f}_{n-1} + \boldsymbol{e} \tag{3.2}$$

where $\boldsymbol{L}$ is the linear transformation and $\boldsymbol{e}$ the "error" comprising the piecewise constant image represented by the last term in (3.1). Note that both $\boldsymbol{L}$ and $\boldsymbol{e}$ are independent of n in this case. In the limit, because of the fixed-point theorem,

$$\boldsymbol{f}_\infty = \lim_{n \to \infty} \boldsymbol{T}^n \boldsymbol{f}_0. \tag{3.3}$$

In practice, convergence is generally obtained in a few tens of iterations, but it is by no means simple to achieve a specified rate-distortion trade-off with the method.

If an orthonormal wavelet transform is used, however, (3.1) is replaced by

$$\boldsymbol{f}_{n,r} = s_{dr}\boldsymbol{R}^i\boldsymbol{f}^j\boldsymbol{f}_{n-1,d} + \boldsymbol{e}_{n,r} \tag{3.4}$$

where $\boldsymbol{R}, \boldsymbol{F}$ are as in (3.1) and the scale factor s_{dr} is no longer constrained to the range $|s| < 1$. Because the WT is adapted to dilations (cf. (2.8)), both range and domain blocks are 4×4 pixels in size; if an orthonormal WT is used, iteration is redundant: errors at level n are orthogonal to the data on level $n-1$ from which the prediction was derived. Consequently, the residuals are coded using the optimal orthonormal basis - the KLT. This reduces the iterated function coder to a form of predictive vector quantisation (VQ) across scales and avoiding the most visible artefacts associated with the original IFS coder, which were due to the block structure it used. In the experiments performed with this coder, 512×512 pixel images have been coded using a Daubechies 16-pt wavelet pair. Levels above the second scale were coded using scalar quantization; 4×4

(a)

(b)

Figure 1. (a) Wavelet fractal coding result using Daubechies 16-pt "symmetric" orthonormal wavelets. Bit rate is 0.16bpp, output PSNR=31.49dB, and (b) bit rate 0.25bpp, PSNR=32.99dB

(a)

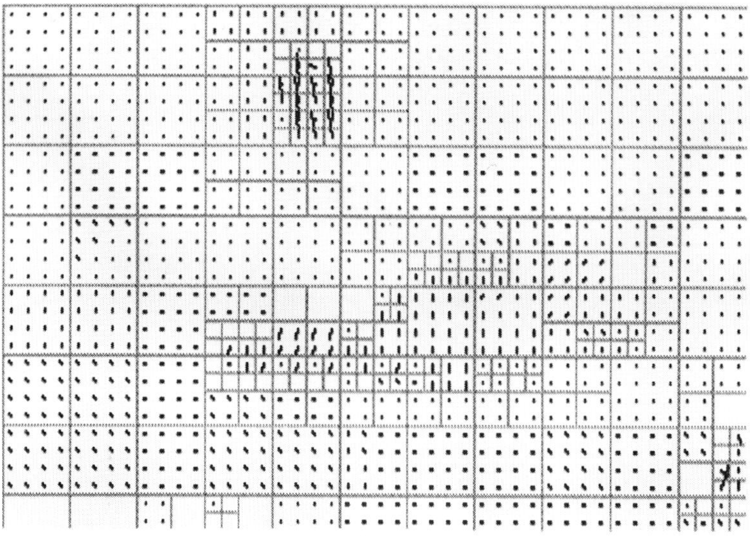

(b)

(c)

Figure 2. (a) Image from "table tennis" sequence, with linear features overlaid, (b) affine motion between frames, showing scale at which model was accepted, and (c) features from first frame, moved and overlaid on second frame quantized blocks on level 3 were used to form the domain pool, from which 256 vectors were selected for tree-structured VQ applied to levels 2 and 1. All three wavelet bands at each scale were coded together, so that the vectors in the VQ were 48-d. Figure 1 shows the coder's performance on the widely used "Lena" image, coded at 0.16 bits per pixel (bpp) and 0.25bpp. The results compare favourably with those reported in [3] and other WT-based coders and are a significant improvement on the standard JPEG algorithm, both in terms of signal-noise ratio and visual quality.

A second application for symmetries in coding is in the compression of image sequences: much of what appears in frame i in a sequence results from movement of what appeared in frame $i-1$. The relevant symmetry group in this case is the full 2-d affine group, which is a linearisation of the projective group

$$f_i(\vec{x}) = f_{i-1}(\boldsymbol{A}^{-1}(\vec{x}-\vec{a})) + \nu_i(\vec{x}). \tag{3.5}$$

The algorithm uses both featureless and feature-based estimation: the featureless component is based on a discrete form of (2.18), in which a coarse-fine search is terminated when the likelihood for the affine model exceeds a threshold; this is corrected by a feature-based local update, designed to reduce errors at occlusion boundaries. The featureless estimates are sufficiently accurate to avoid the usual problem with feature-based approaches - the *correspondence* problem, in other words, identifying which feature in frame $i-1$ corresponds to which in frame i.

Identification of the motion parameters uses the discrete MFT. The use of a form of FT allows the factoring of the estimation into two parts cf. (2.16): the translation is estimated by cross-correlation

$$\hat{\vec{a}} = \arg\max_{\vec{y}} \sum_{\vec{x}} f_i(\vec{x}) f_{i-1}(\boldsymbol{A}^{-1}(\vec{x}-\vec{y})), \tag{3.6}$$

after the linear transform has been found by identifying from the Fourier transformed patches , $Mf_i(\vec{x},\vec{\omega},\sigma)$, a pair of axes based on 2-Means clustering of the spectral magnitudes

$$\{\vec{\chi}_{i,j},\ j=1,2\} = \arg\min_{\vec{\omega}_1,\vec{\omega}_2} \sum_{\vec{\omega}} |Mf_i(\vec{x},\vec{\omega},\sigma)|^2 \min_k \|\vec{\omega}-\vec{\omega}_k\|^2. \tag{3.7}$$

The matrix $\boldsymbol{A}$ is found from the co-ordinate transform between the two

$$\vec{\chi}_{2,j} = \boldsymbol{A}^T \vec{\chi}_{1,j}, \quad j = 1,2. \tag{3.8}$$

These area affine estimates are illustrated in Figure 2(b), for a pair of frames from the "Table Tennis" sequence. Note how the window scale is smaller near occlusion boundaries around the bat and ball, for example. Updating of the featureless estimates uses the linear features estimated and tracked from frame to frame using a modification of the procedures described in [11]. First, any feature in frame $i-1$ (Figure 2(a)) is moved using the appropriate affine estimate, then a local search is performed to correct the motion estimate. Because the features are approximately linear, only the rotation and translation perpendicular to the feature are updated, but this prevents the accumulation of errors at occlusion boundaries, as the updated features in Figure 2(c) show. The last stage in the process is to amend the running feature map by deleting any features which have vanished and adding any new ones. The whole process is implemented using the MFT, to provide an effective and computationally efficient way of both separating the translation from the linear transformation and selecting model scale automatically.

4 Conclusions

Although much work remains to be done, the results shown here illustrate the usefulness of wavelet transforms in dealing with the symmetries associated with visual motions, both real and virtual. The orthonormal wavelet transform allows the iterated function compression scheme to be seen in a new light, which emphasises its connection with conventional image coding methods, while improving its performance significantly. The use of the MFT in video compression has illustrated the importance in general of having a representation which deals adequately with the problem of locality, as well as capturing the full 2-d affine group of symmetries. A novel "featureless/feature-based" estimator using the MFT was briefly described. In both applications, results show the advantages of taking symmetries properly into account when selecting representations for image data compression.

This raises the perplexing question of why such symmetries, "if they really exist", do not show up in image statistics, for in practice they never do. In the first place, to expect symmetries to reveal themselves in statistics such as the autocorrelation is as unreasonable as to expect a fair coin when tossed 100 times to yield precisely 50 heads and 50 tails. What can be said is that unless the number of heads deviates by an unlikely percentage from 50, the coin is fair: if you claim the world lacks a symmetry, then have adequate statistical grounds for that assertion. The visual world has an obvious geometrical structure [15], which ought to be reflected in the representations we use to explore it, even in such low level tasks as data compression, unless the statistics demand otherwise.

Acknowledgment

This work was supported by the UK EPSRC. Thanks to Andrew Calway and Stefan Kruger of the University of Bristol for supplying the motion estimates used in the sequence work.

Bibliography

1. Daubechies, I. (1992). *Ten Lectures on Wavelets.* SIAM Press, Penn.

2. Antonini, M., Barlaud, M., Mathieu, P. and Daubechies, I. (1992). Image coding using wavelet transform. *IEEE Trans. Image Process.*, **1**, 205-220.

3. Shapiro, J. (1993). Embedded image coding using zerotrees of wavelet coefficients. *IEEE Trans. on Signal Proc.*, **41**, 100-110.

4. Coifman, R.R. and Wickerhauser, M.V. (1992). Entropy-based algorithms for best basis selection. *IEEE Trans. Inform. Theory.* **38**, 713-718.

5. Jacquin, A.E. (1992). Image coding based on a fractal theory of iterated contractive image transformations. *IEEE Trans. Image Proc.* **1**, 18-30.

6. Barnsley, M.F. (1988). *Fractals Everywhere.* Academic Press, New York.

7. Jain, A.K. (1989). *Fundamentals of Digital Image Processing.* Academic Press, New York.

8. Lenz, R. (1990). *Group Theoretical Methods in Image Processing.* Springer-Verlag, Berlin.

9. Tekalp, A.M. (1995). *Digital Video Processing.* Prentice Hall, New Jersey.

10. Hsu, T-I. and Wilson, R. (1994). Texture analysis using a generalised wavelet transform. *Proc. IEEE ICIP-94.* **3**, Austin, 436-440.

11. Wilson, R., Calway, A.D. and Pearson, E.R.S. (1992). A generalized wavelet transform for Fourier analysis: the multiresolution Fourier transform and its application to image and audio signal analysis. *IEEE Trans. Inform. Theory.* **38**, 674-690.

12. Antoine, J-P. (1996). Symmetry-adapted wavelet analysis. *Proc. IEEE ICIP-96.* **3**, Lausanne, 177-180.

13. Krüger, S.A. and Calway, A.D. (1996). A multiresolution frequency domain method for estimating affine motion parameters. *Proc. IEEE ICIP-96.* **1**, Lausanne, 113-116.

14. Davenport, W.B. and Root, W.L. (1987). *An Introduction to the Theory of Random Signals and Noise.* IEEE Press, New York, 312-350.

15. Gibson, J.J. (1974). *The perception of the visual world.* Greenwood Press, Conn.

Vanishing Moments and Biorthogonal Wavelet Systems

Jun Tian and Raymond O. Wells, Jr.

Computational Mathematics Laboratory, Rice University Houston, Texas, USA

Abstract

We consider biorthogonal wavelet systems with the vanishing of moments equally distributed between scaling functions and wavelet functions. Such wavelet systems provide good wavelet sampling approximation with exponential decay. Using a time domain design method, the closed form (minimum length) solution is obtained. An attractive feature behind it is that all filter coefficients are dyadic rational, which means that we can implement a very fast multiplication-free discrete wavelet transform, which consists of only addition and shift operations, on digital computers. Moreover these wavelet systems converge to the sinc wavelet system, and half of them are symmetric.

1 Introduction

In a wavelet system, vanishing moments are a necessary condition for the smoothness of the wavelet function [1]. In 1988, I. Daubechies, in her celebrated paper [2], introduced a class of compactly supported orthonormal wavelet systems in general, as well as a family with growing smoothness for increasing support, *Daubechies wavelet systems.* Daubechies wavelet systems put all freedom of parameters into vanishing moments of the wavelet function. One year later, R. Coifman suggested that it might be worthwhile to construct orthogonal wavelet systems with vanishing moments not only for wavelet functions, but also for scaling functions. Such orthogonal wavelet systems with vanishing moments equally distributed between scaling functions and wavelet functions (which are called *Coiflets* by Daubechies in [3]) are closer to being symmetric than Daubechies wavelet systems, which suggests that with vanishing moments on scaling functions, we might be able to design more symmetric wavelet systems. This is important since in signal processing, symmetry corresponds to linear phase. A method to construct Coiflets was proposed in [3] but the general existence problem is still open.

In [4], we presented a sampling approximation theorem. It says that for an optimal wavelet sampling approximation, vanishing moments should be equally distributed between the scaling function and the wavelet function, and the approximation error converges to zero with exponential decay.

Thus vanishing moments of the wavelet function is necessary for the smoothness of the wavelet system; vanishing moments of the scaling function will improve the symmetry; and equal vanishing moments distribution will provide an optimal wavelet sampling approximation with exponential decay. Based on these observations, we introduce *biorthogonal Coifman wavelet systems*, a family of compactly supported biorthogonal wavelet systems with the vanishing of moments equally distributed between scaling functions and wavelet functions. Biorthogonal Coifman wavelet systems provide multiplication-free discrete wavelet transform, and converge to the noncompactly supported sinc wavelet system. We want to thank Howard Resnikoff and Xiaodong Zhou for very helpful discussion concerning the convergence property.

2 A sampling approximation theorem

The sampling approximation theorem is a quite general result. It is not only valid for a wavelet system (orthogonal, semiorthogonal, or biorthogonal), but also holds for an L^2 function which satisfies a two-scale difference equation.

Theorem 1. (Sampling approximation theorem) *If an $L^2(\mathbf{R})$ function $\phi(x)$ is a solution of the two-scale difference equation*

$$\phi(x) = \sum_{k \in \mathbf{Z}} a_k \phi(2x - k),$$

and it is normalized

$$\int_{-\infty}^{\infty} \phi(x)\, dx = 1,$$

where $\{a_k\}$ is a real sequence with only finite nonzero elements, satisfying

$$\sum_{k \in \mathbf{Z}} (2k)^n a_{2k} = \sum_{k \in \mathbf{Z}} (2k+1)^n a_{2k+1} = 0,$$

for $n = 1, \cdots, N$, *and*

$$\sum_{k \in \mathbf{Z}} a_{2k} = \sum_{k \in \mathbf{Z}} a_{2k+1} = 1,$$

for $f(x) \in C_0^{N,1}(\mathbf{R}), j \in \mathbf{N}$, define

$$S^j(f)(x) := 2^{-j/2} \sum_{k \in \mathbf{Z}} f\left(\frac{k}{2^j}\right) \phi_{j,k}(x),$$

where $\phi_{j,k}(x) = 2^{j/2}\phi(2^j x - k)$. Then

$$\| f(x) - S^j(f)(x) \|_{L^2} \le C 2^{-j(N+1)},$$

where C is a constant, depending only on $f(x)$ and $\{a_k\}$.
If in addition, $\phi(x) \in C^m(\mathbf{R})$, where $m \in \mathbf{Z}, 0 \le m \le N$, then

$$||f(x) - S^j(f)(x)||_{H^m} \le C2^{-j(N+1-m)},$$

where C is a constant, depending only on $f(x)$ and $\{a_k\}$.

In the case of a wavelet system, the linear conditions on the sequence $\{a_k\}$ (which is the scaling vector) in the above theorem are exactly the vanishing moments conditions on the scaling function and the wavelet function [5, 6]. Thus, with vanishing moments equally distributed between the scaling function and the wavelet function, if we use sample values of a smooth function as scaling function coefficients at a fine scale, then the resulting wavelet series approximates the underlying smooth function with increasing accuracy in L^2 (or H^m) type norm as the number N of vanishing moments gets larger or the scale j gets larger (which corresponds to a finer scale). In addition, for an optimal wavelet sampling approximation, vanishing moments should be equally distributed between the scaling function and the wavelet function. The proof can be found in [4].

3 Biorthogonal Coifman wavelet systems

A big advantage of biorthogonal wavelet systems [7, 8] over orthonormal wavelet systems is that biorthogonal systems can be symmetric, while orthonormal systems can't, except for the *Haar wavelet system.*

3.1 Definition

Definition 2. *A biorthogonal wavelet system with compact support is called a* biorthogonal Coifman wavelet system *of degree N if the following two conditions are satisfied,*

- *the vanishing moments of the synthesis scaling function $\tilde{\phi}(x)$ and wavelet function $\tilde{\psi}(x)$ are of degree N, i.e.,*

$$\int_{-\infty}^{\infty} x^n \tilde{\phi}(x)\, dx = 0, \quad \text{for } n = 1, \cdots, N,$$

$$\int_{-\infty}^{\infty} x^n \tilde{\psi}(x)\, dx = 0, \quad \text{for } n = 0, \cdots, N.$$

- *the vanishing moment of the analysis wavelet function $\psi(x)$ is of degree N,*

$$\int_{-\infty}^{\infty} x^n \psi(x)\, dx = 0, \quad \text{for } n = 0, \cdots, N.$$

Note that in the definition of the biorthogonal Coifman wavelet system, although there is no vanishing moment requirement on the analysis scaling function $\phi(x)$, it turns out that $\phi(x)$ also has vanishing moments up to degree N, because of the perfect reconstruction condition

$$\sum_{k\in\mathbf{Z}} a_k \tilde{a}_{k+2l} = 2\delta_{0,l}, \quad \forall l \in \mathbf{Z}.$$

Lemma 3. *For a biorthogonal Coifman wavelet system of degree N, the vanishing moments' degree of the analysis scaling function $\phi(x)$ is also N,*

$$\int_{-\infty}^{\infty} x^n \phi(x)\, dx = 0, \quad \text{for}\, n = 1, \cdots, N.$$

The proof of Lemma 3 can be found in [9]. So based on Theorem 1, biorthogonal Coifman wavelet systems will provide very good wavelet sampling approximation.

3.2 Construction

Using a time domain design method, the closed form solutions of the minimum length biorthogonal Coifman wavelet systems of all degrees are obtained.

Theorem 4. *The minimum length biorthogonal Coifman wavelet systems of degree N have the analysis scaling vector $\{a_k\}$ and the synthesis scaling vector $\{\tilde{a}_k\}$ of the form*

$$\tilde{a}_0 = 1, \quad \tilde{a}_{2k} = 0\, \text{when}\, k \neq 0,$$

- *if N is even, $N = 2n$,*

$$a_{2k+1} = \tilde{a}_{2k+1} = \frac{(-1)^k(2n+1)}{2^{4n-1}(2k+1)} \binom{2n-1}{n-1} \binom{2n}{n+k}$$

- *if N is odd, $N = 2n-1$,*

$$a_{2k+1} = \tilde{a}_{2k+1} = \frac{(-1)^k(2n-1)}{2^{4n-3}(2k+1)} \binom{2n-2}{n-1} \binom{2n-1}{n+k}$$

and

$$a_{2k} = 2\delta_{0,k} - \sum_{l\in\mathbf{Z}} \tilde{a}_{2l+1}\tilde{a}_{2l+1-2k}.$$

The scaling vectors of the minimum length biorthogonal Coifman wavelet systems with degrees $N = 0, 1, 2, 3$ and 4 are listed in Table 1.

Table 1. The coefficients of biorthogonal Coifman wavelet systems

N	a_k	$\tilde{a}_k$
$N=0$	$a_0 = 1$	$\tilde{a}_0 = 1$
	$a_1 = 1$	$\tilde{a}_1 = 1$
$N=1$	$a_{-2} = -1/4$	
	$a_{-1} = 1/2$	$\tilde{a}_{-1} = 1/2$
	$a_0 = 3/2$	$\tilde{a}_0 = 1$
	$a_1 = 1/2$	$\tilde{a}_1 = 1/2$
	$a_2 = -1/4$	
$N=2$	$a_{-4} = 3/64$	
	$a_{-3} = 0$	
	$a_{-2} = -3/16$	
	$a_{-1} = 3/8$	$\tilde{a}_{-1} = 3/8$
	$a_0 = 41/32$	$\tilde{a}_0 = 1$
	$a_1 = 3/4$	$\tilde{a}_1 = 3/4$
	$a_2 = -3/16$	$\tilde{a}_2 = 0$
	$a_3 = -1/8$	$\tilde{a}_3 = -1/8$
	$a_4 = 3/64$	
$N=3$	$a_{-6} = -1/256$	
	$a_{-5} = 0$	
	$a_{-4} = 9/128$	
	$a_{-3} = -1/16$	$\tilde{a}_{-3} = -1/16$
	$a_{-2} = -63/256$	$\tilde{a}_{-2} = 0$
	$a_{-1} = 9/16$	$\tilde{a}_{-1} = 9/16$
	$a_0 = 87/64$	$\tilde{a}_0 = 1$
	$a_1 = 9/16$	$\tilde{a}_1 = 9/16$
	$a_2 = -63/256$	$\tilde{a}_2 = 0$
	$a_3 = -1/16$	$\tilde{a}_3 = -1/16$
	$a_4 = 9/128$	
	$a_5 = 0$	
	$a_6 = -1/256$	
$N=4$	$a_{-8} = 15/16384$	
	$a_{-7} = 0$	
	$a_{-6} = -35/2048$	
	$a_{-5} = 0$	
	$a_{-4} = 345/4096$	
	$a_{-3} = -5/128$	$\tilde{a}_{-3} = -5/128$
	$a_{-2} = -405/2048$	$\tilde{a}_{-2} = 0$
	$a_{-1} = 15/32$	$\tilde{a}_{-1} = 15/32$
	$a_0 = 10317/8192$	$\tilde{a}_0 = 1$
	$a_1 = 45/64$	$\tilde{a}_1 = 45/64$
	$a_2 = -405/2048$	$\tilde{a}_2 = 0$
	$a_3 = -5/32$	$\tilde{a}_3 = -5/32$
	$a_4 = 345/4096$	$\tilde{a}_4 = 0$
	$a_5 = 3/128$	$\tilde{a}_5 = 3/128$
	$a_6 = -35/2048$	
	$a_7 = 0$	
	$a_8 = 15/16384$	

3.3 Multiplication-free discrete wavelet transform

As it can be seen in Table 1, the scaling vectors $\{a_k\}$ and $\{\tilde{a}_k\}$ are all dyadic rational, i.e., all the nonzero elements in the scaling vectors are of the form $(2p+1)/2^q$, for some integers p and q. Actually this assertion is true for scaling vectors of the minimum length biorthgonal Coifman wavelet systems of all degrees. Thus we have obtained a family of biorthogonal wavelet systems on which we can implement a very fast multiplication-free discrete wavelet transform [10], which consists of only addition and shift operations, on digital computers. This is one of the main advantage of biorthogonal Coifman wavelet systems over other widely used irrational wavelet systems, like the Cohen-Daubechies-Feauveau 9-7 biorthogonal filters (CDF-9-7) [7].

Theorem 5. *In a biorthogonal Coifman wavelet system of degree N, the scaling vectors are dyadic rationals, i.e., $\forall k \in \mathbf{Z}$, there must exist four integers p_1, p_2, q_1, q_2, such that*

$$a_k = \frac{2p_1+1}{2^{q_1}}, \quad \tilde{a}_k = \frac{2p_2+1}{2^{q_2}},$$

whenever a_k or $\tilde{a}_k$ are nonzero.

Proof. It is sufficient to prove only $\tilde{a}_{2k+1}$ is dyadic rational since the addition, subtraction, and multiplication are closed operations in dyadic rationals.

1. If N is even, $N = 2n$, for nonzero $\tilde{a}_{2k+1}$, we have $-n \leq k \leq n$. When $k = -1, 0$, or n, the proof is trivial. Let's first look at $0 < k < n$. Since

$$\begin{aligned} \tilde{a}_{2k+1} &= \frac{(-1)^k}{2k+1}\binom{2n-1}{n-1}\binom{2n}{n+k}\frac{2n+1}{2^{4n-1}} \\ &= \frac{(-1)^k}{2^{4n-1}(2k+1)} \cdot \frac{(2n-1)!(2n+1)!}{(n-1)!n!(n+k)!(n-k)!}, \end{aligned}$$

(a) If $2k+1$ is a prime number, from number theory, it suffices to show that

$$\left\lfloor \frac{2n-1}{2k+1} \right\rfloor + \left\lfloor \frac{2n+1}{2k+1} \right\rfloor > \left\lfloor \frac{n-1}{2k+1} \right\rfloor + \left\lfloor \frac{n}{2k+1} \right\rfloor + \left\lfloor \frac{n+k}{2k+1} \right\rfloor + \left\lfloor \frac{n-k}{2k+1} \right\rfloor \tag{3.1}$$

where $\lfloor \cdot \rfloor$ is the integer part of a real number. Set $n-k = t(2k+1)+r$, where $t, r \in \mathbf{Z}, t \geq 0, 0 \leq r \leq 2k$.

(i) If $r = 0$, then

$$\begin{aligned} n+k &= t(2k+1)+2k, \\ 2n+1 &= (2t+1)(2k+1). \end{aligned}$$

So

$$\left\lfloor \frac{2n+1}{2k+1} \right\rfloor = 2t+1 = \left\lfloor \frac{n-k}{2k+1} \right\rfloor + \left\lfloor \frac{n+k}{2k+1} \right\rfloor + 1.$$

We know that

$$\left\lfloor \frac{2n-1}{2k+1} \right\rfloor \geq \left\lfloor \frac{n-1}{2k+1} \right\rfloor + \left\lfloor \frac{n}{2k+1} \right\rfloor,$$

Thus (3.1) follows immediately.

(ii) If $1 \le r \le k$, then

$$\begin{aligned} n &= t(2k+1) + (k+r)\,, \\ n-1 &= t(2k+1) + (k+r-1)\,, \\ 2n-1 &= (2t+1)(2k+1) + (2r-2)\,. \end{aligned}$$

Thus

$$\left\lfloor \frac{2n-1}{2k+1} \right\rfloor = 2t+1 = \left\lfloor \frac{n-1}{2k+1} \right\rfloor + \left\lfloor \frac{n}{2k+1} \right\rfloor + 1\,.$$

(iii) If $k+1 \le r \le 2k$, then

$$\begin{aligned} n+k &= (t+1)(2k+1) + (r-1)\,, \\ 2n+1 &= (2t+2)(2k+1) + (2r-2k-1)\,. \end{aligned}$$

So

$$\left\lfloor \frac{2n+1}{2k+1} \right\rfloor = 2t+2 = \left\lfloor \frac{n-k}{2k+1} \right\rfloor + \left\lfloor \frac{n+k}{2k+1} \right\rfloor + 1\,.$$

(b) If $2k+1$ is not a prime number, assume $2a+1$ is a prime factor of $2k+1$, and $2k+1 = (2a+1)(2b+1)$, i.e., $k = 2ab+a+b$. Then it suffices to show that

$$\left\lfloor \frac{2n-1}{2a+1} \right\rfloor + \left\lfloor \frac{2n+1}{2a+1} \right\rfloor > \left\lfloor \frac{n-1}{2a+1} \right\rfloor + \left\lfloor \frac{n}{2a+1} \right\rfloor + \left\lfloor \frac{n+k}{2a+1} \right\rfloor + \left\lfloor \frac{n-k}{2a+1} \right\rfloor \tag{3.2}$$

Since

$$\begin{aligned} \left\lfloor \frac{n+k}{2a+1} \right\rfloor + \left\lfloor \frac{n-k}{2a+1} \right\rfloor &= \left\lfloor \frac{n+2ab+a+b}{2a+1} \right\rfloor + \left\lfloor \frac{n-2ab-a-b}{2a+1} \right\rfloor \\ &= \left\lfloor \frac{n+a}{2a+1} \right\rfloor + \left\lfloor \frac{n-a}{2a+1} \right\rfloor \end{aligned}$$

the inequality (3.2) is exactly (3.1) when replacing k with a.

Now suppose $-n \le k \le -2$. Set $l = -k, 2 \le l \le n$. We want to show that (as before, if $2l-1$ is not prime, it can be converted to the following inequality)

$$\left\lfloor \frac{2n-1}{2l-1} \right\rfloor + \left\lfloor \frac{2n+1}{2l-1} \right\rfloor > \left\lfloor \frac{n-1}{2l-1} \right\rfloor + \left\lfloor \frac{n}{2l-1} \right\rfloor + \left\lfloor \frac{n+l}{2l-1} \right\rfloor + \left\lfloor \frac{n-l}{2l-1} \right\rfloor \tag{3.3}$$

Set $n-l = t(2l-1)+r$, where $t, r \in \mathbf{Z}, t \ge 0, 0 \le r \le 2l-2$.

(i) If $0 \le r \le l-2$, then

$$n = t(2l-1) + (l+r)\,,$$

$$n-1 = t(2l-1) + (l+r-1)\,,$$

$$2n-1 = (2t+1)(2l-1) + 2r\,.$$

So

$$\left\lfloor \frac{2n-1}{2l-1} \right\rfloor = 2t+1 = \left\lfloor \frac{n-1}{2l-1} \right\rfloor + \left\lfloor \frac{n}{2l-1} \right\rfloor + 1\,.$$

Since

$$\left\lfloor \frac{2n+1}{2l-1} \right\rfloor \ge \left\lfloor \frac{n+l}{2l-1} \right\rfloor + \left\lfloor \frac{n-l}{2l-1} \right\rfloor,$$

the inequality (3.3) follows.

(ii) If $l-1 \le r \le 2l-3$, then

$$n+l = (t+1)(2l-1) + (r+1)\,,$$

$$2n+1 = (2t+2)(2l-1) + (2r-2l+3)\,.$$

So

$$\left\lfloor \frac{2n+1}{2l-1} \right\rfloor = 2t+2 = \left\lfloor \frac{n+l}{2l-1} \right\rfloor + \left\lfloor \frac{n-l}{2l-1} \right\rfloor + 1\,.$$

(iii) If $r = 2l-2$, then

$$n+l = (t+2)(2l-1)\,,$$

$$2n+1 = (2t+3)(2l-1)\,.$$

So

$$\left\lfloor \frac{2n+1}{2l-1} \right\rfloor = 2t+3 = \left\lfloor \frac{n+l}{2l-1} \right\rfloor + \left\lfloor \frac{n-l}{2l-1} \right\rfloor + 1\,.$$

2. If N is odd, $N = 2n-1$,

$$\tilde{a}_{2k+1} = \frac{(-1)^k}{2k+1} \binom{2n-2}{n-1} \binom{2n-1}{n+k} \frac{2n-1}{2^{4n-3}}\,.$$

Since $\tilde{a}_{2k+1} = \tilde{a}_{-2k-1}$, it will be sufficient to only prove for the case $1 \le k \le n-2$. Let $n-k-1 = t(2k+1)+r$, where $t, r \in \mathbf{Z}, t \ge 0, 0 \le r \le 2k$.

(a) If $r = 0$, then

$$2n-1 = (2t+1)(2k+1)\,,$$

i.e.,

$$(2k+1) | (2n-1)\,.$$

(b) If $1 \le r \le 2k$, we will prove the inequality (again, if $2k+1$ is not prime, it can be reduced to the following inequality)

$$\left\lfloor \frac{2n-2}{2k+1} \right\rfloor + \left\lfloor \frac{2n-1}{2k+1} \right\rfloor > \left\lfloor \frac{n-1}{2k+1} \right\rfloor + \left\lfloor \frac{n-1}{2k+1} \right\rfloor + \left\lfloor \frac{n-k-1}{2k+1} \right\rfloor + \left\lfloor \frac{n+k}{2k+1} \right\rfloor$$

(i) If $1 \le r \le k$, then

$$\begin{aligned} n-1 &= t(2k+1) + (k+r)\,, \\ 2n-2 &= (2t+1)(2k+1) + (2r-1)\,. \end{aligned}$$

So

$$\left\lfloor \frac{2n-2}{2k+1} \right\rfloor = 2t+1 = \left\lfloor \frac{n-1}{2k+1} \right\rfloor + \left\lfloor \frac{n-1}{2k+1} \right\rfloor + 1\,.$$

(ii) If $k+1 \le r \le 2k$, then

$$\begin{aligned} n+k &= (t+1)(2k+1) + r\,, \\ 2n-1 &= (2t+2)(2k+1) + (2r-2k-1)\,. \end{aligned}$$

So

$$\left\lfloor \frac{2n-1}{2k+1} \right\rfloor = 2t+2 = \left\lfloor \frac{n-k-1}{2k+1} \right\rfloor + \left\lfloor \frac{n+k}{2k+1} \right\rfloor + 1\,.$$

3.4 Convergence to sinc wavelet system

The *sinc wavelet system* is a basic wavelet system whose scaling vector $\{a_k^{sinc}, k \in \mathbf{Z}\}$ is defined by

$$a_{2k}^{sinc} = \delta_{0,k}\,, \quad a_{2k+1}^{sinc} = \frac{(-1)^k 2}{(2k+1)\pi}\,.$$

It had been a problem for some time to find a sequence of scaling functions with compact supports which approximates the function $\mathrm{sinc}(\pi x) = \frac{\sin \pi x}{\pi x}$, which is the scaling function of the sinc wavelet system. This problem is important because of the special relation of sinc function to signal processing applications. The family of biorthogonal Coifman wavelet systems just provides a very suitable candidate.

Theorem 6. *Suppose $\{\tilde{a}_k^N, k \in \mathbf{Z}\}$ to be the synthesis scaling vector of the biorthogonal Coifman wavelet system of degree N. Then*

$$\lim_{N\to\infty} \left|\left|(\tilde{a}^N) - (a^{sinc})\right|\right|_{l^2} = \lim_{N\to\infty} \left(\sum_{k\in\mathbf{Z}} \left(\tilde{a}_k^N - a_k^{sinc}\right)^2 \right)^{1/2} = 0\,.$$

Before proving Theorem 6, we first prove that $\{\tilde{a}^N\}$ converges to $\{a^{sinc}\}$ termwise.

Lemma 7. *Assume the same condition as in Theorem 6, then*

$$\lim_{N\to\infty} \tilde{a}_k^N = a_k^{sinc}.$$

Proof. First let's look at the case when N is even, $N = 2n$. Assume $|k| < n$ (otherwise we can choose a larger N), we have

$$\begin{aligned}\tilde{a}_{2k+1}^N &= \frac{(-1)^k}{2k+1}\begin{pmatrix} 2n-1 \\ n-1 \end{pmatrix}\begin{pmatrix} 2n \\ n+k \end{pmatrix}\frac{2n+1}{2^{4n-1}} \\ &= \frac{(-1)^k}{2k+1}\cdot\frac{(2n-1)!(2n+1)!}{2^{4n-1}(n-1)!n!(n-k)!(n+k)!}.\end{aligned}$$

We want to show that

$$\lim_{n\to\infty}\frac{(2n-1)!(2n+1)!}{2^{4n-1}(n-1)!n!(n-k)!(n+k)!} = \frac{2}{\pi}. \tag{3.4}$$

Recall that Stirling's formula states

$$\lim_{n\to\infty}\frac{n!}{\sqrt{2\pi}n^{n+1/2}e^{-n}} = 1,$$

so

$$\begin{aligned}&\lim_{n\to\infty}\frac{(2n-1)!(2n+1)!}{2^{4n-1}(n-1)!n!(n-k)!(n+k)!} \\ &= \lim_{n\to\infty}\frac{(2n-1)^{2n-1/2}(2n+1)^{2n+3/2}e^{-1}}{2^{4n}\pi(n-1)^{n-1/2}n^{n+1/2}(n-k)^{n-k+1/2}(n+k)^{n+k+1/2}} \\ &= \lim_{n\to\infty}\frac{e^{-1}}{2^{4n}\pi}\left(\frac{2n-1}{n-1}\right)^{n-1/2}\left(\frac{2n+1}{n}\right)^{n+1/2}\left(\frac{2n-1}{n-k}\right)^{n} \\ &\quad\cdot\left(\frac{2n+1}{n+k}\right)^{n+1}\left(\frac{n-k}{n+k}\right)^{k-1/2} \\ &= \lim_{n\to\infty}\frac{e^{-1}}{2^{4n}\pi}\cdot 2^{n-1/2}e^{1/2}\cdot 2^{n+1/2}e^{1/2}\cdot 2^{n}e^{1/2}\cdot 2^{n+1}e^{-1/2}\cdot 1 \\ &= \frac{2}{\pi}.\end{aligned}$$

When N is odd, $N = 2n-1$, the ratio

$$\frac{\tilde{a}_{2k+1}^N}{\tilde{a}_{2k+1}^{N-1}} = \frac{2n-1}{2n+2k} \longrightarrow 1 \quad \text{as } N\to\infty.$$

Thus the result also holds when N is odd.

Proof of Theorem 6. Based on Lemma 7, to apply Lebesgue's Dominated Convergence Theorem, we just need to find an $l^2(\mathbf{R})$ dominating sequence $\{d_k, k \in \mathbf{Z}\}$ such that

$$|\tilde{a}_k^N - a_k^{\mathrm{sinc}}| \leq d_k \text{ for all } k.$$

1. If N is even, $N = 2n$, assume $|k| \leq n$. Note that

$$(n-k)!(n+k)! \geq (n!)^2$$

then

$$\begin{aligned} |\tilde{a}_{2k+1}^N| &= \left| \frac{(2n-1)!(2n+1)!}{(2k+1)2^{4n-1}(n-1)!n!(n-k)!(n+k)!} \right| \\ &\leq \left| \frac{(2n-1)!(2n+1)!}{(2k+1)2^{4n-1}(n-1)!(n!)^3} \right| \end{aligned}$$

Set $k = 0$ in (3.4),

$$\lim_{n\to\infty} \frac{(2n-1)!(2n+1)!}{2^{4n-1}(n-1)!(n!)^3} = \frac{2}{\pi}$$

So for N large enough,

$$\frac{(2n-1)!(2n+1)!}{2^{4n-1}(n-1)!(n!)^3} \leq 1$$

$$|\tilde{a}_{2k+1}^N| \leq \left| \frac{1}{2k+1} \right|$$

2. If N is odd, $N = 2n-1$, since

$$(n+k)!(n-1-k)! \geq (n-1)!n!,$$

we have

$$\begin{aligned} |\tilde{a}_{2k+1}^N| &= \left| \frac{1}{2k+1} \binom{2n-2}{n-1} \binom{2n-1}{n+k} \frac{2n-1}{2^{4n-3}} \right| \\ &= \left| \frac{((2n-1)!)^2}{2^{4n-3}(2k+1)((n-1)!)^2(n+k)!(n-1-k)!} \right| \\ &\leq \left| \frac{((2n-1)!)^2}{2^{4n-3}(2k+1)((n-1)!)^2(n-1)!n!} \right| \\ &= \left| \frac{1}{2k+1} \tilde{a}_1^N \right| \end{aligned}$$

Because

$$\lim_{N\to\infty} \tilde{a}_1^N = a_1^{\mathrm{sinc}} = \frac{2}{\pi},$$

it follows that

$$\tilde{a}_1^N \leq 1 \text{ for } N \text{ large enough}.$$

Then

$$|\tilde{a}_{2k+1}^N| \leq \left|\frac{1}{2k+1}\right|.$$

Set

$$d_{2k+1} = \left|\frac{1}{2k+1}\right| + |a_{2k+1}^{\text{sinc}}| \leq \left|\frac{2}{2k+1}\right|, \quad d_{2k} = 0$$

we have $(d_k)_{k\in\mathbf{Z}} \in l^2(\mathbf{R})$ and

$$|\tilde{a}_k^N - a_k^{\text{sinc}}| \leq d_k,$$

By Lebesgue's Dominated Convergence Theorem and Lemma 7,

$$\lim_{N\to\infty} ||\tilde{\alpha}^N - \alpha^{\text{sinc}}||_{l^2} = 0.$$

The theorem follows.

4 Conclusions

In this paper we present a new family of wavelet systems, the biorthogonal Coifman wavelet systems, which have very nice properties both in the theoretical sense and the application sense. The vanishing moments conditions have played an important role in such systems. These conditions not only give growing smoothness of wavelet systems, improve symmetry, but also provide fast wavelet sampling approximation. Another attractive feature of biorthogonal Coifman wavelet systems is that all the coefficients are dyadic rational. Thus we can have a very fast multiplication-free discrete wavelet transform implemented on digital computers.

Bibliography

1. Daubechies, I. (1992). *Ten lectures on wavelets.* CBMS-NSF regional conference series in applied mathematics. SIAM, Philadelphia, Pennsylvania.

2. Daubechies, I. (1988). Orthonormal bases of compactly supported wavelets. *Comm. Pure Appl. Math.*, **XLI**:906–966.

3. Daubechies, I. (1993). Orthonormal bases of compactly supported wavelets II. Variations on a theme. *SIAM J. Math. Anal.*, **24**(2):499–519.

4. Tian, J. and Wells, Jr., R. O. (1996). A remark on vanishing moments. In *Proc. 30th Asilomar Conf. on Signals, Systems, and Computers*, Pacific Grove, CA.

5. Steffen, P. and Heller, P. N. and Gopinath, R. A. and Burrus, C. S. (Dec. 1993). Theory of regular m-band wavelet bases. *IEEE Trans. on Signal Processing*, **41**(12).

6. Heller, P. N. (1995). Rank m wavelet matrices with n vanishing moments. *SIAM J. Matr. Anal.*, **16**:502–518.

7. Cohen, A. and Daubechies, I. and Feauveau, J.-C. (1992). Biorthogonal bases of compactly supported wavelets. *Comm. Pure Appl. Math.*, **XLV**:485–560.

8. Vetterli, M. and Herley, C. (1992). Wavelets and filter banks: Theory and design. *IEEE Trans. Acoust. Speech Signal Process.*, **40**:2207–2232.

9. Tian, J. and Wells, Jr., R. O. (1996). Dyadic rational biorthogonal Coifman wavelet systems. Technical Report CML TR 96-07, Rice University. (ftp://cml.rice.edu/pub/reports/CML9607.ps.Z).

10. Mallat, S. G. (Sept. 1989). Multiresolution approximation and wavelet orthonormal bases of $L^2(\mathbf{R})$. *Trans. AMS*, **315**(1):69–87.

On Sliding Window Schemes for Discrete Least-Squares Approximation by Trigonometric Polynomials

Heike Fassbender

FB3-Mathematik und Informatik, Universität Bremen, Germany

Abstract

Fast, efficient, and reliable algorithms for up- and downdating discrete least-squares approximations of a real-valued function given at arbitrary distinct nodes in $[0, 2\pi)$ by trigonometric polynomials are presented. A combination of the up- and downdating algorithms yields a sliding window scheme. The algorithms are based on schemes for the solution of (inverse) unitary eigenproblems and require only $\mathcal{O}(mn)$ arithmetic operations as compared to $\mathcal{O}(mn^2)$ operations needed for algorithms that ignore the structure of the problem. Numerical examples show that the proposed algorithms produce consistently accurate results that are often better than those obtained by general QR decomposition methods for the least-squares problem.

Keywords. trigonometric approximation, unitary Hessenberg matrix, Schur parameter, Szegö polynomial, updating, downdating, sliding window scheme

1 Introduction

A problem in signal processing is the approximation of a function known only at some measured points by a trigonometric polynomial. A number of different models for representing the measured points as a finite superposition of sine- and cosine-oscillations are possible. One choice could be to compute the trigonometric interpolating function. Then several numerical algorithms are available [1]. But in general a large number of measured points are given, such that this approach leads to a trigonometric polynomial with a lot of superposed oscillations (and a large linear system to be solved). In practical applications it is often sufficient to compute a trigonometric polynomial with only a small number of superposed oscillations. A different, often chosen approach is the (fast) Fourier transform [1]. In this case the frequencies of the sine- and cosine-oscillations have to be chosen equidistant. The following approach gives more freedom in the choice of the frequencies and the number of superposed oscillations. Given a set of m arbitrary distinct nodes $\{\theta_k\}_{k=1}^{m}$ in the interval $[0, 2\pi)$, a set of m

positive weights $\{\omega_k^2\}_{k=1}^m$, and a real-valued function f whose values at the nodes θ_k are explicitly known. Then the trigonometric function

$$t(\theta) = a_0 + \sum_{j=1}^{\ell}(a_j \cos j\theta + b_j \sin j\theta), \qquad a_j, b_j \in \mathbf{R}, \tag{1.1}$$

of order at most $\ell < m/2$ is sought that minimizes the discrete least-squares error

$$||f - t||_{\mathbf{R}} := \sqrt{\sum_{k=1}^{m} |f(\theta_k) - t(\theta_k)|^2 \omega_k^2}. \tag{1.2}$$

In general, m (the number of measured functional values) is much larger than $n = 2\ell + 1$ (the number of coefficients to be determined).

Let the trigonometric polynomial $t(\theta) = a_0 + \sum_{j=1}^{\ell}(a_j \cos j\theta + b_j \sin j\theta)$ be the optimal solution of the approximation problem (1.2) corresponding to the data $Y_m = \{\theta_k, \omega_k^2\}_{k=1}^m$. Suppose Y_{m+1} is obtained from Y_m by augmenting a new node-weight pair $(\theta_{m+1}, \omega_{m+1}^2)$. Solving the approximation problem for Y_{m+1} assuming the knowledge of its solution for Y_m is called *updating* the least-squares fit. Solving the approximation problem for Y_m assuming the knowledge of its solution for Y_{m+1} is called *downdating* the least-squares fit.

Standard algorithms for solving the approximation problem (1.2) require $\mathcal{O}(mn^2)$ arithmetic operations. It can be observed however, that Szegö polynomials, that is polynomials that are orthogonal with respect to an inner product on the unit circle, arise naturally as a convenient basis for solving the least-squares problem (1.2). Updating and downdating of polynomial approximations when all nodes z_k are real has received a lot of attention in the literature, see [2] and the references therein. A collection of algorithms for updating and downdating based on orthogonal polynomials is presented in [3]. Downdating Szegö polynomials is considered in [4], while the updating process is the topic of [5, 6, 7].

Section 2 reviews fast algorithms for solving (1.2) via updating procedures which make use of the special structure of the problem (1.2). Fast downdating methods are presented in Section 3. The updating and downdating procedures can be combined to yield a sliding window scheme, in which one node is replaced by another.

2 Updating

The problem (1.2) can easily be reformulated as the standard least-squares problem of minimizing

$$||D\widetilde{A}\widetilde{t} - D\widetilde{f}||_2 = \min \tag{2.1}$$

over all coefficient vectors $\widetilde{t}$ in the Euclidean norm, where $\widetilde{f} = (f(\theta_1), \ldots, f(\theta_m))^T$ is a vector of the measured values of the function f, $D = diag(\omega_1, ..., \omega_m) \in \mathbf{R}^{m\times m}$ is a diagonal matrix with the given weights on the diagonal, and

$$\widetilde{A} = \begin{pmatrix} 1 & \sin\theta_1 & \cos\theta_1 & \cdots & \sin\ell\theta_1 & \cos\ell\theta_1 \\ \vdots & \vdots & \vdots & & \vdots & \vdots \\ 1 & \sin\theta_m & \cos\theta_m & \cdots & \sin\ell\theta_m & \cos\ell\theta_m \end{pmatrix} \in \mathbf{R}^{m\times n}.$$

A different approach is used by Reichel, Ammar, and Gragg in [5]. They noted that the problem (1.2) can be reformulated as the following standard least-squares problem: Minimize

$$||DAc - Dg||_2 = \min, \tag{2.2}$$

where A is a transposed Vandermonde matrix

$$A = \begin{pmatrix} 1 & z_1 & \cdots & z_1^{n-1} \\ 1 & z_2 & \cdots & z_2^{n-1} \\ \vdots & \vdots & & \vdots \\ 1 & z_m & \cdots & z_m^{n-1} \end{pmatrix} \in \mathbb{C}^{m\times n}$$

with $z_k = exp(\imath\theta_k), \imath = \sqrt{-1}$. $g = (g(z_1), ..., g(z_m))^T \in \mathbb{C}^m$ is a vector of the values of a complex function $g(z)$ and $c = (c_0, ..., c_{n-1})^T \in \mathbb{C}^n$ is the solution vector. With the proper choice of g ($g = \Lambda^\ell \widetilde{f}$, $\widetilde{f}$ as above, $\Lambda = diag(z_1, \ldots, z_m)$) it is easy to see that the coefficients of the trigonometric polynomial (1.1) that minimizes the error (1.2) can be read off of the least-squares solution $\widehat{c}$ of (2.2) (see [5])

$$a_0 = \widehat{c}_\ell, \qquad a_j = 2Re(\widehat{c}_{j+\ell}), \qquad b_j = -2Im(\widehat{c}_{j+\ell}), \qquad 1 \le j \le \ell.$$

The usual way to solve these least-squares problems is to compute the QR decomposition of DA or $D\widetilde{A}$. Ignoring the special structure of DA or $D\widetilde{A}$ this requires $\mathcal{O}(mn^2)$ arithmetic operations. It can be observed however, that Szegö polynomials, that is polynomials that are orthogonal with respect to an inner product on the unit circle, arise naturally as a convenient basis for solving the above standard least-squares problems. This observation can be used to develop fast, efficient and reliable algorithms for solving the approximation problem (2.2).

Observe that

$$\begin{aligned} DA &= \begin{pmatrix} \omega_1 & \omega_1 z_1 & \omega_1 z_1^2 & \cdots & \omega_1 z_1^{n-1} \\ \vdots & \vdots & \vdots & & \vdots \\ \omega_m & \omega_m z_m & \omega_m z_m^2 & \cdots & \omega_m z_m^{n-1} \end{pmatrix} \\ &= (q, \Lambda q, \Lambda^2 q, ..., \Lambda^{n-1} q) \\ &= \sigma_0(q_0, \Lambda q_0, \Lambda^2 q_0, ..., \Lambda^{n-1} q_0) \end{aligned}$$

with $q = (\omega_1, ..., \omega_m)^T, \sigma_0 = ||q||_2, q_0 := \sigma_0^{-1} q$ and $\Lambda = diag(z_1, ..., z_m)$. Thus, the matrix DA is given by the first n columns of the Krylov matrix $K(\Lambda, q_0, m) = (q_0, \Lambda q_0, ..., \Lambda^{m-1} q_0)$. We may therefore use the following consequence of the Implicit Q Theorem [8] to compute the desired QR decomposition. If there exists a unitary matrix U with $Ue_1 = q_0$ such that $U^H \Lambda U = H$ is a unitary upper Hessenberg matrix with positive subdiagonal elements, then the QR decomposition of $K(\Lambda, q_0, m)$ is given by UR with $R = K(H, e_1, m)$. The construction of such a unitary Hessenberg matrix from spectral data, here contained in Λ and q_0, is an inverse eigenproblem. Hence the best trigonometric approximation to f can be computed via solving this inverse eigenproblem. It can be seen that the elements of U are the values of the Szegö polynomials at the nodes z_k. Thus solving the inverse unitary Hessenberg eigenvalue problem $U^H \Lambda U = H$ is equivalent to computing Szegö polynomials.

From the above observation, an updating formulation for the approximation problem (2.2) in terms of the inverse unitary eigenproblem can easily be given:

Given

$\sigma_0 > 0$	
H_m	unitary upper Hessenberg matrix of size $m \times m$
d_m	a vector of length m
(λ, ν^2)	a node-weight pair

(σ_0, H_m, d_m representing the solution of (2.2) for some data set Y_m) find $\sigma_0 > 0$, a vector d_{m+1} and a unitary upper Hessenberg matrix H_{m+1} such that

1. the eigenvalues of H_{m+1} are $e^{\imath\lambda}$ and those of H_m,

2. the vector d_{m+1} contains the first components of the eigenvectors of H_{m+1}, that is if the entries of d_m are $\delta_1/\sigma_0, \ldots, \delta_m/\sigma_0$ and $\sigma_0 = (\sum_{k=1}^m \delta_k^2)^{\frac{1}{2}}$, then the new σ_0 will be $\sigma_0 = (\sigma_0^2 + \nu^2)^{\frac{1}{2}}$ and the entries of d_{m+1} are δ_k/σ_0 for $k = 1, \ldots, m$ and ν/σ_0.

Hence the approximation problem (1.2) can entirely be solved by updating, starting from the trivial solution for $m = 1$.

Unitary Hessenberg matrices have special properties which allow the development of efficient algorithms for this class of matrices. Any $n \times n$ unitary Hessenberg matrix with positive subdiagonal elements can be uniquely parameterized by n complex parameters, that is

$$H = G_1(\gamma_1) G_2(\gamma_2) \cdots G_n(\gamma_n)$$

for certain complex-valued parameters $|\gamma_k| < 1, 1 \leq k < n$, and $|\gamma_n| = 1$. Here $G_k(\gamma_k)$ denotes the $n \times n$ elementary reflector in the $(k, k+1)$ plane

$$G_k = G_k(\gamma_k) = diag(I_{k-1}, \begin{pmatrix} -\gamma_k & \sigma_k \\ \sigma_k & \overline{\gamma_k} \end{pmatrix}, I_{n-k-1})$$

with $\gamma_k \in \mathbb{C}$, $\sigma_k \in \mathbf{R}^+$, $|\gamma_k|^2 + \sigma_k^2 = 1$, and

$$G_n(\gamma_n) = diag(I_{n-1}, -\gamma_n)$$

with $\gamma_n \in \mathbb{C}$, $|\gamma_n| = 1$. The nontrivial entries γ_k are called *Schur parameters* and the σ_k are called *complementary Schur parameters.* This parameterization can be used to develop an efficient and reliable algorithm for solving the inverse unitary Hessenberg eigenvalue problem/the updating problem.

Such an algorithm was first described by Ammar, Gragg and Reichel in [6]. The idea is to build up the Hessenberg matrix successively by adding node-weight pairs (θ_k, ω_k^2) one at a time, in an updating fashion. The algorithm manipulates the n complex parameters instead of the n^2 matrix elements. An adaption of this scheme to the computation of the least-squares solution $\widehat{c}$ can be given, which requires $\mathcal{O}(mn + n^2)$ arithmetic operations. For details see [5].

The coefficients of the optimal trigonometric polynomial t of (1.2) can be recovered from $\widehat{c}$. This representation of t is convenient if we desire to integrate or differentiate the polynomial or if we wish to evaluate it at many equidistant points on a circle with a center at the origin. If we, on the other hand, only desire to evaluate t at a few points, then we can use the representation of t in terms of Szegö polynomials.

As $D\widetilde{A}$ in (2.1) is a real $m \times n$ matrix with full column rank, there exists a unique "skinny" real QR decomposition $\widetilde{Q}_1\widetilde{R}_1$ of $D\widetilde{A}$ where $\widetilde{Q}_1 \in \mathbf{R}^{m\times n}$ has orthonormal columns and $\widetilde{R}_1 \in \mathbf{R}^{n\times n}$ is upper triangular with positive diagonal entries. This ansatz leads to orthogonal Laurent polynomials and the (generalized) inverse unitary eigenproblem $\widetilde{U}^H(\Lambda - \lambda I)\widetilde{U}G_e = G_o - \lambda G_e$, where G_o and G_e are unitary block diagonal matrices with 1×1 or 2×2-blocks on the diagonal. The nonzero entries of G_o and G_e are just the Schur parameters and the complementary Schur parameters:

$$\begin{aligned} G_o &= G_1(\gamma_1)G_3(\gamma_3)\cdots G_{2[(n+1)/2]-1}(\gamma_{2[(n+1)/2]-1}) \\ &= \begin{pmatrix} -\gamma_1 & \sigma_1 & & & \\ \sigma_1 & \overline{\gamma_1} & & & \\ & & -\gamma_3 & \sigma_3 & \\ & & \sigma_3 & \overline{\gamma_3} & \\ & & & & \ddots \end{pmatrix} \end{aligned}$$

is the product of the odd numbered elementary reflectors and

$$\begin{aligned} G_e^H &= G_2(\gamma_2)G_4(\gamma_4)\cdots G_{2[n/2]}(\gamma_{2[n/2]}) \\ &= \begin{pmatrix} 1 & & & \\ & -\gamma_2 & \sigma_2 & \\ & \sigma_2 & \overline{\gamma_2} & \\ & & & \ddots \end{pmatrix} \end{aligned}$$

is the product of the even numbered elementary reflectors. The (generalized) inverse eigenproblem $\widetilde{U}^H(\Lambda - \lambda I)\widetilde{U}G_e = G_o - \lambda G_e$, where a *Schur parameter pencil* is constructed from spectral data, is equivalent to the inverse unitary Hessenberg eigenproblem $U^H \Lambda U = H = G_1(\gamma_1)\ldots G_n(\gamma_n)$ [9].

Observe that with

$$F = diag(2, \begin{pmatrix} -\imath & 1 \\ \imath & 1 \end{pmatrix}, \ldots, \begin{pmatrix} -\imath & 1 \\ \imath & 1 \end{pmatrix})$$

we have

$$\begin{aligned} D\widetilde{A} &= \frac{1}{2}(q, \Lambda q, \Lambda^H q, \Lambda^2 q, (\Lambda^H)^2 q, \ldots, \Lambda^\ell q, (\Lambda^H)^\ell q)F \\ &= \frac{1}{2}\sigma_0(q_0, \Lambda q_0, \Lambda^H q_0, \Lambda^2 q_0, (\Lambda^H)^2 q_0, \ldots, \Lambda^\ell q_0, (\Lambda^H)^\ell q_0)F \\ &= \frac{1}{2}\sigma_0 \kappa(\Lambda, q_0, \ell)F \end{aligned}$$

with q, σ_0, q_0 and Λ as before. A QR-like decomposition of $D\widetilde{A}$ can be obtained using the following result [10, 11] : If there exists a unitary matrix V such that $V(\Lambda - \lambda I)V^H G_e = G_o - \lambda G_e, V^H e_1 = q_0$, then the QR decomposition of $\kappa(\Lambda, q_0, \ell)$ is given by VR with $R = \kappa(G_o G_e^H, e_1, \ell)$. Hence $D\widetilde{A} = \frac{\sigma_0}{2} VRF$ and the optimal solution of (2.1) is given by $\widetilde{t} = 2\sigma_0^{-1} F^{-1} R^{-1} V^H D\widetilde{f}$. The construction of such a Schur parameter pencil from spectral data is a (generalized) inverse eigenproblem. Thus the best trigonometric approximation to f can be computed via solving this inverse eigenproblem. As explained in [7], the elements of V are the values of orthogonal Laurent polynomials at the nodes θ_k. Thus solving the inverse unitary eigenproblem $V(\Lambda - \lambda I)W = G_o - \lambda G_e$ is equivalent to computing orthogonal Laurent polynomials.

An updating formulation for the approximation problem (2.1) in terms of the generalized inverse unitary eigenproblem can be given analogous to the one for problem (2.2). The special structure of the inverse eigenproblem $V(\Lambda - \lambda I)V^H G_e = G_o - \lambda G_e$ can be used to develop an efficient and reliable algorithm for computing V, γ_j and $\sigma_j, j = 1, \ldots, n$. The advantage of this approach over (2.2) is that here an algorithm can be given which solves the real-valued problem (1.2) using only real arithmetic. For details see [11].

The two algorithms sketched above are updating procedures in the sense that the least-squares fit is obtained by incorporating the nodes of the inner product one at a time. In certain applications it may be desirable to replace certain node-weight pairs (θ_k, ω_k^2). This can be carried out by successively removing a node-weight pair from the current approximation, and then adding a new node-weight pair. Downdating Szegö polynomials/orthogonal Laurent polynomials and a given least-squares fit when one node is deleted from the inner product can easily be implemented solving unitary eigenproblems. The updating and downdating procedures, based on solving inverse unitary eigenproblems and unitary eigenproblems, can be combined to yield a sliding window scheme, in which one node is replaced by another.

3 Downdating

Assume the solution t of (1.2) corresponding to the data set Y_{m+1} is obtained by the updating method based on the inverse unitary Hessenberg eigenproblem discussed in the Section 2. The problem of downdating the optimal trigonometric approximation t of (1.2) can then be expressed as follows:

Given

$\sigma_0 > 0$	
H_{m+1}	unitary upper Hessenberg matrix of size $(m+1) \times (m+1)$
d_{m+1}	a vector of length $m+1$
(λ, ν^2)	a node-weight pair from Y_{m+1}

($\sigma_0, H_{m+1}, d_{m+1}$ representing the solution of (1.2) for some data set $Y_{m+1} = \{\theta_k, \omega_k^2\}_{k=1}^{m+1}$) find $\sigma_0 > 0$, a vector d_m and a unitary upper Hessenberg matrix H_m such that

1. the eigenvalues of H_m are $\{e^{\imath\theta_k}\}_{k=1}^{m+1} \setminus e^{\imath\lambda}$,
2. the vector d_m contains the first components of the eigenvectors of H_m, that is if the entries of d_{m+1} are $\delta_1/\sigma_0, \ldots, \delta_{m+1}/\sigma_0$ and $\sigma_0 = (\sum_{k=1}^{m+1} \delta_k^2)^{\frac{1}{2}}$, then the new σ_0 will be $\sigma_0 = (\sigma_0^2 - \nu^2)^{\frac{1}{2}}$ and the entries of d_m are $\{\delta_k\}_{k=1}^{m+1} \setminus \nu$ normalized by σ_0.

Such a downdating procedure was first described by Ammar, Gragg and Reichel in [4]. They present an $\mathcal{O}(m)$ algorithm which is based on the unitary QR algorithm introduced by Gragg in [12]. Here we will develop a downdating procedure for the generalized unitary eigenproblem approach using these ideas. Assume the optimal least-squares solution t of (1.2) corresponding to the data $Y_{m+1} = \{\theta_k, \omega_k^2\}_{k=1}^{m+1}$ has been computed via the updating algorithm based on the generalized inverse unitary eigenproblem discussed in Section 2. Then a unitary matrix Q and a Schur parameter pencil $G_o^{m+1} - \lambda G_e^{m+1}$ is known such that

$$Q(\Lambda - \lambda I)Q^H G_e^{m+1} = G_o^{m+1} - \lambda G_e^{m+1}, \quad Q^H e_1 = \sigma_0^{-1}(\omega_1, \ldots, \omega_{m+1})^T \quad (3.1)$$

where

$$\sigma_0 = (\sum_{k=1}^{m+1} \omega_k^2)^{\frac{1}{2}}, \qquad \Lambda = diag(\lambda_1, \ldots, \lambda_{m+1}), \qquad \lambda_k = e^{\imath\theta_k}.$$

Instead of (3.1) we can just as well consider the equivalent equation

$$Q\Lambda Q^H = G_o^{m+1}(G_e^{m+1})^H.$$

Let (θ_j, ω_j^2) be the node-weight pair to be deleted from the solution. Using the knowledge of the above solution, we wish to construct a Schur parameter pencil $G_o^m - \lambda G_e^m$ or equivalently a matrix $G_o^m(G_e^m)^H$ such that

$$\widehat{W}\widehat{\Lambda}\widehat{W}^H = G_o^m(G_e^m)^H,$$

where

$$\begin{aligned}\widehat{\Lambda} &= diag(\lambda_1, \ldots, \lambda_{j-1}, \lambda_{j+1}, \ldots, \lambda_{m+1}), \\ \widehat{W}^H e_1 &= \widehat{\sigma}_0^{-1}(\omega_1, \ldots, \omega_{j-1}, \omega_{j+1}, \ldots, \omega_{m+1})^T, \\ \widehat{\sigma}_0 &= (\sigma_0^2 - \omega_j^2)^{\frac{1}{2}}\end{aligned}$$

or some suitable permutation of $\widehat{\Lambda}$ and $\widehat{W}^H e_1$. In other words, we wish to determine a unitary matrix V such that

$$\begin{pmatrix} 1 & 0 \\ 0 & V \end{pmatrix} \begin{pmatrix} \delta & \sigma_0 e_1^T \\ \sigma_0 e_1 & G_o^{m+1}(G_e^{m+1})^H \end{pmatrix} \begin{pmatrix} 1 & 0 \\ 0 & V^H \end{pmatrix} = \begin{pmatrix} \delta & \widehat{\sigma}_0 \widehat{e}_1^T & \omega_j \\ \widehat{\sigma}_0 \widehat{e}_1 & G_o^m (G_e^m)^H & 0 \\ \omega_j & 0 & \lambda_j \end{pmatrix}.$$

Then, as $\begin{pmatrix} \widehat{\Lambda} & \\ & \lambda_j \end{pmatrix} = P_j^H \Lambda P_j$ where $P_j = (e_1, \ldots, e_{j-1}, e_{j+1}, \ldots, e_{m+1}, e_j)$

$$\begin{pmatrix} \widehat{W} & \\ & 1 \end{pmatrix} = P_j^H Q^H V^H.$$

Analogous to the ideas of Ammar, Gragg and Reichel in [4] we want to compute $\widehat{W}$ via a QR-type step. In [13] a QR-like algorithm for Schur parameter pencils is introduced. The method is based on the standard QR algorithm applied to a matrix of the form $G_o G_e^H$. No initial reduction to Hessenberg form is performed. It is shown that each iterate is then of the same form as $G_o G_e^H$ again. Hence, applying one QR-step with the exact shift λ_j to the matrix $G_o^{m+1}(G_e^{m+1})^H$ determines a unitary matrix $\widetilde{V}$ such that

$$\widetilde{V} G_o^{m+1}(G_e^{m+1})^H \widetilde{V}^H = \begin{pmatrix} X & \\ & \lambda_j \end{pmatrix}$$

because λ_j is an eigenvalue of $G_o^{m+1}(G_e^{m+1})^H$. X is a unitary matrix of the same form as $G_o^m(G_e^m)^H$. $\widetilde{V}$ is a matrix of the form $G_1(G_3G_2G_3)(G_5G_4G_5)\cdots$. Hence, the vector $\sigma_0 e_1$ will not be transformed as required as $\widetilde{V}e_1$ is a full vector.

Applying one step of an RQ algorithm with the exact shift λ_j to $G_o^{m+1}(G_e^{m+1})^H$ determines an upper triangular matrix R and a unitary U of the form

$$G_1(G_3G_2G_3)(G_5G_4G_5)\cdots = G_1G_3G_5\cdots G_2G_3G_4G_5\cdots$$

that is, $U = G_o H$ for some upper Hessenberg matrix H and a matrix of the form G_o. R and U are determined such that $G_o^{m+1}(G_e^{m+1})^H - \lambda_j I = RU$ and

$$U G_o^{m+1}(G_e^{m+1})^H U^H = UR + \lambda_j I = \begin{pmatrix} \widehat{X} & \\ & \lambda_j \end{pmatrix}.$$

The vector $\sigma_0 e_1$ is transformed such that only the first two entries are nonzero

$$\sigma_0 U e_1 = (x, x, 0, \ldots, 0)^T.$$

Observe that with the reversal matrix $J = (e_{m+1}, e_m, \ldots, e_1)$

$$\begin{aligned} G_o^{m+1}(G_e^{m+1})^H = RU \quad &\Leftrightarrow \quad JG_o^{m+1}(G_e^{m+1})^H J = (JRJ)(JUJ) \\ &\Leftrightarrow \quad J(G_o^{m+1}(G_e^{m+1})^H)^T J = (JU^T J)(JR^T J). \end{aligned}$$

Let $G^P := J(G_o^{m+1}(G_e^{m+1})^H)^T J$ and let us assume for simplicity that $m+1$ is odd (a similar argumentation can be given in the case that $m+1$ is even). Then

$$G_o^{m+1}(G_e^{m+1})^H = G_1(\gamma_1)G_3(\gamma_3)\cdots G_{m+1}(\gamma_{m+1})G_2^H(\gamma_2)G_4^H(\gamma_4)\cdots G_m^H(\gamma_m)$$

and

$$G^P = G_1(\overline{\gamma_m})G_3(\overline{\gamma_{m-2}})\cdots G_{m-1}(\overline{\gamma_2})G_0(\gamma_{m+1})G_2^H(\gamma_{m-1})G_4^H(\gamma_{m-3})\cdots G_m^H(\gamma_1)$$

where

$$G_0(\gamma) = diag(-\gamma, I).$$

That is, $G_o^{m+1}(G_e^{m+1})^H$ and G^P are of the same form. Further, as U is a matrix of the form $G_o H$ for some upper Hessenberg matrix H and a matrix of the form G_o, $JU^T J$ is a matrix of the same form. Hence, as $JR^T J$ is upper triangular again, applying the RQ algorithm to $G_o^{m+1}(G_e^{m+1})^H$ is equivalent to applying the QR algorithm to G^P. One iteration of the QR algorithm with the exact shift λ_j applied to G^P generates a unitary matrix V

$$G_1(\beta_1)G_3(\beta_3)\cdots G_{m-1}(\beta_{m-1})G_2(\delta_2)G_3(\delta_3)\cdots G_{m+1}(\delta_{m+1})$$

such that

$$V^H G^P V = \begin{pmatrix} X' & 0 \\ 0 & \lambda_j \end{pmatrix}. \tag{3.2}$$

Moreover, δ_{m+1} can be taken to be an arbitrary unimodular number, because deflation has taken place.

Only the last two components of $\sigma_0 V^H e_{m+1}$ are nonzero; they are given by

$$\begin{aligned} &\begin{pmatrix} 1 & 0 \\ 0 & -\overline{\delta_{m+1}} \end{pmatrix} \begin{pmatrix} -\overline{\delta_m} & (1-|\delta_m|^2)^{\frac{1}{2}} \\ (1-|\delta_m|^2)^{\frac{1}{2}} & \delta_m \end{pmatrix} \begin{pmatrix} 0 \\ \sigma_0 \end{pmatrix} \\ &= \begin{pmatrix} (1-|\delta_m|^2)^{\frac{1}{2}}\,\sigma_0 \\ -\overline{\delta_{m+1}}\,\delta_m\,\sigma_0 \end{pmatrix}. \end{aligned}$$

We can choose $\delta_{m+1} = -\delta_m/|\delta_m|$ to obtain

$$\begin{pmatrix} (1-|\delta_m|^2)^{\frac{1}{2}}\sigma_0 \\ |\delta_m|\sigma_0 \end{pmatrix}.$$

Transforming (3.2) by similarity using $\widehat{J} = (e_m, e_{m-1}, \ldots, e_1, e_{m+1})$ and transposing the result, we obtain

$$W G_o^{m+1} (G_e^{m+1})^H W^H = \begin{pmatrix} X'' & 0 \\ 0 & \lambda_j \end{pmatrix}, \quad \text{where} \quad W = J\overline{V}\widehat{J}.$$

Moreover,

$$\sigma_0 W^H e_1 = \begin{pmatrix} (1 - |\delta_m|^2)^{\frac{1}{2}} \sigma_0 \widehat{e}_1 \\ |\delta_m| \sigma_0 \end{pmatrix},$$

and by the uniqueness of the reduction, $X'' = G_o^m (G_e^m)^H$, $\sigma_0 (1 - |\delta_m|^2)^{\frac{1}{2}} = \widehat{\sigma}_0$, and $|\delta_m|\sigma_0 = \omega_j$.

Note that the downdating process requires knowledge of the node θ_j to be deleted, but not of the corresponding weight ω_j^2. In an implementation of the process the computed weight can therefore be used to assess the accuracy of the computation.

If the optimal least-squares solution t of (1.2) corresponding to the data Y_{m+1} has been computed via the algorithm based on the generalized inverse unitary eigenproblem discussed in Section 2, the least-squares solution $\widetilde{t} = \widetilde{R}_1^{-1} \widetilde{Q}_1^H D \widetilde{f} \in \mathbb{C}^{m+1}$ of (2.1) is known. The optimal least-squares solution t of (1.2) corresponding to the data Y_m is then obtained by applying $\widehat{W}^H$ to $t' = \widetilde{Q}_1^H D \widetilde{f}$ incrementally

$$\widehat{W}^H t' = \begin{pmatrix} t'' \\ \omega_j f(z_j) \end{pmatrix}.$$

In a second step a new $\widetilde{t} \in \mathbb{C}^m$ has to be computed from t'' using the Schur parameters of $G_o^m (G_e^m)^H$ via a simplified Levinson algorithm. This is analogous to the second step of the updating procedure. There first $\widetilde{Q}_1^H D \widetilde{f}$ is computed via solving the generalized inverse unitary eigenproblem. Then $\widetilde{R}_1^{-1}$ is computed using the Schur parameters of G_o and G_e via a simplified Levinson algorithm. Details are given in [11].

A QR step has to be applied to a matrix X of the form $G_o G_e^H$. We get a unitary matrix $X' = V^H X V$, where X' can be written as $G_o'(G_e')^H$ again. If X corresponds to an unreduced Schur parameter pencil $G_o - \lambda G_e$, then X' will correspond to an unreduced Schur parameter pencil $G_o' - \lambda G_e'$. The transformation to an unreduced Schur parameter pencil is uniquely determined, up to unitary scaling, if the first column of the transformation matrix is given. Therefore one can derive $G_o' - \lambda G_e'$, up to scaling from $G_o - \lambda G_e$ by any unitary transformation $Q^H (G_o - \lambda G_e) P$ to Schur parameter pencils, for which the first column of Q coincides with a scalar multiple of the first column of V. This was used by Bunse-Gerstner and Elsner in [13] to derive an implicit single shifted QR step.

In a preparatory step, a matrix $V_1 = G_1(\alpha_1)$ is determined such that

$$V_1^H (G_o - \lambda_j G_e) e_1 = \rho e_1.$$

The pencil $V_1^H(G_o - \lambda G_e)$ differs from a Schur parameter pencil only by three additional entries

$$\begin{pmatrix} x & x & & & \\ x & x & & & \\ & & x & x & \\ & & x & x & \\ & & & & \ddots \end{pmatrix} - \lambda \begin{pmatrix} x & + & + & & \\ + & x & x & & \\ & x & x & & \\ & & & x & x \\ & & & x & x \end{pmatrix}.$$

This "bulge" is then chased down along the diagonal to restore the Schur parameter pencil form again. This is an $\mathcal{O}(m)$ process.

4 Concluding remarks

Due to space limitation, we will refrain from giving numerical examples for the updating and downdating procedures. For numerical examples and a detailed discussion on the updating process see [5, 10, 11]. For numerical examples and a detailed discussion on downdating Szegö polynomials/the downdating process in terms of the unitary eigenproblem see [4].

Bibliography

1. P. Henrici. *Applied and Computational Analysis*, volume 3. Wiley, 1986.
2. L.B. Scott and L.R. Scott. Efficient methods for data smoothing. *SIAM J. Numer. Anal.*, 26:681 – 692, 1989.
3. S. Elhay, G.H. Golub, and J. Kautsky. Updating and downdating of orthogonal polynomials with data fitting applications. *SIAM J. Matrix Anal. Appl.*, 12:327 – 353, 1991.
4. G.S. Ammar, W.B. Gragg, and L. Reichel. Downdating of Szegö Polynomials and Data Fitting Applications. *Lin. Alg. and its Applic.*, 172:315 – 336, 1992.
5. L. Reichel, G. S. Ammar, and W. B. Gragg. Discrete Least Squares Approximation by Trigonometric Polynomials. *Math. Comp.*, 57:273 – 289, 1991.
6. G. S. Ammar, W. B. Gragg, and L. Reichel. Constructing a Unitary Hessenberg Matrix from Spectral Data. In G.H. Golub and P. Van Dooren, editors, *Numerical Linear Algebra, Digital Signal Processing, and Parallel Algorithms*, pages 385–396. Springer-Verlag, Berlin, 1991.
7. H. Faßbender. Inverse unitary eigenproblems and related orthogonal functions. *Numerische Mathematik*, 77:323 – 345, 1997.

8. G. H. Golub and C. F. Van Loan. *Matrix Computation.* The John Hopkins University Press, second edition, 1989.

9. G. S. Ammar, W. B. Gragg, and L. Reichel. On the Eigenproblem for Orthogonal Matrices. In *Proc. 25th IEEE Conference on Decision and Control*, pages 1963 – 1966, 1986.

10. H. Faßbender. *Numerische Verfahren zur diskreten trigonometrischen Polynomapproximation.* Dissertation Universität Bremen, 1993.

11. H. Faßbender. On Numerical Methods for Discrete Least-Squares Approximation by Trigonometric Polynomials. *Math. Comp.*, 66:719 – 741, 1997.

12. W. B. Gragg. The QR algorithm for unitary Hessenberg matrices. *J. Comp. Appl. Math.*, 16:1 – 8, 1986.

13. A. Bunse-Gerstner and L. Elsner. Schur Parameter Pencils for the Solution of the Unitary Eigenproblem. *Lin. Alg. and its Appl.*, 154 - 156:741 – 778, 1991.

Jump Detection by Padé Approximation

Riccardo March and Piero Barone

Istituto per le Applicazioni del Calcolo, CNR, Roma, Italy

Abstract

We propose a method based on the Padé approximation to reconstruct a piecewise constant function from a finite number of noisy Fourier coefficients. The method is based on the remark that the poles of the Padé approximants to the Z-transform of the Fourier coefficients show, asymptotically, clusters in the complex plane which allow the identification of the discontinuities of the function. This property is relevant for the design of an effective noise filter to recover the piecewise constant function.

1 Introduction

In this paper we consider the problem of reconstructing a piecewise constant function from a finite number of its Fourier coefficients perturbed by noise. A relevant example of such a problem is provided by the reconstruction of magnetic resonance images [1]. It is well known that the reconstruction provided by the truncated Fourier series is affected by the Gibbs oscillations that may hide important features of the true signal.

Given a real interval $[A, B]$ and $N+1$ numbers $A \leq l_1 < l_2 \ldots < l_{N+1} \leq B$, let $\mathcal{F}$ be the class of functions defined as

$$F(t) = \sum_{j=1}^{N} w_j \chi_j(t), \qquad \chi_j(t) = \begin{cases} 1 & \text{if } t \in [l_j, l_{j+1}] \\ 0 & \text{otherwise} \end{cases} \qquad w_j \in \mathbb{R}.$$

In the following we assume, without loss of generality, $[A, B] = [-\pi, \pi]$. The problem is to find a function $F(t) \in \mathcal{F}$ from a finite number of its Fourier coefficients

$$a_k = \frac{1}{2} \int_{-\pi}^{\pi} F(t) e^{itk} dt + \varepsilon_k = s_k + \varepsilon_k \ , \quad k = 0, \ldots, m+n \ , \tag{1.1}$$

where ε_k is a noisy perturbation that will be specified in the following, and m and n are positive integers. The unperturbed Fourier coefficients s_k are

$$s_k = \frac{1}{2} \int_{-\pi}^{\pi} F(t) e^{itk} dt = \sum_{j=1}^{N} \frac{w_j}{k} \sin\left(k \frac{l_{j+1} - l_j}{2}\right) \exp\left(ik \frac{l_{j+1} + l_j}{2}\right) . \tag{1.2}$$

The reconstruction problem then reduces to the estimation of w_j and l_j from a_k, $k = 0, \ldots, m+n$, which is a highly nonlinear problem. After the l_j are determined, the w_j can be computed by solving a linear problem.

In order to estimate the jump points l_j of $F(t)$ we consider the Padé approximants to the Z–transform of the sequence $\{a_k\}$ of the noisy Fourier coefficients. By using a theory due to Stahl [2] we study the relation between the asymptotic location of the poles of the Padé approximants and the l_j. By exploiting the properties of the Padé poles a reconstruction method for recovering a stable solution which is not affected by Gibbs artifact is designed. The case $N = 1$ has been considered in [3, 4]. Some experimental results showing the denoising performance of the method are discussed. We also present an experiment which shows that our method can have some super–resolution properties too.

2 The Padé approximation method

We study the asymptotic location of the poles of the Padé approximants to the Z–transform of the sequence $\{a_k\}$ given by the formal power series

$$f(z) = \sum_{k=0}^{\infty} a_k z^{-k} = \sum_{k=0}^{\infty} s_k z^{-k} + \sum_{k=0}^{\infty} \varepsilon_k z^{-k} = s(z) + \eta(z) \ . \tag{2.1}$$

The Z–transform $s(z)$ of the unperturbed Fourier coefficients s_k is

$$s(z) = \sum_{j=1}^{N} w_j \left(\frac{l_{j+1} - l_j}{2} + \frac{1}{2i} \ln \frac{z - e^{il_j}}{z - e^{il_{j+1}}} \right) , \tag{2.2}$$

which is a multiple–valued function in the complex plane.

The perturbation $\eta(z)$ is represented by

$$\eta(z) = \sum_{r=1}^{\infty} A_r / (1 - \alpha_r z^{-1}) , \tag{2.3}$$

where $|\alpha_r| = 1$, and $|A_r| <$ const. $\exp(-r^{1+\delta})$, with $\delta > 0$. The function $\eta(z)$ has the unit circle as a curve of singularities. Functions of this type were proposed by Gammel [5] to model the effect of the noise, in the terms of the power series, on the properties of the Padé approximants. In order to apply the Stahl's results we assume that the series in the Equation 2.3 has a countable compact set of singularities denoted by S

$$S = \bigcup_{r=1}^{\infty} \{\alpha_r\} ,$$

with the α_r uniformly distributed on the unit circle so as to represent a white noise.

The Padé approximant $[m,n]_f$ of orders m and n to the function $f(z)$ is given by [6]

$$[m,n]_f = Q_m(z^{-1})/P_n(z^{-1}) ,$$

where $Q_m(z)$ and $P_n(z)$ are polynomials in z of degree m and n, respectively, which satisfy

$$P_n(z^{-1})f(z) - Q_m(z^{-1}) = O(z^{-(m+n+1)}) , \quad \text{as } |z| \to \infty . \qquad (2.4)$$

In the next section we study the asymptotic location of poles of the $[m,n]_f$ Padé approximant for large orders m and n.

3 Distribution of the Padé poles

The Z–transform $s(z)$ given by the Equation 2.2 is a multiple–valued function with branch points on the unit circle at $c_j = \exp(il_j)$, $j = 1, \ldots, N+1$, if l_j is a point of discontinuity of $F(t)$. If a set of arcs joining the branch points is introduced a single–valued function in the complex plane cut along the arcs is obtained. Since the resulting $s(z)$ is discontinuous across the cuts, $[m,n]_f$, which is continuous except at poles, can represent a discontinuous function only if the poles cluster along the cuts. The other poles have to approach the unit circle in order to approximate the noise. There is an infinity of ways to choose the cuts, however, it is found that the branch points always determine a specific set of cuts. Furthermore Stahl [2] proved that the Padé poles are asymptotically distributed along the cuts according to the equilibrium measure of the logarithmic potential theory.

In the following we assume that $s(z)$ has an even number $2M$ of branch points at finite points in the complex plane, so that $f(z)$ is analytic at infinity. If $F(t)$ has an odd number of discontinuities in the interval $[-\pi,\pi]$ then $s(z)$ has a branch point at infinity, and we may resort to the previous case by means of the transformation $z \to 1/z$. To determine the asymptotic distribution in the complex plane of the poles of $[m,n]_f$ as $m,n \to \infty$, Stahl [2] introduces a domain D defined by the three properties:

(i) $\infty \in D$ and $f(z)$ has a single–valued analytic continuation in D.

(ii) The complement K of D is of minimal logarithmic capacity.

(iii) D is the union of all domains satisfying the assertions (i) and (ii).

The domain D is uniquely determined by the branch points of $f(z)$ and the set S of singularities. Stahl then proved that [2] the complement K of the domain D consists of analytic Jordan arcs connecting the branch points and a compact set $S_0 \subseteq S$. Hence the function $f(z)$ has a single–valued analytic continuation in the complex plane cut along a set of arcs that belong to K.

The distribution of the poles of the Padé approximants is described by means of the pole counting measure. If $Q_m(z)$ and $P_n(z)$ have no common factors the *pole counting measure* ν_{mn} for $[m,n]_f$ is

$$\nu_{mn} = \frac{1}{n} \sum_{P_n(z^{-1})=0} \delta_z \ ,$$

where δ_z denotes the Dirac's measure for the point z, and poles are repeated according to multiplicity. Furthermore we denote by μ_K the *equilibrium measure* for the set K according to the logarithmic potential theory. The asymptotic distribution of the poles of $[m,n]_f$ is given by the following result which is a particular case of a theorem proved by Stahl [2].

Theorem 1. *Let $f(z)$ be a function with branch points and the set S of singularities, and let $m/n \to 1$ for $m+n \to \infty$. Then the sequence of the pole counting measures for the Padé approximants $[m,n]_f$ converges weakly to the equilibrium measure for K:*

$$\nu_{mn} \overset{\star}{\to} \mu_K \ , \quad \textit{for } m+n \to \infty \ , \tag{3.1}$$

where $\overset{\star}{\to}$ denotes the weak convergence of measures.

Hence the branch points always determine the asymptotic distribution of the Padé poles, which is the distribution of electrostatic equilibrium on the set K. The theorem involves the weak convergence of measures which describes how the poles behave on average when m and n are large, so that poles outside of K may exist.

We denote by L the set of analytic cuts which connect the branch points, so that $K = L \cup S_0$. We then write the equilibrium measure $\mu_K(z)$ on L as

$$\mu_K(z) = \rho(z)|dz| \ , \quad z \in L \ ,$$

so that $\rho(z)$ represents the (normalized) linear density of the poles along L. From an analysis of the condition of electrostatic equilibrium on the set K of minimal capacity, we prove [7] that the density $\rho(z)$ satisfies the singular integral equation of Cauchy type

$$\int_L \frac{\rho(t)}{t-z}|dt| = -\int_{S_0} \frac{1}{t-z} d\mu_K(t) \ , \quad z \in L \ . \tag{3.2}$$

If L consists of a set of arcs ending at the branch points c_j, by using a method of Nuttall [8], the Cauchy integral in the left hand side of the Equation 3.2 may be inverted yielding the expression

$$\rho(z) = \left| \frac{\xi(z)}{X(z)} \right| \ , \quad z \in L \ , \tag{3.3}$$

where

$$X(z) = \left[\prod_{j=1}^{2M} (z - c_j) \right]^{1/2} ,$$

and $\xi(z)$ is a function analytic in the complex plane cut along L. If at a branch point c_j, $j = 1, \ldots, 2M$, the condition $\xi(c_j) \neq 0$ holds, then from the Equation 3.3 it follows that, for z close to c_j, the density $\rho(z)$ behaves as

$$\rho(z) \sim |z - c_j|^{-1/2} , \quad z \in L . \tag{3.4}$$

Furthermore in [7] we show that this condition holds at most of the branch points.

Thus a subset of the Padé poles is distributed along a set of arcs so as to give rise to clusters at the branch points on the unit circle. From an analysis of the condition of electrostatic equilibrium it follows [7] that such clusters of poles are stable by varying the orders m and n. The other poles lie close to the unit circle and model the noise. We find that these poles are unstable, so that an analysis of the distribution of the poles in the complex plane at different orders allows the identification of the branch points, and hence of the jumps of $F(t)$.

4 Numerical experiments

From the discussion in the previous section the following qualitative features of the poles of the Padé approximants can be exploited:

- A subset of the poles clusters at the branch points allowing the identification of the discontinuities l_j.
- The other poles are distributed in a neighbour of the unit circle modeling the effect of the noise.

These results on the asymptotic behaviour of the Padé poles can be used to build a numerical procedure for reconstructing $F(t)$. The reconstruction process is splitted in two parts. First the branch points of $s(z)$ are identified from the noisy data. Then the weights w_j are estimated. We locate the branch points by means of a "statistical" approach [1] based on the computation of the two–dimensional empirical distribution of the computed Padé poles.

The pole distribution is computed by using a grid in the neighbourhood of the unit circle with cells of the same area delimited by a set of equispaced rays starting in zero and a set of concentric circles with origin in zero. By using a_k, $k = 0, \ldots, m+n$, we compute the $[m, n]_f$ Pade' approximants of $f(z)$ for several values of m and n. We compute a two dimensional empirical distribution of the computed poles in a neighbour of the unit circle. Then we select the cells where the pole distribution has clear peaks and we take the average of the poles in such cells as an estimate of the branch points.

After the jump points l_j of $F(t)$ are determined, the weights w_j, $j = 1, \ldots, N$, are estimated by solving a linear problem. We compute the discrete inverse Fourier transform $\hat{F}(t)$ of a_k, $k = 0, \ldots, m+n$. In each interval determined by two consecutive arguments of the estimated branch points, we take a trimmed average of $\hat{F}(t)$ as an estimate of the weights w_j.

Figure 1 reports a numerical example of denoising. The top left part of the figure shows the function $F(t)$; the top right shows the Fourier reconstruction with a signal–to–noise ratio (SNR, measured as the ratio between the standard deviations of signal and noise) equal to 1; the bottom left shows the reconstruction by the Padé method; the bottom right shows the reconstruction by the Haar wavelet shrinkage method [9] for a comparison: some discontinuities are lost in this case.

Numerical experiments show that the proposed procedure has some super–resolution abilities too. This fact was exploited when the procedure was applied to the class of functions whose Fourier transform is a linear combination of complex exponentials [10, 11]. We find that if the SNR is large enough the procedure is able to estimate branch points whose arguments are closer than the

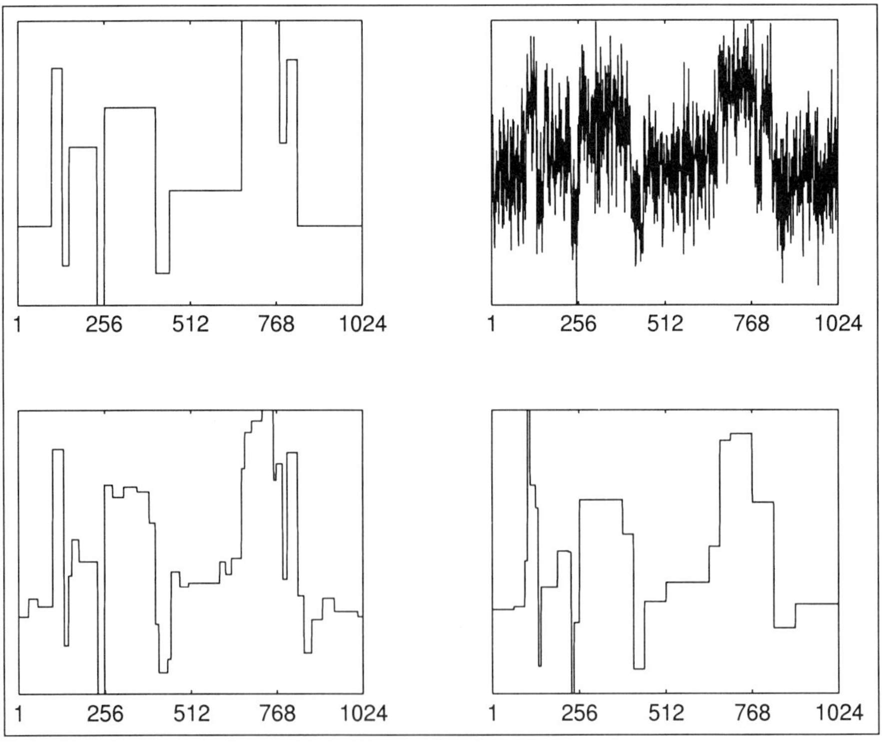

Figure 1. Numerical example of denoising

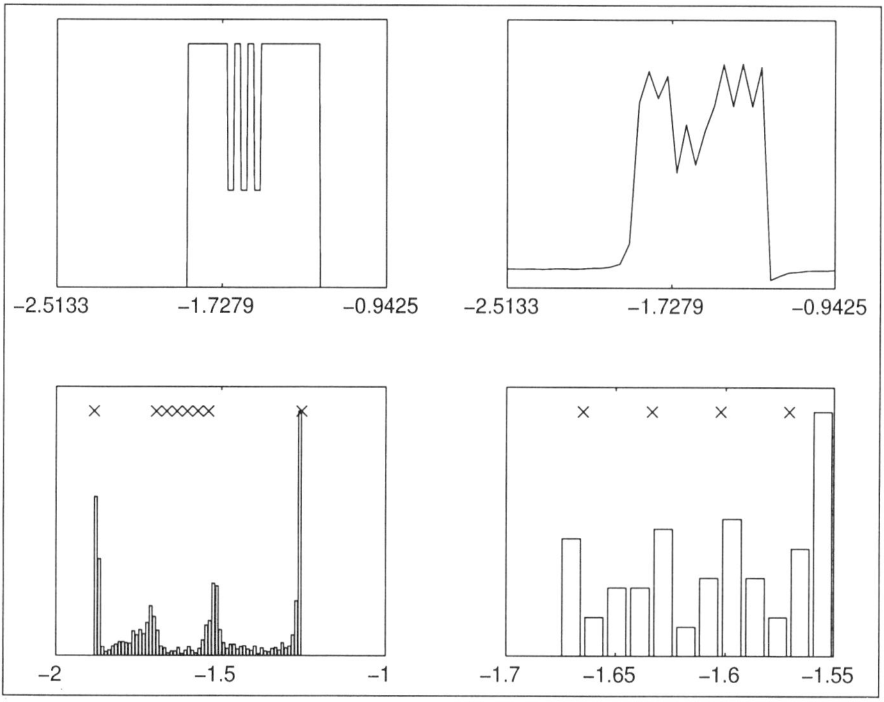

Figure 2. Numerical example of super-resolution

Nyquist frequency. Figure 2 reports a numerical example of super–resolution. The top left part of the figure shows the true signal: the spacing between the closest branch points is 0.4 times the used Nyquist rate and the SNR is equal to 200; the top right shows the Fourier reconstruction; the histogram of the pole arguments in a neighbour of the unit circle is plotted in the bottom left, and the crosses indicate the arguments of the branch points of the true signal; the bottom right shows a zoom of the part of the histogram corresponding to the two narrow boxes in the middle: all the branch points are clearly identified by the peaks in the distribution. At the moment we are not able to state exactly under which conditions super–resolution takes place. However, when it works, the localization of the jumps is much better than that provided by standard out–of–band extrapolation methods, like for example alternate projections.

5 Conclusion

The problem of reconstructing a piecewise constant function from noisy Fourier coefficients is considered. A new method is proposed, based on the asymptotic properties of the poles of the Padé approximants to the Z–transform of the observed Fourier coefficients. Numerical experiments compare favourably with recently appeared methods. However some problems require further research, such as a statistical representation of the noise in the Padé approximation process, and the quantification of the quality of the results to be expected for a given SNR.

Bibliography

1. Barone, P. and Sebastiani, G. (1992). A new method of magnetic resonance image reconstruction with short acquisition time and truncation artifact reduction. *IEEE Trans. Med. Imaging*, **2**, 250–259.
2. Stahl, H. (1986). Orthogonal polynomials with complex–valued weight function, I. *Constr. Approx.*, **2**, 225–240.
3. Barone, P. and March, R. (1996). On the super–resolution properties of Prony's method. *ZAMM: Z. angew. Math. Mech.*, **76 S2**, 177–180.
4. March, R. and Barone, P. (1998). Application of the Padé method to solve the noisy trigonometric moment problem: some initial results. *SIAM J. on Appl. Math.*, **58**, 324–343.
5. Gammel, J.L. (1972). Effect of random errors (noise) in the terms of a power series on the convergence of the Padé approximants. *Padé approximants*, Editor P. Graves–Morris, The Institute of Physics, London and Bristol, 132–133.
6. Baker, G.A. Jr. (1975). *Essentials of Padé approximants.* Academic Press, New York.
7. March, R. and Barone, P. (1996). Reconstruction of a piecewise constant function from noisy Fourier coefficients by using the Padé method. *Quaderno IAC–CNR*, **10**, submitted to *SIAM J. on Appl. Math.*
8. Nuttall, J. (1984). Location of poles of Padé approximants to Entire Functions. *Rational Approximation and Interpolation*, Editors: P.R. Graves-Morris, E.B. Saff and R.S. Varga, Springer Verlag, Berlin, 354–363.
9. Donoho, D.L., Johnstone, I.M., Kerkyacharian, G. and Picard, D. (1995). Wavelet Shrinkage: Asymptopia? *J.R. Statist. Soc. B*, **57**, 301–369.
10. Barone, P., Guidoni, L., Ragona, R., Viti, V. and Degani, H. (1994). Modified Prony method to resolve and quantify in vivo 31P NMR spectra of tumors. *Journal of Magnetic Resonance, Series B*, **104** 137–146.
11. Viti, V., Ragona, R., Guidoni, L. Barone, P., Furman, E. and Degani, H. (1997). Hormonally induced modulation in the phosphate metabolites of breast cancer: analysis of in vivo 31P MRS signals with a modified Prony method. *Magnetic Resonance in Medicine*, **38**, 285–295.

Discrete Simulation Models for Multidimensional Systems Based on Functional Transformations

Rudolf Rabenstein

Lehrstuhl für Nachrichtentechnik I, Universität Erlangen-Nürnberg, Germany

Abstract

A method for the discrete simulation of continuous multidimensional systems is presented. It is based on a transfer function description of the continuous system, which is obtained by application of appropriate functional transformations for each independent variable. Digital signal processing analog-to-discrete transformations turn the transfer function into a discrete simulation model. It is suitable for computer implementation and provides simulation algorithms, which are competitive with customary numerical methods. A comparison with a commercial software package demonstrates the advantage of the proposed simulation models.

1 Introduction

Multidimensional systems describe relations between signals depending on two or more independent variables. In many physical and technical applications these are the time and space coordinates. Systems of this kind are also called systems with distributed parameters. They are conventionally modelled by partial differential equations. Typical applications are wave propagation or heat and mass transfer.

In order to simulate the behaviour of these continuous multidimensional systems on a digital computer, one has to turn the partial differential equation description into suitable discrete models containing shift and delay operations instead of derivatives. Traditionally, these models are created by replacing differential operators by difference operators (finite difference methods) or by solving a variational problem on a set of finite elements (finite element methods). Recently, also discrete models based on signal processing methods have been proposed, like multidimensional wave-digital filters [1, 2, 4, 5, 6, 15] and functional transformation methods [9, 10, 11, 12, 13]. The discrete simulation models based on functional transformations are the subject of this contribution. The procedure for their derivation is briefly described by two key points:

- employ functional transformations for the independent variables to describe the continuous system by multidimensional transfer functions,
- derive from these transfer functions a discrete model in the form of a multidimensional difference equation suitable for computer implementation.

These steps will be discussed in detail using a parabolic and a hyperbolic system as examples. A comparison with a commercial software package shows the reduced numerical expense of the proposed discrete simulation models.

2 Partial differential equation description of multidimensional systems

We consider multidimensional systems with a partial differential equation description of the form

$$\begin{aligned} D\{y(\mathbf{x},t)\} + L\{y(\mathbf{x},t)\} &= v(\mathbf{x},t), \quad \mathbf{x} \in V\,, \\ \mathbf{y}(\mathbf{x},0) &= \mathbf{y}_i(\mathbf{x}), \quad \mathbf{x} \in V\,, \\ f_b\{y(\mathbf{x},t)\} &= \phi(\mathbf{x},t), \quad \mathbf{x} \in S\,, \end{aligned} \tag{2.1}$$

with the time variable t, the vector $\mathbf{x}$ of space variables in the domain V, the linear operator D of time derivatives, the self-adjoint operator L of spatial derivatives, and the boundary operator f_b defined on the surface S. f_b may describe boundary conditions of the first, second, or third kind. The excitation function $v(\mathbf{x},t)$, the vector of initial values $\mathbf{y}_i(\mathbf{x})$, and the boundary values $\phi(\mathbf{x},t)$ are assumed to be known.

This general form of a multidimensional description comprises many special cases of practical relevance, as illustrated by two examples.

2.1 Example: Parabolic problem

For heat flow and diffusion problems, (2.1) specializes to a parabolic partial differential equation with the operators for time and space derivation

$$D\{y(\mathbf{x},t)\} = \dot{y}(\mathbf{x},t)\,, \tag{2.2}$$

$$L\{y(\mathbf{x},t)\} = -\frac{1}{c(\mathbf{x})}\left(\operatorname{div}(\lambda(\mathbf{x})\operatorname{grad} y(\mathbf{x},t))\right) \tag{2.3}$$

where c and λ are material constants. The general form of the initial and boundary conditions is in this case

$$\mathbf{y}(\mathbf{x},0) = y(\mathbf{x},0) = y_i(\mathbf{x}) \tag{2.4}$$

$$f_b\,\{y(\mathbf{x},t)\} = p(\mathbf{x})\,y(\mathbf{x},t) + q(\mathbf{x})\lambda(\mathbf{x})\frac{\partial}{\partial \mathbf{n}}y(\mathbf{x},t) \tag{2.5}$$

The quantities p and q describe general boundary conditions of the third kind.

2.2 Example: Hyperbolic problem

The voltage distribution for an electrical transmission line is given by a hyperbolic equation. It follows from (2.1) for a onedimensional space variable x along the line and with the derivation operators

$$D\{y(x,t)\} = lc\,\ddot{y}(x,t) + (lg + rc)\dot{y}(x,t) + rg\,y(x,t)\,, \tag{2.6}$$

$$L\{y(x,t)\} = -y''(x,t)\,. \tag{2.7}$$

The electrical properties of the transmission line are given in terms of l, c, r, and g. The initial conditions are

$$\mathbf{y}(x,0) = \begin{bmatrix} y(x,0) \\ \dot{y}(x,0) \end{bmatrix} = \begin{bmatrix} y_{i0}(x) \\ y_{i1}(x) \end{bmatrix} \tag{2.8}$$

while the boundary conditions depend on the termination of the line at the coordinates x_0 and x_1. For boundary conditions of the first kind, they are given by

$$y(x_0,t) = \phi(t), \qquad y(x_1,t) = 0\,. \tag{2.9}$$

3 Transfer function description of multidimensional systems

A necessary requirement for the transfer function description of multidimensional systems is the choice of suitable functional transformations for the time and space variables. Their task is to remove the differentiation operators D and L in Equation (2.1) and to replace them by expressions in terms of the frequency variables. This turns the partial differential Equation (2.1) into an algebraic equation, from which a transfer function description can be derived. How can we find transformations with these properties?

It is well known, that the Laplace transformation turns linear ordinary differential equations with constant coefficients into algebraic equations. The key to this property is its differentiation theorem. Applying the Laplace transformation with respect to time to a partial differential equation of the form (2.1) results in a boundary value problem for the space variable. Now we need a transformation with respect to the space variable, which turns the boundary value problem into an algebraic equation. It is shown in [3, 12] that this transformation for the space variable is given in terms of a Sturm-Liouville transformation. Its transformation kernel follows from a Sturm-Liouville problem derived from the partial differential equation of the multidimensional system. The resulting transfer functions are of first order with respect to time for parabolic problems and of second order for hyperbolic problems. Details of this procedure are presented below.

3.1 Example: Parabolic problem

The Laplace transformation and its differentiation theorem are given by

$$\mathcal{L}\{y(\mathbf{x},t)\} = Y(\mathbf{x},s) = \int_0^\infty y(\mathbf{x},t)\mathrm{e}^{-st}\,dt, \qquad (3.1)$$

$$\mathcal{L}\{\dot{y}(\mathbf{x},t)\} = sY(\mathbf{x},s) - y(\mathbf{x},0)\,. \qquad (3.2)$$

Its application to the parabolic problem (2.2) and (2.4) results in the boundary value problem

$$\begin{aligned} sY(\mathbf{x},s) + L\{Y(\mathbf{x},s)\} &= V(\mathbf{x},s) + y_i(\mathbf{x}), & \mathbf{x} \in V \\ f_b\{Y(\mathbf{x},s)\} &= \Phi(\mathbf{x},s), & \mathbf{x} \in S\,. \end{aligned} \qquad (3.3)$$

The transformation with respect to the space variable is given by the Sturm-Liouville transformation [3, 9, 11]

$$\mathcal{T}\{Y(\mathbf{x},s)\} = \int_V c(\mathbf{x})Y(\mathbf{x},s)K(\mathbf{x},\beta)dV\,. \qquad (3.4)$$

with $c(\mathbf{x})$ from (2.2). The transformation kernel $K(\mathbf{x},\beta)$ is the solution of the Sturm-Liouville problem

$$L\{K(\mathbf{x},\beta_\mu)\} = \beta_\mu^2 K(\mathbf{x},\beta_\mu)\,, \quad f_b\{K(\mathbf{x},\beta_\mu)\} = 0 \qquad (3.5)$$

with the discrete eigenvalues β_μ^2 and the associated eigenfunctions $K(\mathbf{x},\beta_\mu)$. This choice of the transformation kernel yields a differentiation theorem for the operator L according to (2.2) which is analogous to the differentiation theorem (3.2) of the Laplace-transformation. In order to derive this differentiation theorem, we rewrite

$$\mathcal{T}\{L\{Y(\mathbf{x},s)\}\} = \int_V c(\mathbf{x})L\{Y(\mathbf{x},s)\}K(\mathbf{x},\beta)\,dV \qquad (3.6)$$

with the help of Green's formula and obtain (arguments are omitted and derivation is denoted by a prime to simplify the notation)

$$\mathcal{T}\{L\{Y\}\} = \int_V cL\{Y\}K\,dV = \int_V cY L\{K\}dV - \int_S \lambda[KY' - YK']\,dS \qquad (3.7)$$

For the volume integral follows from Equations (3.4) and (3.5)

$$\int_V cY L\{K\}dV = \beta_\mu^2 \mathcal{T}\{Y\} \qquad (3.8)$$

The surface integral can be expressed by the boundary values Φ from (3.2). This is shown separately for boundary conditions of the first kind ($p(\mathbf{x}) = 1,\ q(\mathbf{x}) = 0$)

and of the second or third kind ($q(\mathbf{x}) \neq 0$). For boundary conditions of the first kind we write with (3.3) and (3.5) (see [14])

$$\lambda[KY' - YK'] = \begin{vmatrix} f_b\{K\} & \lambda K' \\ f_b\{Y\} & \lambda Y' \end{vmatrix} = -\lambda K' \Phi \tag{3.9}$$

and for the second and third kind

$$\lambda[KY' - YK'] = \frac{1}{q} \begin{vmatrix} K & f_b\{K\} \\ Y & f_b\{Y\} \end{vmatrix} = \frac{1}{q} K \Phi \,. \tag{3.10}$$

We can combine these results into a function $\bar{G}_b(\mathbf{x}, \beta_\mu)$ (returning to full notation)

$$\bar{G}_b(\mathbf{x}, \beta_\mu) = \begin{cases} -\lambda(x) \dfrac{\partial}{\partial \mathbf{n}} K(\mathbf{x}, \beta_\mu) & \text{first kind} \\ \\ \dfrac{1}{q(x)} K(\mathbf{x}, \beta_\mu) & \text{second and third kind} \end{cases} \tag{3.11}$$

and express the surface integral in Equation (3.7) by

$$\bar{\Phi}_b(\beta_\mu, s) = \int\limits_S \bar{G}_b(\mathbf{x}, \beta_\mu) \Phi(\mathbf{x}, s) \, dS \,. \tag{3.12}$$

The differentiation theorem of the Sturm-Liouville transformation $\mathcal{T}$ for the operator L follows from Equations (3.7), (3.8) and (3.12)

$$\mathcal{T}\{L\{Y(\mathbf{x}, s)\}\} = \beta_\mu^2 \bar{Y}(\beta_\mu, s) - \bar{\Phi}_b(\beta_\mu, s) \,. \tag{3.13}$$

Note the formal similarity between the differentiation theorems of the Laplace-transformation $\mathcal{L}$ (3.2) and of the Sturm-Liouville transformation $\mathcal{T}$ (3.13). In both cases, the transform of a derived quantity is obtained from the transform itself multiplied by the frequency variable and an expression which follows directly from the initial or boundary values, respectively.

Application of $\mathcal{T}$ to the boundary value problem (3.3) gives an algebraic equation for the transform of the solution $\bar{Y}(\beta_\mu, s) = \mathcal{T}\{\mathcal{L}\{y(\mathbf{x}, t)\}\}$

$$\bar{Y}(\beta_\mu, s) = \bar{G}(\beta_\mu, s) \left[\bar{V}(\beta_\mu, s) + \bar{y}_i(\beta_\mu) + \bar{\Phi}_b(\beta_\mu, s) \right] \,. \tag{3.14}$$

It is a frequency domain input-output description of a multidimensional system. The input quantities $\bar{V}(\beta_\mu, s)$, $\bar{y}_i(\beta_\mu)$, $\bar{\Phi}_b(\beta_\mu, s)$ are the transforms of the right-hand side signals of the partial differential Equation (2.1). The output $\bar{Y}(\beta_\mu, s)$ is the transform of the solution $y(\mathbf{x}, t)$ of (2.1). The physical behaviour of the multidimensional system is described by the transfer function

$$\bar{G}(\beta_\mu, s) = \frac{1}{s + \beta_\mu^2} \,. \tag{3.15}$$

in terms of the frequency variables s for the time coordinate and β_μ for the space coordinate.

3.2 Example: Hyperbolic problem

The boundary value problem for the hyperbolic example problem follows from (2.1) with (2.6) and vanishing initial conditions by Laplace transformation

$$\gamma^2(s)Y(x,s) - Y''(x,s) = V(x,s)\,, \tag{3.16}$$

$$Y(x_0,s) = \Phi(s) \qquad Y(x_1,s) = 0 \tag{3.17}$$

with the propagation function of the transmission line

$$\gamma(s) = \sqrt{(sl+r)(sc+g)}\,. \tag{3.18}$$

The Sturm-Liouville transformation for this problem has the form

$$\mathcal{T}\{Y(x,s)\} = \int_{x_0}^{x_1} Y(x,s)K(x,\beta)\,dx\,. \tag{3.19}$$

The transformation kernel $K(x,\beta)$ follows from the Sturm-Liouville problem according to (3.16)

$$-K''(x,\beta_\mu) = \beta_\mu^2 K(x,\beta_\mu), \qquad K(x_0,\beta_\mu) = 0, \qquad K(x_1,\beta_\mu) = 0 \tag{3.20}$$

with the closed form solution

$$K(x,\beta_\mu) = \sin\beta_\mu(x-x_0), \qquad \beta_\mu = \pi\frac{\mu-1}{x_1-x_0}\,,\ \mu = 1,2,3,\ldots \tag{3.21}$$

such that the spatial transformation (3.19) is a finite sine transformation [14]. Integration by parts gives the differentiation theorem

$$\mathcal{T}\{L\{Y(x,s)\}\} = \beta_\mu^2\bar{Y}(x,\beta_\mu) - \beta_\mu\Phi(s)\,. \tag{3.22}$$

Application of this transformation to the boundary value problem (3.16) and (3.17) yields the transfer function desription of this system

$$\bar{Y}(\beta_\mu,s) = \bar{G}_e(\beta_\mu,s)\bar{V}(\beta_\mu,s) + \bar{G}_b(\beta_\mu,s)\Phi(s) \tag{3.23}$$

with the transfer functions $\bar{G}_e$ for the excitation $v(x,t)$ and $\bar{G}_b$ for the boundary value $\phi(x,t)$

$$\bar{G}_e(\beta_\mu,s) = \frac{1}{\gamma^2(s)+\beta_\mu^2}\,, \qquad \bar{G}_b(\beta_\mu,s) = \frac{\beta_\mu}{\gamma^2(s)+\beta_\mu^2}\,. \tag{3.24}$$

3.3 Block diagram of a multidimensional system

The examples show, that multidimensional systems with a partial differential equation description in the form of (2.1) can also be represented by transfer functions in the temporal and spatial frequency domain. This transfer function description allows to separate the effects of excitation, initial, and boundary values by regarding them as separate inputs and assigning to them the transfer functions $\bar{G}_e(\beta_\mu,s)$, $\bar{G}_i(\beta_\mu,s)$, and $\bar{G}_b(\beta_\mu,s)$ as shown in Figure 1.

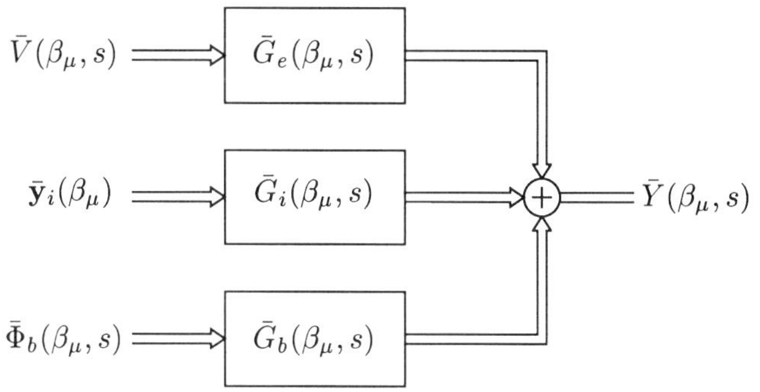

Figure 1. Transfer function model of a multidimensional system in the frequency domain

4 Discrete simulation models

From the transfer function description follows a discrete simulation model by time and space discretization:

- Time discretization turns the system with continuous time and space coordinates into a discrete-time, continuous-space system (*hybrid system*). Well-known analog-to-discrete transformations from onedimensional signal processing like impulse, step, ramp invariant or bilinear transformation can be applied [7].

- Space discretization turns the hybrid system into a discrete-time, discrete-space system or simply a *discrete system*. The discrete system is derived either with useful convolution properties of the Sturm-Liouville transformation or by performing the inverse transformation numerically.

The resulting discrete models are well suited for computer implementation since they require only addition, multiplication and delay elements and are free of implicit loops. Details of the construction of these discrete simulation models are given in [10, 11, 13].

4.1 Example: Parabolic problem

The application of time and space discretization to the parabolic problem according to the aforementioned principles is shown in [11]. We give here the details for a boundary value problem which results from (2.1) and (2.4) by the simplifications

$$v(\mathbf{x}, t) = 0, \quad y_i(\mathbf{x}) = 0, \quad \phi(\mathbf{x}, t) = \phi_1(\mathbf{x})\phi_2(t)\,. \tag{4.1}$$

The transfer function description of this case is

$$\bar{Y}(\beta_\mu, s) = \bar{G}(\beta_\mu, s)\bar{\phi}_{b1}(\beta_\mu)\Phi_2(s) \quad \text{with} \quad \bar{\phi}_{b1}(\beta_\mu) = \int_S \bar{G}_b(\mathbf{x}, \beta_\mu)\phi_1(\mathbf{x})\, dS\,. \tag{4.2}$$

Time discretization is carried out by the time response invariant transformation of the transfer function $\bar{G}(\beta_\mu, s)$ according to [7]

$$\bar{G}^h(\beta_\mu, z) = \frac{\mathcal{Z}\left\{\mathcal{L}^{-1}\left\{V(s)\,\bar{G}(\beta_\mu, s)\right\}\big|_{t=kT}\right\}}{\mathcal{Z}\left\{\mathcal{L}^{-1}\left\{V(s)\right\}\big|_{t=kT}\right\}}\,. \tag{4.3}$$

$\mathcal{Z}$ denotes z-transformation of a sequence, which is obtained by sampling a time signal at $t = kT$, where T is the sampling interval and k is the discrete time variable. With $V(s) = 1/s$ results the transfer function of the hybrid system according to the step invariant transformation

$$\bar{G}^h(\beta_\mu, z) = \frac{1}{\beta_\mu}\frac{1 - \bar{c}(\beta_\mu)}{z - \bar{c}(\beta_\mu)} \quad \text{with} \quad \bar{c}(\beta_\mu) = \exp(-\beta_\mu^2 T)\,. \tag{4.4}$$

The transfer function description of the hybrid system (denoted by the superscript h) is now

$$\bar{Y}^h(\beta_\mu, z) = \bar{G}^h(\beta_\mu, z)\bar{\phi}_{b1}(\beta_\mu)\Phi_2^d(z) = \frac{\bar{b}(\beta_\mu)}{z - \bar{c}(\beta_\mu)}\Phi_2^d(z) \tag{4.5}$$

with

$$\bar{b}(\beta_\mu) = \frac{1 - \bar{c}(\beta_\mu)}{\beta_\mu}\bar{\phi}_{b1}(\beta_\mu) \tag{4.6}$$

and the z-transform of the sampled boundary value signal $\Phi_2^d(z) = \mathcal{Z}\{\phi_2(kT)\}$. Reordering of the terms and inverse z-transformation $\bar{y}^h(\beta_\mu, k) = \mathcal{Z}^{-1}\{\bar{Y}^h(\beta_\mu, z)\}$ gives the description of the hybrid system in form of a difference equation

$$\bar{y}^h(\beta_\mu, k) = \bar{c}(\beta_\mu)\bar{y}^h(\beta_\mu, k-1) + \bar{b}(\beta_\mu)\phi_2(kT)\,. \tag{4.7}$$

The last step in order to obtain the output signal $y^h(\mathbf{x}, k)$ is the inverse Sturm-Liouville transformation $\mathcal{T}^{-1}$. It is given by the orthogonal eigenfunction expansion

$$\mathcal{T}^{-1}\{\bar{y}^h(\beta_\mu, k)\} = y^h(\mathbf{x}, k) = \sum_{\mu=1}^{\infty}\frac{1}{N_\mu}\bar{y}^h(\beta_\mu, k)\,K(\mathbf{x}, \beta_\mu)\,. \tag{4.8}$$

N_μ is the squared L_2-norm of the eigenfunctions. The series representation (4.8) can be used for a numerical evaluation. Space discretization is most simply performed by evaluating $y^h(\mathbf{x}_n, k)$ from (4.8) only at points $\mathbf{x}_n$ of interest. Possible convergence problems in the evaluation of an infinite series can be avoided by measures described in [11].

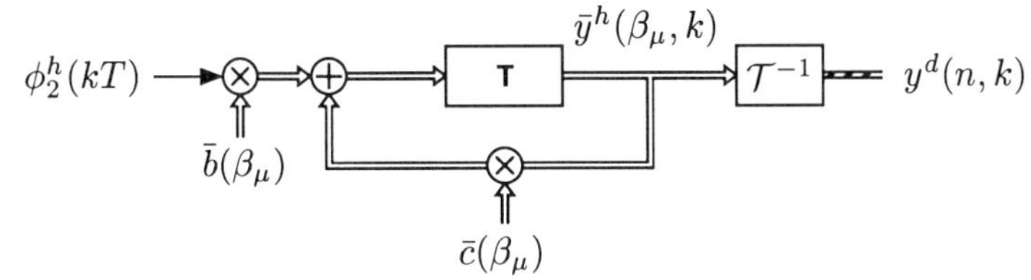

Figure 2. Discrete model for the parabolic problem

The structure of the discrete simulation model is shown in Figure 2. It consists of a first order recursive system with respect to time for the calculation of the hybrid solution $\bar{y}^h(\beta_\mu, k)$ and of an inverse Sturm-Liouville transformation evaluated at the discrete spatial points $\mathbf{x}_n$. The empty double lines indicate a parallel arrangement of all discrete spatial frequency coefficients. The hatched double lines denote the disrete-time, discrete-space solution $y^d(n,k) = y^h(\mathbf{x}_n, k)$.

4.2 Example: Hyperbolic problem

The discrete model for the hyperbolic system with zero excitation is shown in Figure 3. Instead of performing the inverse Sturm-Liouville transformation numerically, the convolution properties of the finite sine transformation (3.21) have been exploited [14]. It can be expressed in terms of a Discrete Fourier Transformation (DFT), performed on equidistant spatial samples x_n of x, where $n = 0, \ldots, N$ is the discrete space index. The coefficients in the feedback loop depend on the polynomial $\gamma^2(s)$ in (3.16). c_0 is a constant and $c_1(n)$ is a spatial sequence representing the effects of wave propagation and dispersion through a circular convolution of length $2N$. The derivation of these sequences is presented in detail in [13].

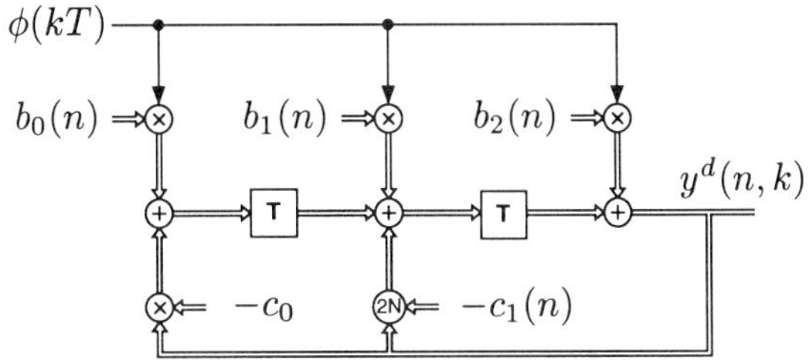

Figure 3. Discrete model for the hyperbolic problem

5 Numerical Results

A twodimensional heat flow problem was chosen in order to compare the efficiency of the described discrete simulation models with conventional methods for the numerical solution of partial differential equations. It is a demonstration example from a commercial numerical mathematics program [8]. Figure 4 shows a rectangular domain with two different heat conducting materials. The material within the diamond is heated by an internal source (for example electrical current), while the material outside of the diamond is free of heat sources. The domain is surrounded by a well-stirred fluid with spatially constant temperature, which varies sinusoidally with time. A sample solution for a certain point in time is shown in Figure 4 (top view) and Figure 5 (3D view). All coordinates are in normalized units.

The resulting temperature variations within the domain were caluclated for certain ranges of the normalized simulation time t. The methods used were a Finite Element Method (FEM) from [8] and the discrete simulation model according to Figure 2 with proper measures to avoid slow convergence of the inverse Sturm-Liouville transformation (Functional Transformation Method). The comparison of the required computer time for these two methods in Figure 6 shows a clear advantage of the presented discrete simulation models derived by functional transformations.

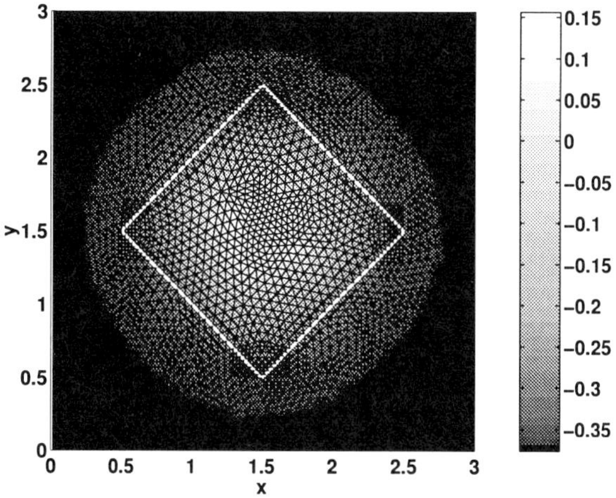

Figure 4. Problem geometry

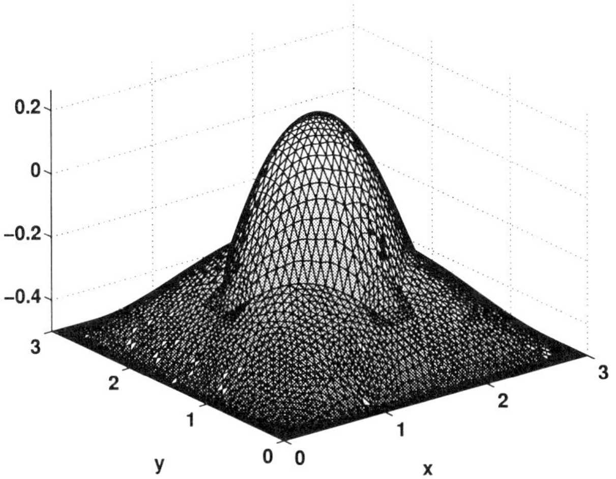

Figure 5. Sample solution for parabolic problem

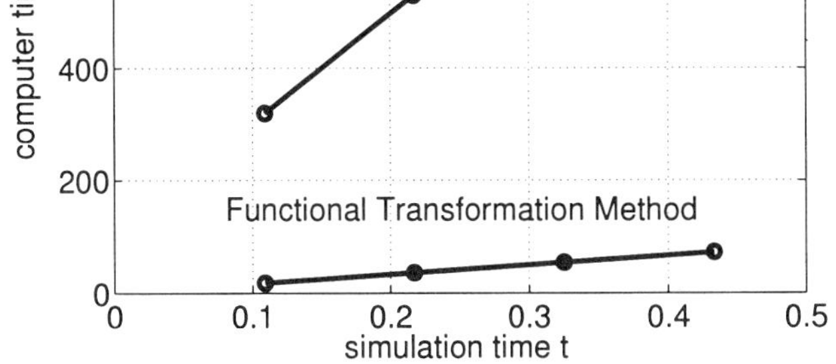

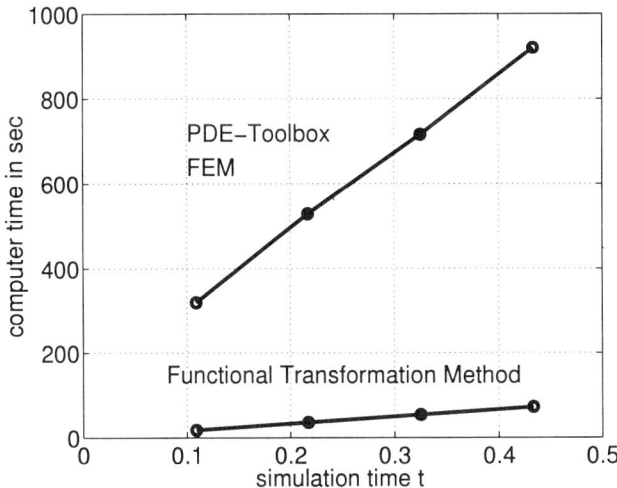

Figure 6. Comparison of computer time for parabolic problem: functional transformation method and finite element method (FEM) from MATLAB PDE-Toolbox

6 Conclusion

Discrete simulation models for continuous multidimensional systems can be constructed with signal processing methods. Appropriate functional transformations for the independent variables turn a partial differential equation into a transfer function description. Standard analog-to-discrete transformations and orthogonal series expansions complete the construction of the discrete models. They consist of adders, multipliers, and scalar products and they do not contain any delay free loops, which makes these structures suitable for direct processor implementation. The resulting simulation algorithms are well competitive with conventional numerical methods. For stable continuous systems, also the discrete simulation models derived here will be stable.

Bibliography

1. R. Bernhardt and D. Dahlhaus. Numerical integration of the Euler equations by means of wave digital filters. In *Proc. Int. Conf. Acoustics, Speech, and Signal Processing (ICASSP 94)*, pages VI-1 – VI-4. IEEE, April 1994.
2. M. Bolle. Wave digital filters for the migration of seismic data. In *Proc. Int. Conf. Acoustics, Speech, and Signal Processing (ICASSP 94)*, pages VI-5 – VI-8. IEEE, April 1994.
3. R.M. Cotta. *Integral Transforms in Computational Heat and Fluid Flow*. CRC Press, Boca Raton, 1993.
4. A. Fettweis. Multidimensional wave digital filters for discrete-time modelling of maxwell's equations. *Int. Journal of Numerical Modelling, Electronic Networks, Devices, and Fields*, 5:183–201, 1992.
5. A. Fettweis. Multidimensional wave-digital principles: From filtering to numerical integration. In *Proc. Int. Conf. Acoustics, Speech, and Signal Processing (ICASSP 94)*, pages VI-173 – VI-181. IEEE, April 1994.
6. A. Fettweis and G. Nitsche. Numerical integration of partial differential equations using principles of multidimensional wave digital filters. *Journal of VLSI Signal processing*, 3:7–24, 1991.
7. Z. Kowalczuk. Discrete approximation of continuous-time systems: a survey. *IEE Proceedings-G*, 140(4):264–278, Aug. 1993.
8. Partial Differential Equation Toolbox For Use with MATLAB. The MathWorks, Inc., Natick, Mass., USA, Aug. 1995.
9. R. Rabenstein. Discrete simulation of dynamical boundary value problems. In F. Breitenecker and I. Husinsky, editors, *Proc. of the 1995 EUROSIM Conference (EUROSIM'95)*, pages 177–182, Amsterdam, 1995. Elsevier.
10. R. Rabenstein. Discrete models for multidimensional system simulation. In G. Ramponi et al., editor, *Proc. of VIII European Signal Proc. Conf. (EUSIPCO 96)*, pages 2125–2128. EURASIP, Sept. 1996.
11. R. Rabenstein. Multidimensional system simulation with functional transformations. In *Proc. Int. Conf. Acoustics, Speech, and Signal Proc. (ICASSP 96)*, pages 2377–2380. IEEE, 1996.

12. R. Rabenstein. Transfer function models for multidimensional systems. In I. Troch and F. Breitenecker, editors, *Proc. IMACS Symposium on Mathematical Modelling*, pages 1047–1052, 1997.

13. R. Rabenstein and H. Krauß. Discrete simulation of uniform transmission lines by multidimensional digital filters. *Int. Journal of Numerical Modelling, Electronic Networks, Devices, and Fields*, 9:271–294, 1996.

14. I.N. Sneddon. *The Use of Integral Transforms.* Tata McGraw-Hill, New Delhi, 1974.

15. T. Utsunomiya and A. Fettweis. Discrete modelling of plasma equations with ion motion using technique of wave digital filters. In *Proc. Int. Conf. Acoustics, Speech, and Signal Processing (ICASSP 94)*, pages VI–21 - VI–24. IEEE, April 1994.

Analytic Performance Evaluation of Third-Order Statistics based Estimators for Allpass Identification

Thomas Kaiser[1]

Department of Communication Engineering, Gerhard-Mercator-Universität Duisburg, Germany

Abstract

In the past decade there has been an increasing interest in identifying the parameters of a non-minimum phase system using higher-order statistics, where only a realization $x(k)$, $k = 0(1)N-1$ of the output sequence $X(k)$ is known. In contrast, a realization $z(k)$ of the input sequence $Z(k)$, which has some special properties described later, is not available. This model is applied in many different signal processing areas, for example in seismic signal processing, in speech processing and also in blind equalization (deconvolution). Since a non-minimum phase system can be divided in a series connection of a minimum phase system and an allpass system, the parameters of the minimum phase part can be estimated with well-known algorithms based on second-order statistics, whereas for the allpass part higher-order statistics are useful.

In this article we consider the problem of estimating the third-order moment γ_Z of the input sequence $Z(k)$ of an allpass system since γ_Z is an important parameter for allpass identification. In the allpass case, we are faced with some unusual facts. For example, it is quite amazing that we can estimate γ_Z by

$$\hat{\gamma}_Z = \sum_{n=-L}^{L} \sum_{m=-L}^{L} \hat{c}_X(n,m)$$

$$\text{or} \quad \hat{\gamma}_Z = \sum_{n=-L}^{L} \hat{c}_X(n,n)$$

$$\text{or} \quad |\hat{\gamma}_Z| = \sqrt{\sum_{n=-L}^{L} \sum_{m=-L}^{L} \hat{c}_X^2(n,m)},$$

where $\hat{c}_X(n,m)$ is an estimator of the third-order moment sequence of the output sequence $X(k)$. We must carefully investigate these estimators,

[1]This work was supported by a scholarship of the NATO Scientific Board and co-ordinated by the Deutscher Akademischer Austauschdienst.

since some of them use the bispectrum of $X(k)$ on the boundaries of its principal region, which exhibits a much larger variance than the values inside the principal region.

Keywords: Higher-order statistics, Allpass identification, Non-minimum phase systems, Performance evaluation

1 Introduction

Recently, Chi [1] has proposed a new algorithm for allpass identification based on higher-order statistics. Of course, various cumulant based Least Squares (LS) methods (see for example Giannakis and Mendel [3] or Swami and Mendel [8]) can be used for estimating the Auto-Regressive (AR) parameters (and therefore all allpass parameters, too) by treating the allpass model as a general Auto-Regressive-Moving-Average (ARMA) model. Chi stated that his algorithm outperforms cumulant based LS methods simply due to the more accurate model. However, the new algorithm is based on nonlinear optimization and therefore global convergence to the true parameters cannot be guaranteed.

Fortunately, the maximum of the objective function is equal to the third-order input moment γ_Z, which can be estimated by the known realization $x(k)$ *without* knowledge of the allpass parameters. So we can use this estimate to check the global convergence. For example, if the global maximum is not reached, we can adjust the starting values. As indicated in the abstract, there are many ways to estimate γ_Z. Therefore this article deals with an analytic performance evaluation about the estimation of γ_Z.

After the necessary definitions in Section 2, we derive some estimators and their theoretical properties. Note that although the proposed new algorithm is suitable for any higher-order moment (more precisely *cumulant* instead of *moment*), we restrict ourselves to third-order moment for reasons of simplicity.

2 Definitions

Consider a linear shift-invariant stable allpass system with a real third-order stationary *input* random sequence $Z(k)$ and an *output* random sequence $X(k)$ described by the difference equation

$$X(k) = \sum_{\kappa=-\infty}^{\infty} h(\kappa) Z(k-\kappa),$$

with impulse response $h(k)$ and transfer function

$$H(z) = \sum_{k=-\infty}^{\infty} h(k)\, z^{-k} = \frac{A(z^{-1})z^{-p}}{A(z)}$$

where $A(z) = 1 + a_1 z^{-1} + ... + a_p z^{-p}$. The input sequence $Z(k)$ is non-Gaussian distributed, zero mean with second- and third-order moment sequence

$$\begin{aligned} r_Z(n) &= \sigma_Z^2 \delta(n), \\ c_Z(n,m) &= \gamma_Z \delta(n,m), \quad \gamma_Z \neq 0, \end{aligned}$$

respectively, where γ_Z is the third-order moment of the *random variable* $Z(k)$, and $\delta(n)$ and $\delta(n,m)$ denote the 1-dimensional and 2-dimensional Kronecker delta, respectively. The bispectrum of the output sequence $X(k)$ is given by

$$B_X(\mathrm{e}^{\mathrm{j}\Omega_1}, \mathrm{e}^{\mathrm{j}\Omega_2}) = \gamma_Z \, H(\mathrm{e}^{\mathrm{j}\Omega_1}) H(\mathrm{e}^{\mathrm{j}\Omega_2}) H(\mathrm{e}^{-\mathrm{j}(\Omega_1+\Omega_2)}), \tag{2.1}$$

whereas the power spectrum $S_X(\mathrm{e}^{\mathrm{j}\Omega})$ is independent of the frequency Ω in the allpass case. Since the absolute value of the transfer function is equal to one we find that

$$|\gamma_Z| = |B_X(\mathrm{e}^{\mathrm{j}\Omega_1}, \mathrm{e}^{\mathrm{j}\Omega_2})|. \tag{2.2}$$

Moreover, if $A(1) \neq 0$ an evaluation of Equation (2.1) for some pairs of frequencies leads to

$$\begin{aligned} \gamma_Z &= B_X(\mathrm{e}^{\mathrm{j}\Omega}, \mathrm{e}^{-\mathrm{j}\Omega}) \\ &= B_X(\mathrm{e}^{\mathrm{j}\Omega}, 1). \end{aligned}$$

This means that not only the absolute value of the bispectrum is constant in the allpass case for all $\Omega_1 \in \mathbf{R}$ and $\Omega_2 \in \mathbf{R}$, but also the phase is constant for $\Omega_1 = -\Omega_2$ (and due to the symmetry properties (see Nikias and Petropulu [7], p. 23) for $\Omega_1 = \Omega$, $\Omega_2 = 0$, etc., too). Based on these two equations a large variety of estimators for γ_Z can be given by using well-known non-parametric bispectrum estimation methods (parametric methods require estimation of the allpass parameters *prior* to the estimation of γ_Z which is in contradiction to the object of this article).

In the following we will consider the *indirect* bispectrum estimator

$$\hat{B}_X(\mathrm{e}^{\mathrm{j}\Omega_1}, \mathrm{e}^{\mathrm{j}\Omega_2}) = \sum_{n=-L}^{L} \sum_{m=-L}^{L} w(\frac{n}{L}, \frac{m}{L}) \, \hat{c}_X(n,m) \, \mathrm{e}^{-\mathrm{j}(\Omega_1 n + \Omega_2 m)}, \tag{2.3}$$

where $\hat{c}_X(n,m)$ is an estimator of the third-order moment sequence of the random sequence $X(k)$ (see Nikias and Mendel [6], p. 20), $L = N^\alpha$, $0 < \alpha < 0.5$ (see Nikias and Petropulu [7], p. 125) and $w(n,m)$ is a two-dimensional window which should satisfy some constraints (see Van Ness [9], p. 1260). Note that $w(n,m)$ is defined for $n, m \in \mathbf{R}$ instead of $n, m \in \mathbf{Z}$. The reciprocal of L is usually referred as the bandwidth Δ. The indirect bispectrum estimator is asymptotically unbiased (see Van Ness [9], p. 1262) with variance

$$\operatorname{var}\left\{\operatorname{Re}\{\hat{B}_X(e^{j\Omega_1},e^{j\Omega_2})\}\right\} = \frac{L^2}{N} S_X(e^{j\Omega_1})\, S_X(e^{j\Omega_2})\, S_X(e^{j(\Omega_1+\Omega_2)})$$

$$\Big[V_1(8\delta(\Omega_1)+\delta(\Omega_2)) + V_2(9\delta(\Omega_1)+\delta(\Omega_2)[1+\delta(\Omega_1-\pi)]$$

$$+[1+\delta(\Omega_1-\Omega_2)]\,[1+\delta(\Omega_1+2\Omega_2-2\pi)+\delta(2\Omega_1+\Omega_2-2\pi)])\Big] \quad (2.4)$$

for $0 \le \Omega_1 \le \pi$, $0 \le \Omega_2 \le \Omega_1$, $2\Omega_1+\Omega_2 \le 2\pi$ and $\operatorname{var}\left\{\operatorname{Im}\{\hat{B}_X(e^{j\Omega_1},e^{j\Omega_2})\}\right\}$ is similiar to Equation (2.4) (see also Lii and Rosenblatt [5]). Here we have used the abbreviations

$$V_1 = \left(\int_{-\infty}^{\infty} w(0,m)\, dm\right)^2$$

$$V_2 = \frac{1}{2}\int_{-\infty}^{\infty}\int_{-\infty}^{\infty} w^2(n,m)\, dn\, dm.$$

Furthermore, $\hat{B}_X(e^{j\Omega_1},e^{j\Omega_2})$ is asymptotically complex Gaussian distributed with independent real- and imaginary part.

Equation (2.4) shows that the variance on the three boundaries

$$\Omega_2 = 0 \quad \text{and} \quad 0 \le \Omega_1 \le \pi$$

$$\Omega_2 = \Omega_1 \quad \text{and} \quad 0 \le \Omega_1 \le \frac{2\pi}{3}$$

$$2\Omega_1+\Omega_2 = 2\pi \quad \text{and} \quad \frac{2\pi}{3} \le \Omega_1 \le \pi$$

of the principal region (for an explanation of the former notion see Hinich and Wolinsky [4], p. 499-500) is clearly increased compared to the variance inside the principal region. Consequently, the estimators for γ_Z

$$\hat{\gamma}_Z = \operatorname{Re}\left\{\hat{B}_X(e^{j\Omega},e^{-j\Omega})\right\}$$

$$= \operatorname{Re}\left\{\hat{B}_X(e^{j\Omega},1)\right\}$$

must be carefully investigated.

3 Some estimators for γ_Z

Although $\gamma_Z = \mathrm{E}\{Z^3(k)\}$ is independent of Ω the first estimator for γ_Z

$$\hat{\gamma}_Z^{(1)}(\Omega) = \sum_{n=-L}^{L}\sum_{m=-L}^{L} w(\frac{n}{L},\frac{m}{L})\, \hat{c}_X(n,m)\cos(\Omega n) \quad (3.1)$$

is dependent on Ω. $\hat{\gamma}_Z^{(1)}(\Omega)$ is asymptotically unbiased and asymptotically Gaussian distributed with variance

$$\mathrm{var}\left\{\hat{\gamma}_Z^{(1)}(\Omega)\right\} = \frac{L^2}{N}\sigma_X^6\left(V_1[1+8\delta(\Omega)] + V_2[2+10\delta(\Omega)+2\delta(\Omega-\pi)]\right) \quad (3.2)$$

where $\sigma_X^6 = (\mathrm{E}\{X^2(k)\})^3 = \sigma_Z^6$ and $0 \leq \Omega \leq \pi$. Observe that the variance is lowest for $0 < \Omega < \pi$ and highest for $\Omega = 0$.

A different kind of estimator can be derived by using the fact that $\hat{B}_X(\mathrm{e}^{\mathrm{j}\Omega_1}, \mathrm{e}^{\mathrm{j}\Omega_2})$ and $\hat{B}_X(\mathrm{e}^{\mathrm{j}\Omega_3}, \mathrm{e}^{\mathrm{j}\Omega_4})$ can be treated as independent random variables for $\Omega_1 \neq \Omega_3$ or for $\Omega_2 \neq \Omega_4$ over the grid in the principal domain if the grid width is larger than or equal to the bandwidth $\Delta = 1/L$ (Nikias and Petropulu [7], p. 143). Therefore, to reduce the variance of $\hat{\gamma}_Z^{(1)}(\Omega)$, the average of the independent bispectrum values along the line $\Omega_1 = \Omega$ and $\Omega_2 = 0$ can be used. Since the real part of the bispectrum exhibits a large variance for $\Omega = 0$ and $\Omega = \pi$, the following three estimators should be investigated

$$\hat{\gamma}_Z^{(2)} = \frac{1}{L}\sum_{\mu=0}^{L-1} \mathrm{Re}\left\{\hat{B}_X(\mathrm{e}^{\mathrm{j}\frac{2\pi\mu}{L}}, 1)\right\} \quad (3.3)$$

$$\hat{\gamma}_Z^{(3)} = \frac{1}{L-1}\sum_{\mu=1}^{L-1} \mathrm{Re}\left\{\hat{B}_X(\mathrm{e}^{\mathrm{j}\frac{2\pi\mu}{L}}, 1)\right\} \quad (3.4)$$

$$\hat{\gamma}_Z^{(4)} = \frac{1}{L+1}\sum_{\mu=0}^{L} \mathrm{Re}\left\{\hat{B}_X(\mathrm{e}^{\mathrm{j}\frac{2\pi\mu}{L}}, 1)\right\}. \quad (3.5)$$

Focusing on $\hat{\gamma}_Z^{(2)}$ yields the time domain estimator

$$\begin{aligned}\hat{\gamma}_Z^{(2)} &= \sum_{n=-L}^{L}\sum_{m=-L}^{L} w(\frac{n}{L},\frac{m}{L})\,\hat{c}_X(n,m)\frac{1}{L}\sum_{\mu=0}^{L-1}\cos\left(\frac{2\pi\mu n}{L}\right) \\ &= \sum_{n=-L}^{L} w(\frac{n}{L},\frac{n}{L})\,\hat{c}_X(n,n).\end{aligned}$$

By some calculation we find that

$$\begin{aligned}\hat{\gamma}_Z^{(3)} &= \frac{L}{L-1}\hat{\gamma}_Z^{(2)} - \frac{1}{L-1}\hat{\gamma}_Z^{(1)}(0) \\ \hat{\gamma}_Z^{(4)} &= \frac{L}{L+1}\hat{\gamma}_Z^{(2)} + \frac{1}{L+1}\hat{\gamma}_Z^{(1)}(0).\end{aligned}$$

Of course, all these estimators $\hat{\gamma}_Z^{(2,3,4)}$ are asymptotically unbiased and asymptotically Gaussian distributed. By using the symmetry properties of the bispectrum (see Nikias and Petropulu [7], p. 23) the variance of $\hat{\gamma}_Z^{(2)}$ is given by

$$\begin{aligned}\mathrm{var}\left\{\hat{\gamma}_Z^{(2)}\right\} &= \frac{1}{L^2}\left(\mathrm{var}\left\{\mathrm{Re}\{\hat{B}_X(1,1)\}\right\} + 4\sum_{\mu=1}^{L/2-1}\mathrm{var}\left\{\mathrm{Re}\{\hat{B}_X(\mathrm{e}^{\mathrm{j}\frac{2\pi\mu}{L}},1)\}\right\}\right.\\ &\qquad \left.+\mathrm{var}\left\{\mathrm{Re}\{\hat{B}_X(-1,1)\}\right\}\right)\\ &= \frac{\sigma_X^6}{N}\Big(V_1(10+2L)+V_2(16+4L)\Big).\end{aligned}$$

Equivalently, we find that

$$\begin{aligned}\mathrm{var}\left\{\hat{\gamma}_Z^{(3)}\right\} &= \frac{L^2\sigma_X^6}{(L-1)^2N}\Big(V_1(1+2L)+V_2(4+4L)\Big)\\ \mathrm{var}\left\{\hat{\gamma}_Z^{(4)}\right\} &= \frac{L^2\sigma_X^6}{(L+1)^2N}\Big(V_1(37+2L)+V_2(52+4L)\Big).\end{aligned}$$

Now, by choosing a window $w(n,m)$ all these estimators can be easily compared by their variances. However, observe that the ratio $\hat{\gamma}_Z^{(2,3,4)}/\hat{\gamma}_Z^{(1)}(\Omega)$ is proportional to $1/L$ for large L.

Another idea to obtain an estimator for γ_Z is based on Equation (2.2). Squaring on both sides and integrating yields

$$|\gamma_Z|^2 = \frac{1}{4\pi^2}\int_{-\pi}^{\pi}\int_{-\pi}^{\pi}|B_X(\mathrm{e}^{\mathrm{j}\Omega_1},\mathrm{e}^{\mathrm{j}\Omega_2})|^2\mathrm{d}\Omega_1\,\mathrm{d}\Omega_2$$

and applying Parseval's equation we obtain

$$|\gamma_Z|^2 = \sum_{n=-\infty}^{\infty}\sum_{m=-\infty}^{\infty}c_X^2(n,m). \tag{3.6}$$

Hence, a further estimator could be

$$\hat{\gamma}_Z = \pm\sqrt{\sum_{n=-L}^{L}\sum_{m=-L}^{L}c_X^2(n,m)}. \tag{3.7}$$

The calculation of the statistics of this estimator is very tedious due to the symmetry properties of the bispectrum. However, to show the basic procedure we will instead consider the estimator

$$\hat{\gamma}_Z^{(5)} = \pm\sqrt{\frac{1}{R}\sum_{\mu=1}^{L/2-1}\sum_{\substack{\nu=1\\ \nu<L-2\mu}}^{\mu}|\hat{B}_X(\mathrm{e}^{\mathrm{j}2\pi\mu/L},\mathrm{e}^{\mathrm{j}2\pi\nu/L})|^2}, \tag{3.8}$$

where the summation includes the principal region without the boundaries and R is the number of summations ($R \approx L^2/3$). The sign of $\hat{\gamma}_Z^{(5)}$ can be taken from the other estimators. However, to check the global maximum of the objective function, only the absolute value is necessary (see Chi [1], p. 242). Let

$$Z = \sum_{\mu=1}^{L/2-1} \sum_{\substack{\nu=1 \\ \nu<L-2\mu}}^{\mu} \frac{|\hat{B}_X(\mathrm{e}^{\mathrm{j}2\pi\mu/L}, \mathrm{e}^{\mathrm{j}2\pi\nu/L})|^2}{\sigma^2_{\hat{B}_X}} \tag{3.9}$$

where

$$\begin{aligned}
\sigma^2_{\hat{B}_X} &= \mathrm{var}\left\{\mathrm{Re}\{\hat{B}_X(\mathrm{e}^{\mathrm{j}2\pi\mu/L}, \mathrm{e}^{\mathrm{j}2\pi\nu/L})\}\right\} \\
&= \mathrm{var}\left\{\mathrm{Im}\{\hat{B}_X(\mathrm{e}^{\mathrm{j}2\pi\mu/L}, \mathrm{e}^{\mathrm{j}2\pi\nu/L})\}\right\} \\
&= \frac{L^2\sigma_X^6 V_2}{N},
\end{aligned}$$

then Z is asymptotically a noncentral chi-square statistic with density function

$$f_Z(z) = \begin{cases} \mathrm{e}^{-\frac{(z+\beta^2)}{2}} \displaystyle\sum_{\nu=0}^{\infty} \frac{(\beta^2/2)^\nu \, z^{\nu+R-1}}{\nu! \, 2^{\nu+R}\,\Gamma(\nu+R)} & z > 0 \\ 0 & z \le 0 \end{cases}$$

with $2R$ degrees of freedom and noncentrality parameter

$$\beta = \sqrt{\frac{R\gamma_Z^2}{\sigma^2_{\hat{B}_X}}}. \tag{3.10}$$

Finally, the density function of $\hat{\gamma}_Z^{(5)}$ is given by

$$f_{\hat{\gamma}_Z^{(5)}}(x) = \begin{cases} \dfrac{2Rx}{\sigma^2_{\hat{B}_X}} f_Z\left(\dfrac{Rx^2}{\sigma^2_{\hat{B}_X}}\right) & x > 0 \\ 0 & x \le 0 \end{cases} \tag{3.11}$$

with variance

$$\mathrm{var}\left\{\hat{\gamma}_Z^{(5)}\right\} = \frac{2L^2\sigma_X^6 V_2}{N}.$$

Observe that not only the variance of $\hat{\gamma}_Z^{(5)}$ is similiar to $\mathrm{var}\left\{\hat{\gamma}_Z^{(1)}(\Omega)\Big|_{\Omega\neq 0,\pi}\right\}$ but also proportional to L^2 instead of L in case of $\hat{\gamma}_Z^{(2,3,4)}$. Therefore, it is unlikely that the inclusion of the boundaries of the principal region in Equation (3.8) would lead to a better estimator than $\hat{\gamma}_Z^{(2,3,4)}$.

4 Conclusions

In this article many estimators for the maximum of an objective function for allpass identification are analytically investigated. After choosing a window $w(n,m)$, all these estimators can be easily compared by their variances as a function of the data length N, the second-order moment σ_X^2 and the cut-off bound L. Since these variances are independent of the allpass parameters a *model-independent* decision among all estimators is feasible.

The most remarkable result is that $\hat{\gamma}_Z^{(2)}$, which uses *only a slice* of the third-order moment sequence, exhibits in general a lower variance than estimators which use the *whole* third-order moment sequence.

Bibliography

1. C.Y. Chi, J.Y. Kung, "A new identification algorithm for allpass systems by higher-order statistics", *Signal Proc.*, Vol. 41, 1995, pp. 239-256.
2. J.A.R. Fonollosa, "Sample Cumulants of Stationary Processes: Asymptotic Results", *IEEE Transactions on Signal Processing*, Vol. 43, No. 4, April 1995, pp. 967-977.
3. G.B. Giannakis, J.M. Mendel, "Identification of non-minimum phase systems using higher-order statistics", *IEEE Transactions on Acoustics, Speech and Signal Processing*, Vol. 37, No. 3, March 1996, pp. 360-377.
4. M.J. Hinich, M.A. Wolinsky, "A Test for Aliasing Using Bispectral Analysis", *Journal of the American Statistical Association*, Vol. 83, No. 402, June 1988 pp. 499-502.
5. K.S. Lii, M. Rosenblatt, "Cumulant spectral estimates: Bias and Covariance", *Proceedings of 3rd Hungarian Colloquium on Limit Theorems*, Pecs, Hungary, 1989.
6. C.L. Nikias, J.M. Mendel, "Signal Processing with Higher-Order Spectra", *IEEE Signal Processing Magazine*, July 1993, pp. 10-37.
7. C.L. Nikias, A.P. Petropulu, "Higher-Order Spectra Analysis", *Prentice Hall Signal Processing Series*, 1993.
8. A. Swami, J.M. Mendel, "ARMA parameter estimation using only output cumulants", *IEEE Transactions on Acoustics, Speech and Signal Processing*, Vol. 38, No. 7, July 1990, pp. 1257-1265.
9. J.W. Van Ness, "Asymptotic Normality of Bispectral Estimates", *Ann. Math. Stat.*, Vol. 37, 1966, pp. 1257-1272.
10. I.G. Žurbenko, "The Spectral Analysis Of Time Series", *North-Holland Series in Statistics and Probability*, Vol. 2, 1986.

Detection of Sleep Apnoea Using Time–Frequency Analysis of Heart Rate Variability

R.A. Bates*, **M.F. Hilton****, **K.R. Godfrey*** and **M.J. Chappell***

**Department of Engineering, University of Warwick, Coventry, and*
***Department of Respiratory Physiology, Birmingham Heartlands Hospital, Birmingham*

Abstract

Spectral analysis of heart rate variability data requires non-standard treatment as the data are nonequispaced in time. Two methods of analysis are compared using specially generated test signals and extended for use with the Discrete Harmonic Wavelet Transform for time–frequency analysis. Two sets of patient data are then analysed, one normal subject and one with sleep apnoea. The results show the potential of time–frequency analysis in the assessment of disorders such as sleep apnoea which involve alterations in autonomic activity.

1 Introduction

Spectral analysis of heart rate variability (HRV) is a useful non-invasive tool for quantifying the neurohumeral autonomic inputs to the heart [1].

The heart rate of a patient is determined from an electrocardiogram (ECG). The features of each period of the ECG waveform are labelled with the letters PQRS and T (see Figure 1), and the time elapsed between successive R peaks, the RR interval, is taken as a measure of heart rate. The RR intervals are then combined to form a nonequispaced RR signal.

One way of analysing HRV data is to fit a curve to the RR signal using a method of interpolation and then to resample at equally spaced time points to obtain a regular time series. The Discrete Fourier Transform (DFT) can then be used to obtain the power spectrum of the interpolated RR signal. This provides information on the frequency content of the signal with a particular resolution in time and frequency. For a signal with N samples, sampled at a frequency f_s, the frequency resolution is f_s/N and the time resolution is N/f_s. There is therefore a trade-off between time and frequency resolution; increasing the length of the series increases frequency resolution but decreases the ability to locate the computed frequencies in time.

Obstructive sleep apnoea (OSA) is a condition where respiration during sleep is punctuated with repetitive cyclical pauses [2] causing transient alterations in both sympathetic and vagal activity, linked to frequencies in the power spectrum

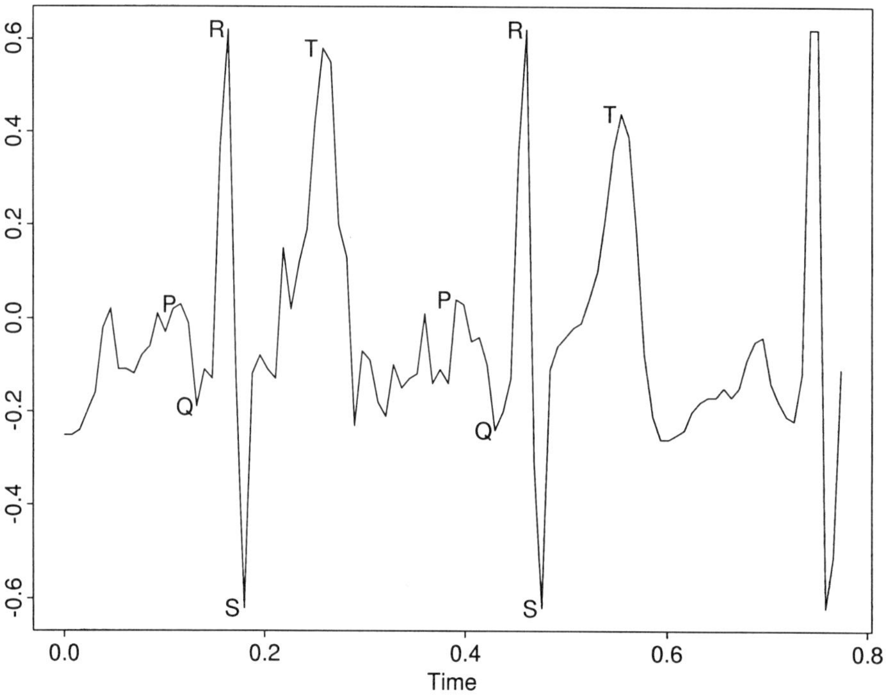

Figure 1. ECG waveform

of the RR signal of 0.04-0.15Hz and 0.15-0.4Hz respectively [3]. These frequency bands are referred to as Low Frequency (LF) and High Frequency (HF) power respectively. As these cycles range between 10 and 90 seconds, calculating the power spectrum of the RR signal, as outlined above, produces Fourier coefficients which represent frequencies over the mean of the whole signal, typically a 5 minute data segment. Thus signal amplitude changes occurring in this 5-minute segment cannot be determined. The ability to do this would provide a useful diagnostic tool, since the cyclical nature of OSA events introduces frequencies to the power spectrum which, using the DFT, are difficult to distinguish from LF power.

The DFT method outlined above is compared with the Nonequispaced Fourier Transform (NeFT), which avoids the use of interpolation methods, in two applications. In the first, it is used to obtain the power spectrum of a test signal and, in the second, it is used as the initial step for time–frequency analysis of both the test signal and real patient data using the Discrete Harmonic Wavelet Transform (DHWT).

The increased time resolution of the DHWT, compared with the DFT, increases the ability to detect transient autonomic alterations, allowing the cyclical alterations in autonomic tone to be more accurately distinguishable from LF power. Improving the spectral analysis of HRV data in this way will greatly add to our understanding of the dynamic function of the human autonomic nervous system especially in disease states where brief alterations in neural tone are prevalent.

The advantage of harmonic wavelets over ordinary wavelets is that each level of the DHWT represents a distinct frequency band. The penalty for this is poorer time resolution because, unlike ordinary wavelets, the harmonic wavelet is not compactly supported in time.

2 Analysis of HRV data

The first step is to generate equispaced Fourier coefficients from a nonequispaced time series. For Fourier analysis the series is assumed to be stationary, therefore for N samples the series is represented by:

$$x_j = x(t_j), \quad j = 0, 1, 2, \ldots, N-1. \tag{2.1}$$

where $t_0 = 0$, $x_0 = x_N$ and the period of the signal, $T = t_N$.

2.1 Discrete Fourier Transform

One solution to the problem of analysing nonequispaced data is to interpolate the data (for example using a cubic spline), resample at equispaced points and use the DFT to compute the Fourier coefficients. For N equispaced data points the series $\{x_j\}$ has Fourier coefficients $\{X_k\}$ defined by

$$X_k = \frac{1}{N} \sum_{j=0}^{N-1} x_j e^{-i(2\pi kj/N)}, \quad k = 0, 1, 2, \ldots, (N-1) \tag{2.2}$$

where $i = \sqrt{-1}$. Equation 2.2 is evaluated using the Fast Fourier Transform (FFT) [4] when N is a power of 2 ($N = 2^m, m \in \mathbb{N}$).

2.2 Nonequispaced Fourier Transform

Dutt and Rokhlin [5] presented a group of algorithms which generalised the DFT to deal with nonequispaced data. The method of computing equispaced frequency coefficients from nonequispaced data points (Problem 1 in [5]) is summarised as:

$$X_k = \sum_{j=0}^{N-1} x_j e^{-i(2\pi k t_j/N)}, \quad k = 0, 1, 2, \ldots, (N-1) \tag{2.3}$$

where $t_j \in [-N/2, N/2]$, $x_j = x(t_j)$ for $j = 0, \ldots, N-1$. The nonequispaced nature of the time series does not provide an orthogonal basis for calculation of the Fourier coefficients, so to compute the Fourier coefficients the x_j's are interpolated to the nodes of an integer grid (the time-frequency grid) using a Fourier series. In this implementation we follow the procedure employed in [5] where the grid is chosen to be $2N$ square, each $e^{i(2\pi k t_j/N)}$ term is approximated by a 28-term Fourier series, and each Fourier coefficient is approximated by interpolation of the values at the nearest 28 equispaced nodes on the grid. In addition the integer grid is scaled so that the Fourier coefficients obtained represent a fixed set of frequencies. This is equivalent to fixing the sampling frequency of the signal.

2.3 Discrete Harmonic Wavelet Transform

The DHWT, described by Newland [6], has the desirable property that the wavelet levels represent distinct frequency bands and can therefore be used for spectral analysis. In general a wavelet at level j, $j \in \mathbb{N}$, translated by k steps of size $1/2^{j-1}$ is defined by:

$$\begin{aligned} W(\omega) &= \frac{1}{2^j\pi} e^{-i2\omega k/2^j} \quad \text{for} \quad \pi 2^j \le \omega \le 2\pi 2^j, \\ &= 0 \quad \text{elsewhere.} \end{aligned} \tag{2.4}$$

The elements of this wavelet family are mutually orthogonal and their coefficients can be computed using the FFT. The DHWT coefficients are used to construct a time-frequency map (described in [6]) which shows how the frequency spectrum evolves with time. For a signal of length $N = 2^m, m \in \mathbb{N}$ there are $m+1$ wavelet levels $(0, \ldots, m)$ with a single coefficient at levels 0 and m and 2^{j-1} coefficients at each subsequent level $j = 1, \ldots, m-1$. The band of frequencies f_j represented by level j is given by

$$\frac{f_s}{2^{(m-j)+1}} \le f_j < \frac{f_s}{2^{(m-j)}} \tag{2.5}$$

where f_s is the sampling frequency of the original time series.

3 R-R test signals

The test signals used in this paper to evaluate the performance of the different methods are based on the Integral Pulse Frequency Modulation (IPFM) model [7] which produces a nonequispaced time series similar to the R-R interval data obtained from an ECG analysis. The IPFM model is a physiologically plausible way of developing an event pulse series from known input signals.

We represent two inputs to the IPFM model as sinusoids of frequencies $f_1 = 0.1$Hz and $f_2 = 0.28$Hz, with amplitudes a and b respectively, using the following equation:

$$m(t) = 1 + a \sin 2\pi f_1 t + b \sin 2\pi f_2 t + c\xi \tag{3.1}$$

where c is a scaling factor and $\xi \sim N(0, \sigma^2)$ represents noise as an independent random number chosen from a normal distribution with mean zero and variance σ^2. The equation for the IPFM model is

$$I_j = \int_{t_j}^{t_{j+1}} m(t)dt. \tag{3.2}$$

where I_j represents the value of integration over the interval $[t_j, t_{j+1}]$. If no noise term is present, there is an analytical solution for Equation 3.2:

$$I_j = \left[t - \frac{a}{2\pi f_1} cos 2\pi f_1 t - \frac{b}{2\pi f_2} cos 2\pi f_2 t \right]_{t_j}^{t_{j+1}} \tag{3.3}$$

where $t_j = j\Delta t$, $j = 0, 1, 2, \ldots$ and Δt is the sample period. If there is noise present, solving Equation 3.2 requires special treatment of the noise factor ξ as $\int \xi(t)dt = 0$, $\xi \sim N(0, \sigma^2)$, over any interval. To include the noise term in the formulation of the RR test signal, Equation 3.3 is evaluated from $t = 0$, with $\Delta t = \frac{1}{128}$s (see below) and a noise term ξ_j is added to the value I_j obtained. The process is repeated at intervals of $\Delta t = \frac{1}{128}$s until the sum of the integral $I_s = \sum_j I_j$ exceeds the threshold level $\overline{I}$ at which point the time of the event is recorded as the nearest edge of the current integration section. I_s is then reset to zero and the process repeated until the required number of events is recorded.

When the ECG is recorded the signal is sampled at 128Hz which determines the accuracy in locating the time of the R peaks in the signal. Choosing an interval $\Delta t = \frac{1}{128}$s mimics the sampling error of the ECG recorder as this also determines the accuracy in locating the time of an event from the IPFM model. Four different test signals are generated by varying the noise scaling factor c in Equation 3.1 where the standard deviation of the noise, $\sigma = 2.5e^{-04}$. Setting $a = 0.24$, $b = 0.16$ and $\overline{I} = 0.9169$ generates a signal which represents an average value of heart rate determined from data collected at Birmingham Heartlands Hospital from 6 healthy patients while asleep between the hours of 24:00 and 05:00. The four signals, summarised in Table 1, are:

1. A: no noise, $c = 0$
2. B: low noise level, $c = 25$
3. C: medium noise level, $c = 50$
4. D: high noise level, $c = 100$.

For a signal S_x the Signal–to–Noise ratios listed in Table 1 are calculated as

$$\text{var}(S_x)/(\text{var}(S_x) - \text{var}(S_A) + \text{var}(S_{fs})) \tag{3.4}$$

where $\text{var}(S_{fs})$ is the sampling frequency measurement error of the ECG recorder calculated as $1/(6 \times (128)^2)$ following [8]. For the DFT method it is necessary to fit a cubic spline to each signal and resample at $f_s = 1.2$Hz to obtain an equispaced series. The frequency bands associated with each level of the DHWT for a series of length $N = 512$, sampled at $f_s = 1.2$Hz are shown in Table 2.

Table 1. RR Test signal data

RR signal	Beats	Mean	Variance	S/N Ratio
No noise (A)	464	0.9204	0.0343	3371.8
Min. noise (B)	463	0.9216	0.0390	8.2799
Med. noise (C)	460	0.9298	0.0542	2.7222
Max. noise (D)	454	0.9408	0.1131	1.4350

Table 2. Wavelet level frequency bands for $f_s = 1.2\text{Hz}$, $N = 512$

Level	Frequency (Hz)		
0	0.00000		
1	0.00234		
2	0.00468	-	0.07031
3	0.00937	-	0.01641
4	0.01875	-	0.03516
5	0.0375	-	0.07266
6	0.075	-	0.1477
7	0.15	-	0.2977
8	0.3	-	0.5977
9	0.6		

4 Results

The periodograms of each of the four RR test signals are given in Figure 2 with two curves for each test signal. The dashed line represents the DFT method and the solid line represents the NeFT method. Each periodogram shows two distinct peaks, the LF peak at 0.1Hz and the HF peak at 0.28Hz. The frequency resolution for the four test signals is 0.002344Hz.

Image plots describe the results of each time–frequency analysis. The plots have been rescaled in the range $[0, 64]$ to obtain a good contrast in order to highlight the main features of the plots. This involves dividing each element in the map by its highest value, δ, and multiplying by 64. In order that the image plot is not dominated by a single peak which may obscure relevant features, δ is chosen to be the average of the 10 highest values in the map. Time-frequency maps for the four RR test signals are shown in Figure 3 for the NeFT method and Figure 4 for the DFT method. In both cases the LF and HF peaks of the test signals fall in Levels 6 and 7 respectively.

An analysis of two sets of patient data collected during sleep using the NeFT method is also presented in Figures 5 and 6. One set of data is from a normal patient, the other set is from a patient wih OSA.

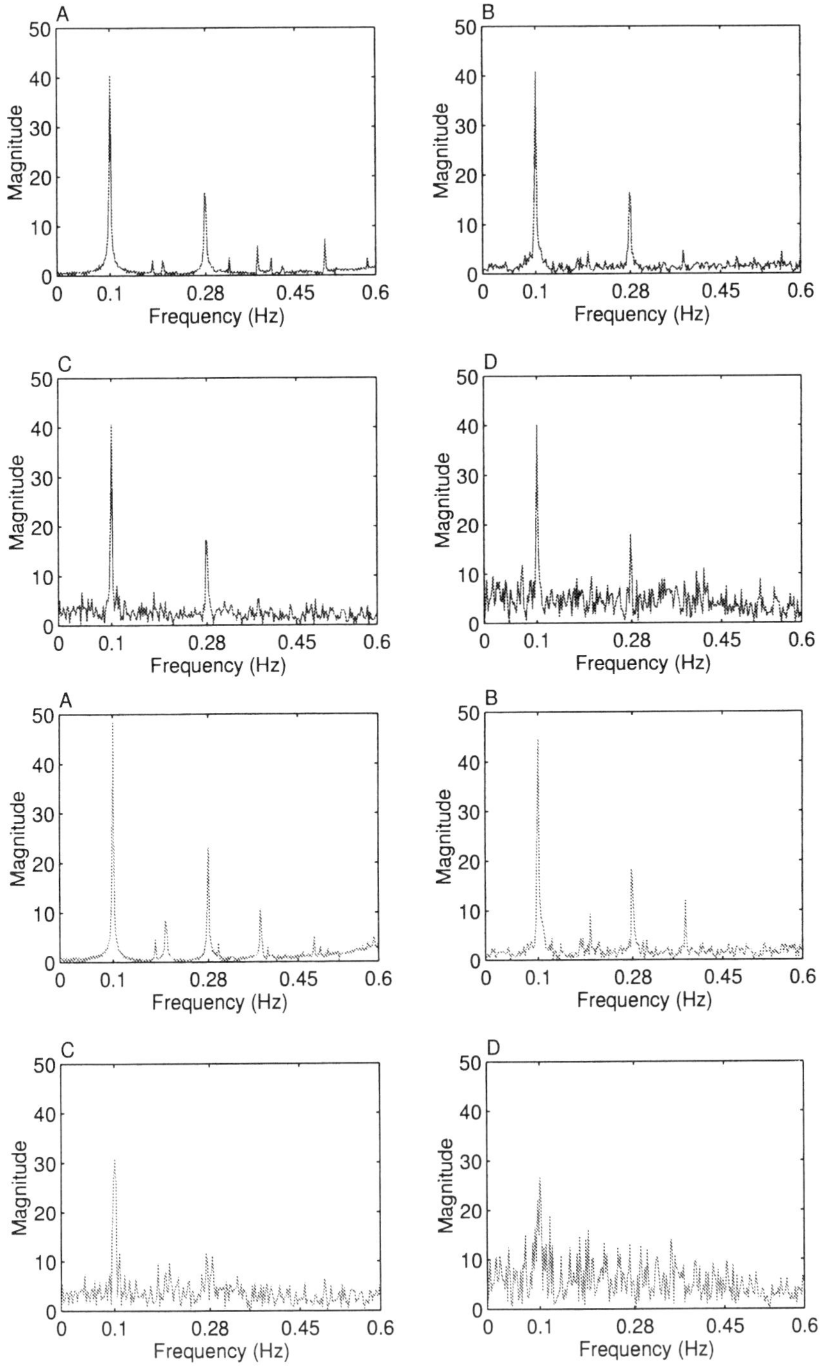

Figure 2. Periodogram of RR test signals A to D; Top: NeFT, Bottom: DFT

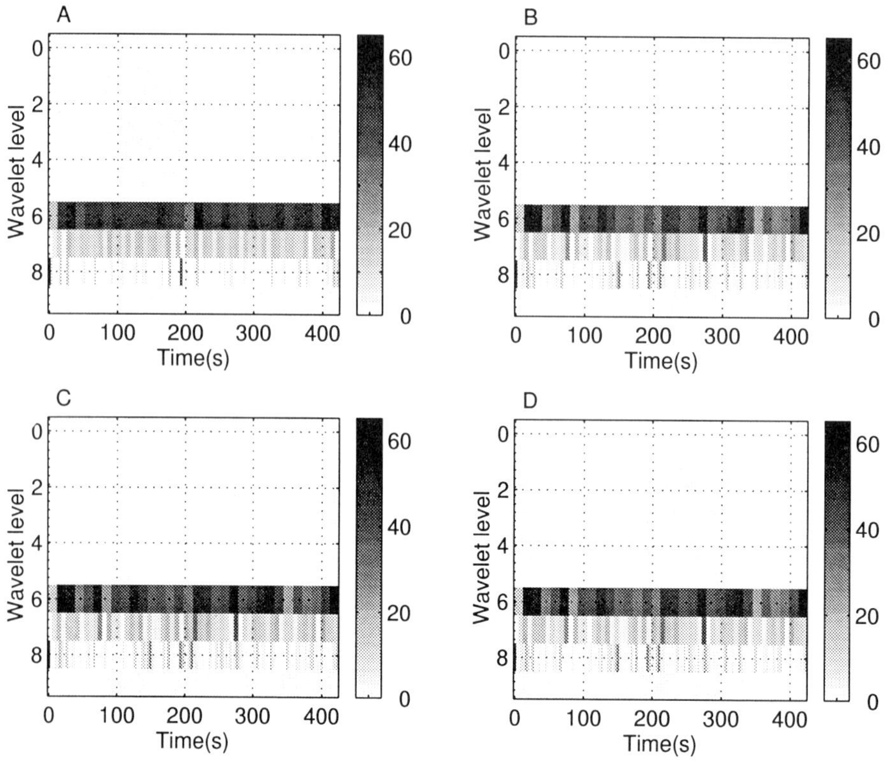

Figure 3. Image map for RR test signals A (no noise) to D (high noise) - NeFT method

5 Discussion

Two methods of calculating Fourier coefficients from RR test signals generated using the IPFM model have been compared. The periodogram plots of Figure 2 show that the NeFT and DFT methods work equally well for test signals A and B but the NeFT method works better for the higher noise signals (C and D).

Analysis using the DHWT also shows a difference between the DFT and NeFT methods. When considering the four RR test signals there is a definite improvement in the quality of analysis using NeFT over the DFT Method with the two frequencies in the test signals being represented more strongly within wavelet Levels 6 and 7. Both methods work equally well when no noise is present in the signal but as more noise is introduced (increasing the signal variance) the NeFT Method distinguishes the frequencies better over time than the DFT Method. As the signal becomes more noisy the HF component of the test signal disperses more into Level 8 and it becomes increasingly difficult to locate the frequency component correctly (as one would anticipate). This can be explained

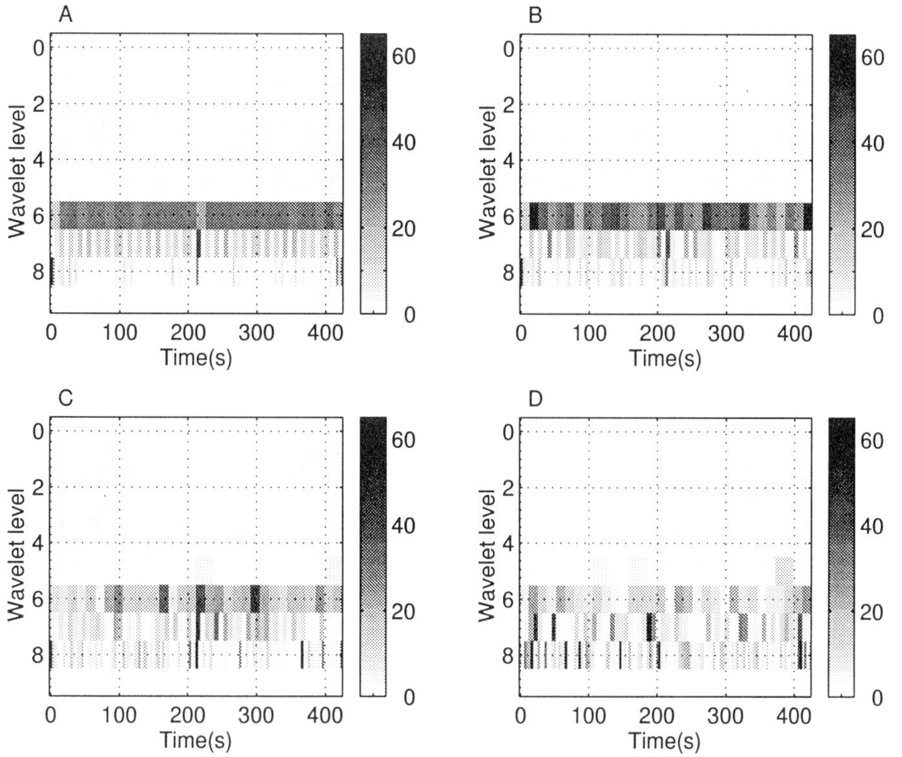

Figure 4. Image map for RR test signals A (no noise) to D (high noise) - DFT method

by noting the closeness of the HF component (0.28Hz) to the boundary between Levels 7 (0.15-0.2977Hz) and 8 (0.3-0.5977Hz).

The Signal–to–Noise ratios for the four test signals are given in Table 1. The test signal A represents an RR signal with a single noise component due to the sampling frequency of the ECG recorder (usually 128Hz) and no white noise. This is likely to be the signal most similar to real patient data which is dependent solely on when the R–peak of the ECG waveform occurs. The other test signals (B, C and D) represent signals with additional white noise and it should be noted that the NeFT method copes well with these signals even though they are excessively noisy.

Analysis of the two sets of patient data using the NeFT method shows a clear difference in the image maps in Figures 5 and 6. The data from the normal patient produces an image map showing clear bands of power in Levels 6 and 7 representing LF and HF power. The image map generated from the OSA patient data is markedly different with the majority of the signal power concentrated in Levels 4 and 5, some in Level 6 and virtually none in Level 7. Power in Level 4 is at the frequency of apnoeic events. This indicates that the DHWT may

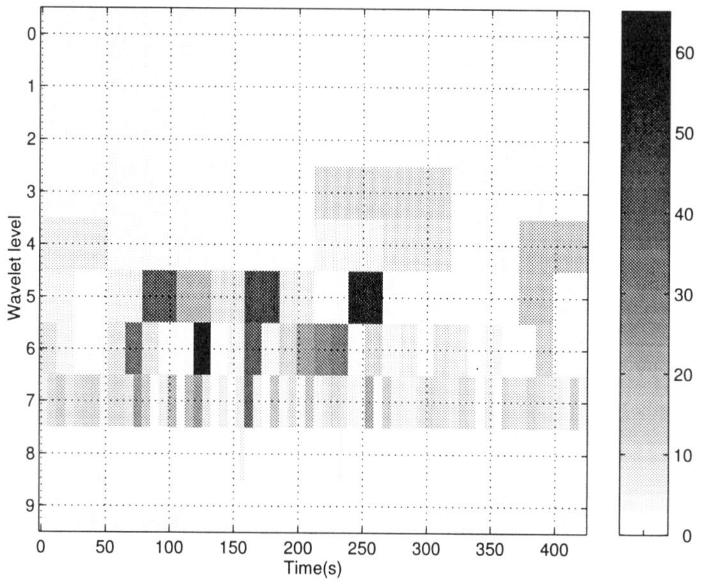

Figure 5. Normal patient

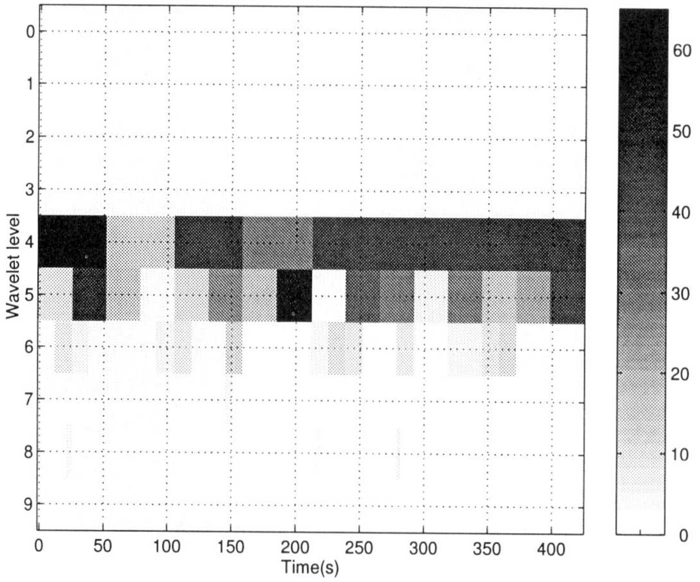

Figure 6. Osa patient

prove useful in distinguishing periods of sleep apnoea from normal sleep by the presence or absence of power in this level.

Acknowledgement

The work described in this paper was carried out as part of Grant GR/J67130 "Identifiability and Indistinguishability of Nonlinear Dynamic Systems" from the U.K. Engineering and Physical Sciences Research Council.

Bibliography

1. S. Akselrod, D. Gordon, F.A. Ubel, D.C. Shannon, A.C. Barger and R.J. Cohen (1981). "Power spectrum analysis of heart rate fluctuation: A quantitative probe of beat-to-beat cardiovascular control". *Science*, **213**, pp. 220–222.

2. C. Guilleminault, A. Tilkian and W.C. Dement (1976). "The sleep apnoea syndromes". *Ann. Rev. Med.*, **27**, pp. 465–484.

3. R.J.O. Davies, P.J. Belt, S.J. Roberts, N.J. Ali and J.R. Stradling (1993). "Arterial blood pressure responses to graded transient arousal from sleep in normal humans". *J. Appl. Physiol*, **74**, pp. 1123–1130.

4. J.W. Cooley and J.W. Tukey (1965). "An algorithm for the machine computation of complex Fourier series". *Math. Comp.*, **19(1)**, pp. 297–301.

5. A. Dutt and V. Rokhlin (1993). "Fast Fourier transforms for nonequispaced data". *SIAM J. Sci. Stat. Comp.*, **14(6)**, pp. 1368–1393.

6. D.E. Newland (1993). *An Introduction to Random Vibrations, Spectral and Wavelet Analysis*, Longman Scientific and Technical.

7. U. Niklasson, U. Wiklund, P. Bjerle and B.O. Olofsson (1993). "Heart-rate variation - what are we measuring?". *Clinical Physiology*, **13(1)**, pp. 71–79.

8. M. Merri, D.C. Farden, J.G. Mottley and E.L. Titlebaum (1990). "Sampling frequency of the electrocardiogram for spectral analysis of heart rate variability". *IEEE Trans. Biomed. Eng.*, **37**, pp. 99–106.

Robust Estimation of Oceanic Background Spectrum

T.S.T. Leung and P.R.White

Institute of Sound and Vibration Research, University of Southampton

Abstract

This paper discusses the estimation of the power spectral density (psd) of oceanic background noise. This is traditionally done by the segment averaging method which may become biased when transients (outliers) are present in the data record. Alternative approaches including the use of order statistics and trimmed mean, have been proposed to minimise such bias. We shall compare the performance of different estimators in terms of their variances and robustness via simulations and present novel theoretical expression for the statistical behaviour of the trimmed mean estimator.

1 Introduction

In the signal processing of oceanic signals, we often minimise the effect of the coloured background noise by using pre-whitening. Optimal detection performance, in coloured Gaussian noise scenario, requires one to normalise signals with respect to the psd of the background noise [1,2].

The most common way of estimating the psd of background noise is to use segment averaging. This spectral estimate will, however, become biased when occasional transients, for example, biological noises, are present in some of the segments. This paper discusses alternative ways to estimate the psd of the background noise so that the estimation will be more robust to the presence of such transients. The techniques discussed herein include the use of order statistics and trimmed mean.

2 Variances of the mean, the order statistic and the trimmed mean estimators

In the following discussion, the normalised power spectral density (psd) estimate will be used so that we can make use of the results of some standard statistical properties:

$$\text{The normalised psd,} \quad S_n(f) = 2\frac{\hat{S}(f)}{S(f)} \tag{2.1}$$

where $\hat{S}(f)$ is the estimator and $S(f)$ is the true psd.

When the estimator is the raw psd $|X(f)|^2/T$ (where $X(f)$ is the Fast Fourier transform (FFT) of the time series $x(t)$ and T is the length of the FFT), the normalised psd (2.1) of a Gaussian input is known to have a chi-squared distribution with 2 d.o.f. (χ_2^2). The spectral estimates discussed in the paper are based on sections of non-overlapping input data using rectangular windowing.

2.1 Mean (segment averaging)

The spectral estimate based on taking the mean value (segment averaging) is defined as follows [5]:

$$\hat{S}_{mn}(f) = \frac{1}{N}\sum_{i=1}^{N} S_i(f) \tag{2.2}$$

where N is the number of sections of signal being averaged, and $S_i(f) = |X_i(f)|^2/T$.

The normalised psd is therefore,

$$\frac{2\hat{S}_{mn}(f)}{S(f)} = \frac{1}{N}\sum_{i=1}^{N}\frac{2\hat{S}_i(f)}{S(f)}. \tag{2.3}$$

The right hand side of (2.3) is a summation of $N\chi_2^2$ random variables and can be said to have a χ^2 distribution with $2N$ d.o.f. [6]. The variance can therefore be expressed as follows (recall: the variance of a χ_N^2 distribution $= 2N$):

$$\mathrm{Var}\left[\frac{2N\hat{S}_{mn}(f)}{S(f)}\right] = 4N \tag{2.4}$$

$$\mathrm{Var}[\hat{S}_{mn}(f)] = \frac{S^2(f)}{N} = \sigma_{mn}^2(f). \tag{2.5}$$

2.2 Order statistics

Suppose there are N data samples $\{Z_i, i = 1, 2, ...N\}$ being sorted into ascending order $\{Y_i, i = 1, 2, ...N\}$, then Y_i is known as the order statistic. When $i = (N+1)/2$ (N is an odd number), the order statistic is called the median. The spectral estimate based on order statistics is defined as follows:

$$\hat{S}_i(f) = k_d U_i \tag{2.6}$$

where U_i is the ith order statistic of $\{|X_j(f)|^2/T, j = 1, 2, ...N\}$ in ascending order.

In general, U_i is not the true mean and we need to introduce a scaling factor k_d to correct the discrepancy so that the expected value is unbiased.

$$k_d = \frac{2}{E[U_{n,i}]} \tag{2.7}$$

where $U_{n,i}$ denotes the normalised value of U_i. The derivation of $E[U_{n,i}]$ is shown in the appendix. This ensures that the normalised psd estimate (2.6) is unbiased.

The variance can be calculated as follows:

$$\text{Var}\left[\frac{2\hat{S}_i(f)}{S(f)}\right] = k_d^2 \text{Var}[U_{n,i}] \tag{2.8}$$

$$\text{Var}[\hat{S}_i(f)] = \frac{k_d^2}{4}\text{Var}[U_{n,i}]S^2(f) \tag{2.9}$$

where $\text{Var}[U_{n,i}]$ is given by (A5). Taking $i = (N+1)/2$ where N is an odd number, we obtain the variance of the median estimate as

$$\text{Var}[\hat{S}_{med}(f)] \approx 2\frac{S^2(f)}{N} = 2\sigma^2_{mn}f. \tag{2.10}$$

By minimising (2.9) with respect to i, the order statistic with the lowest variance can be found, it turns out to be the $U_{0.8N}$, i.e. the 80% value in the sequence arranged in ascending order.

$$\text{Var}[\hat{S}_{80}(f)] \approx 1.3\frac{S^2(f)}{N} = 1.3\sigma^2_{mn}f. \tag{2.11}$$

We call this estimator the 80th order statistic.

2.3 Trimmed mean

A trimmed mean is constructed by sorting the N data samples $\{Z_i, i = 1, 2, ...N\}$ into ascending order $\{Y_i, i = 1, 2, ...N\}$, then discarding the largest 100(1-b)% ($\beta = bN$) of the points and the smallest (100a)% ($\alpha = aN$), and only averaging the remaining (central) points [4]:

$$Y_{trmn} = \frac{1}{n}\sum_{i=\alpha}^{\beta} Y_i \tag{2.12}$$

where $n = \beta - \alpha + 1$.

The trimmed mean spectral estimate is defined as follows:

$$\hat{S}_{trmn}(f) = \frac{k_{tr}}{n}\sum_{i=\alpha}^{\beta} U_i(f). \tag{2.13}$$

If the distribution is symmetric, then selecting $a = (1-b)$ ensures that the average of the trimmed mean equals the true mean. Unfortunately, the spectral values of Gaussian input follow a (scaled) chi-squared distribution and hence

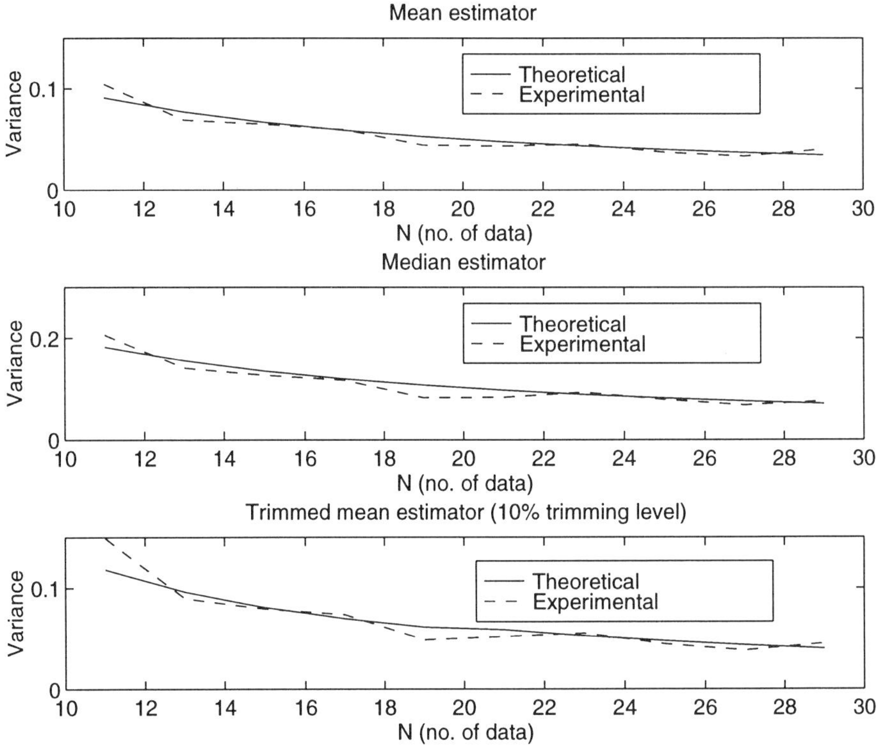

Figure 1. Comparison between theoretical variances and experimental variances

are asymmetric. Therefore, it is necessary to introduce a scaling factor (k_{tr}) in the trimmed mean estimate to correct the discrepancy so that the estimator is unbiased:

$$k_{tr} = \frac{2n}{\sum_{i=\alpha}^{\beta} E[U_{n,i}(f)]}. \tag{2.14}$$

The variance can be calculated as follows:

$$\mathrm{Var}\left[\frac{2\hat{S}_{trmn}(f)}{S(f)}\right] = \left(\frac{k_{tr}}{n}\right)^2 \mathrm{Var}\left[\sum_{i=\alpha}^{\beta} U_{n,i}(f)\right] \tag{2.15}$$

$$\mathrm{Var}[\hat{S}_{trmn}(f)] = \left(\frac{k_{tr}}{2n}\right)^2 \mathrm{Var}\left[\sum_{i=\alpha}^{\beta} U_{n,i}(f)\right] S^2(f) \tag{2.16}$$

where $\text{Var}\left[\sum_{i=\alpha}^{\beta} U_{n,i}(f)\right]$ is derived in the appendix.

In order to check the validity of the expressions (2.5), (2.9) and (2.16), we have performed the following test. We first generated 500 sets of Gaussian white noise. The estimators all with $N = 11$ were applied to estimate their spectral values. A particular frequency bin from each set of data was taken and then used to work out the experimental variances for the estimators. The value N was then increased by 2 and the procedure repeated itself until we had a record of variances from $N = 11$ to $N = 29$. These experimental results are plotted together with the theoretical results according to (2.5), (2.9) and (2.16) in Figure 1. The results show that the expressions so derived conform with the experimental values.

3 Comparison of the estimators

In the following, we shall compare the spectral estimators proposed in terms of their variances and robustness to outliers.

3.1 Variance

The variances of different estimators have been calculated according to (2.5), (2.9) and (2.16). The median and the 80th order statistic were shown in Figure 2 as well as the trimmed means with 10%, 20% and 30% trimming levels.

It can be seen that the variances tend to zero as N increases. The estimators appear to be consistent. Figure 2 demonstrates that in the absence of outliers, the mean estimator is the least variable of the estimators discussed here. This is in fact expected because the mean estimator can be shown to be the optimal (maximum likelihood) estimator.

The median is the most variable of the estimators as expected. The trimmed means become more variable as the trimming level increases. The 80th order statistic, which has the lowest variance for the order statistics, is slightly less variable than the 30% trimmed mean.

3.2 Robustness to outliers

A clip of underwater signal has been selected from our sea trial database. Initially, the ocean was quiet and we estimated the psd of the background noise using segment averaging with a rather large number of segments. This will be used as a template for the comparison. We then estimate the psd of segments with: I. a moderate amount, and II. a large amount, of biological noise (see Figures 5 and 6) using the proposed estimators. The resulting psds for record II are depicted in Figures 3 and 4. The mean squared error (m.s.e.) between the true psd (the template) and the estimated psd has also been calculated and shown in Table 1.

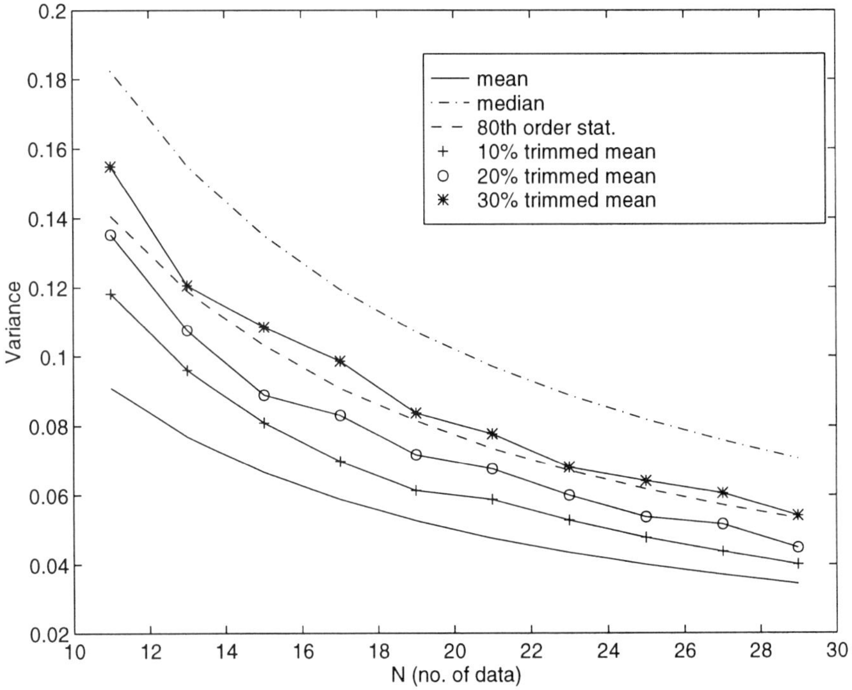

Figure 2. Variance $(xS^2(f))$ of estimators as N increases

In the presence of transients, it can be seen that the mean estimator is the least robust for both records. For record I, the proposed estimators demonstrate similar robustness. For record II, their performances are different. Although the 80th order statistic is optimised for variance, it is not robust for a large number of outliers as occurs in record II. As the trimming level increases, the trimmed mean estimator becomes more robust because more outliers have been discarded, in light of this it is not surprising that the median is the most robust estimator.

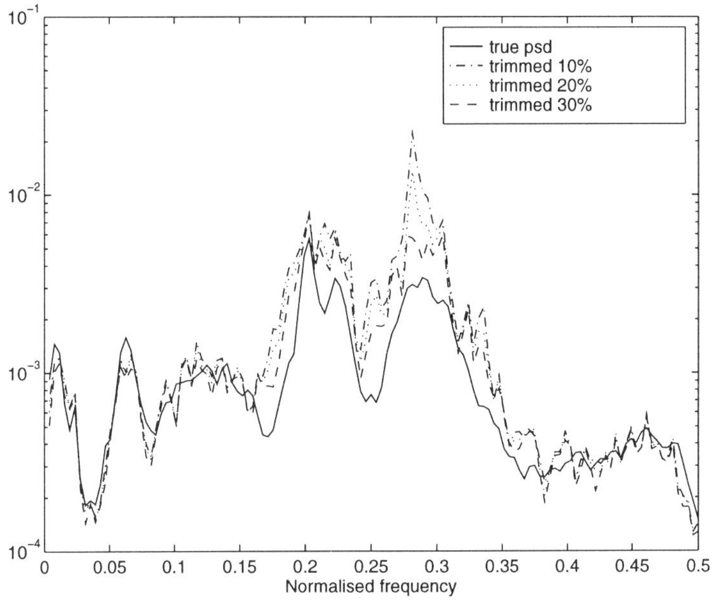

Figure 3. psd estimated by the trimmed mean estimator for record II

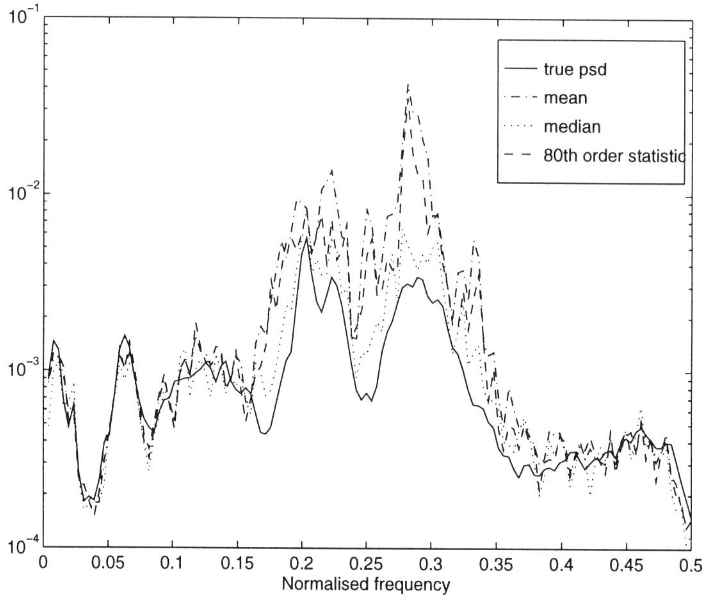

Figure 4. psd estimated by the mean, median and the 80th order statistic estimators for record II

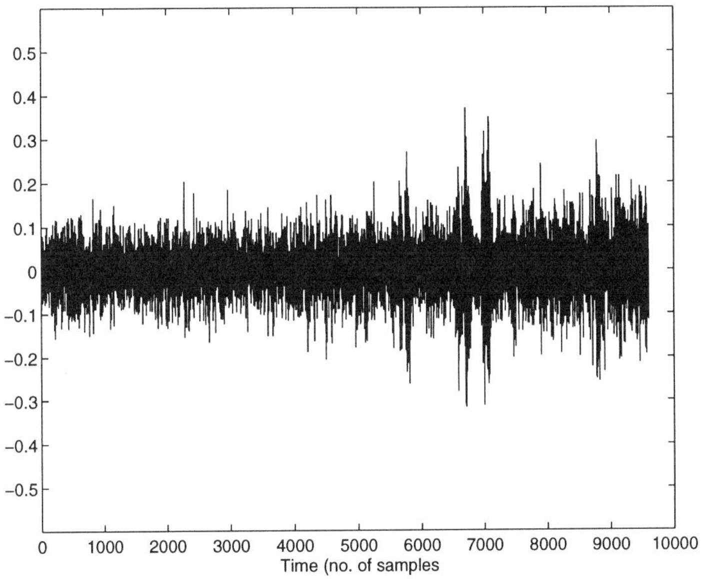

Figure 5. Signal record I: contaminated with a moderate amount of biological noise

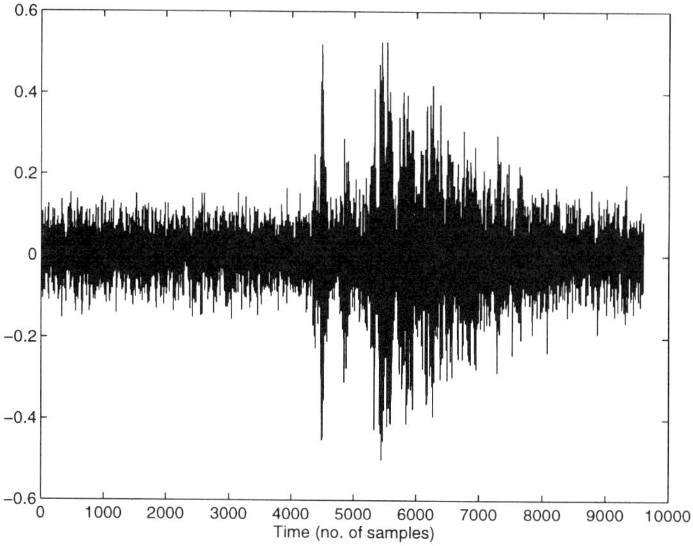

Figure 6. Signal record II: contaminated with a large amount of biological noise

Table 1. m.s.e. of the estimated psd

m.s.e. ($\times 10^{-7}$)	Mean	Trimmed 10%	Trimmed 20%
Record I	26.11	3.34	3.03
Record II	344.57	64.50	22.14

m.s.e. ($\times 10^{-7}$)	Trimmed 30%	Median	80th Order Statistic
Record I	2.77	3.58	3.56
Record II	7.21	6.11	145.23

4 Conclusion

We have investigated the performance of the proposed estimators with regard to their variances in outlier-free scenarios and their robustness in the presence of outliers. The results show that generally there is a trade-off between the variance and the robustness. The classical mean estimator provides the lowest variance but is rather susceptible to outliers. The median estimator is very robust when a large number of outliers exist but exhibits large variability in normal conditions. A good compromise seems to be the trimmed mean which is able to give a reasonably low variance and a significant robustness to outliers. Varying the trimming level allows one to trade variance for robustness.

Acknowledgements

The authors would like to thank ADAC Defence Evaluation and Research Agency, Farnborough, United Kingdom, for their continued financial support for this work, their technical assistance and the supply of data set.

Bibliography

1. White, P.R. (1993). Detection of underwater acoustic transients using adaptive filters and time-frequency methods. *Proceedings of the Institute of Acoustic*, **15(3)**, pp. 881–890.

2. Leung, T.S.T and White, P.R. (1996). A fuzzy logic based underwater acoustic transient classifier. *Proceedings of the 7th IEEE Digital Signal Processing Workshop*, pp. 494–497.

3. Boes, M.G. (1974). *Introduction to the theory of statistics*, McGraw-Hill Book Company, Singapore.

4. Hoaglin, D.C. (1983). *Understanding robust and exploratory data analysis*, John Wiley and Sons Inc., USA.

5. Kay, S.M. (1988). *Modern Spectral Estimation - theory and application*, Prentice Hall, New Jersey.

6. Jenkins, G.M. and Watts, D.G. (1968). *Spectral Analysis and its Applications*, Holden Day, California.

Appendix

1. Derivation of $E[U_{n,i}(f)]$.

$S_{n,i}(f)$ is the ith order statistic with the χ_2^2 distribution before ordering and is denoted by Y_i in the following. The probability density function (pdf) of such an order statistic is given by:

$$f(y) = \frac{N!}{2(i-1)!(N-i)!}\exp(-y/2)[1-\exp(-y/2)]^{i-1}[\exp(-y/2)]^{N-i}. \tag{A1}$$

The expected value is hence calculated by

$$\mu_i = E[Y_i] = \int_{-\infty}^{\infty} yf(y)dy \tag{A2}$$

yielding,

$$E[S_{n,i}(f)] = \frac{2N!}{(i-1)!(N-i)!}\sum_{k=0}^{r}\binom{r}{k}\frac{(-1)^k}{(k+N-i+1)^2}. \tag{A3}$$

2. Derivation of $\text{Var}\left[\sum_{i=\alpha}^{\beta} U_{n,i}(f)\right]$.

The variance of the summation of random variables [3] is given as follows:

$$\text{Var}\left[\sum_{i=\alpha}^{\beta} Y_i\right] = \sum_{i=\alpha}^{\beta}\text{Var}[Y_i] + 2\sum_{i<}\sum_{j}\text{cov}[Y_i, Y_j] \tag{A4}$$

where

$$\text{Var}[Y_i] = \frac{F(\mu_i)[1-F(\mu_i)]}{N[f(\mu_i)]^2} \tag{A5}$$

$F(.)$ and $f(.)$ here are the cumulative distribution function and the pdf of a χ_2^2 distribution.

And the covariance of two order statistics is as follows:

$$\text{cov}[Y_i, Y_j] = E[Y_iY_j] - \mu_i\mu_j \tag{A6}$$

where

$$E[Y_iY_j] = \int_0^\infty \int_{y_i}^\infty (y_iy_j) f_{Y_i,Y_j}(y_i,y_j)dy_jdy_i \tag{A7}$$

where $f_{Y_i,Y_j}(y_i,y_j)$ is the joint pdf of Y_i and Y_j.

$$E[Y_iY_j] = \frac{4N!}{r!m!(p-1)!} \sum_{q=0}^{m}\sum_{k=0}^{r} (-1)^{q+k} \begin{pmatrix} M \\ q \end{pmatrix} \begin{pmatrix} r \\ k \end{pmatrix} \cdot \left(\frac{1}{q+p}\right)\left(\frac{1}{k+h}\right)^2 \left(\frac{2}{k+h}+\frac{1}{q+p}\right) \tag{A8}$$

where $r = i-1$, $m = j-i-1$, $h = N-i+1$, $p = N-j+1$.

Substituting (A5)–(A8) into (A4), the variance of the summation of order statistics can be calculated.